BIOMAGNETICS
Principles and Applications of
Biomagnetic Stimulation and Imaging

BIOMAGNETICS

Principles and Applications of
Biomagnetic Stimulation and Imaging

Edited by

Shoogo Ueno
Professor Emeritus, The University of Tokyo

Masaki Sekino
Associate Professor, The University of Tokyo

CRC Press
Taylor & Francis Group
Boca Raton London New York

CRC Press is an imprint of the
Taylor & Francis Group, an **informa** business

Contents

Foreword

Biomagnetics covers a vast field in the multidisciplinary world of physics, chemistry, biology, and medicine, touching many aspects of our lives. To start with, we ourselves generate some degree of a magnetic field in our bodies, although not strongly. In biological systems, there are interactions with magnetism present in the natural environment as well as with those imposed by modern-age technology where highly developed electromagnetic devices are part of everyday conversation.

In his long and successful academic career, Professor Shoogo Ueno has contributed significantly to the scientific and technological development of the biomagnetics field. He and his associates have worked in many areas of the biological effects of the magnetism field, some of which, such as the Moses effect, have been revelatory. Professor Ueno is well known as the inventor of the "figure-eight" TMS coil, which has been "the device" for noninvasive stimulation of the human brain because of its capacity for stimulus localization.

This book, *Biomagnetics: Principles and Applications of Biomagnetic Stimulation and Imaging*, covers the whole range of the field and builds upon the editors' past work. Many observable phenomena induced by the interaction of magnetic fields with biological materials are described here together with the underlying basic physics of magnetism and/or electromagnetics so that readers can better understand these phenomena. For those whose work is highly specialized in the field, this book will be very useful to widen their scope of understanding biomagnetics. Magnetic fields can permeate normal biological material, and, therefore, it is possible to build noninvasive devices to measure magnetic phenomena inside the body. From the observable interaction of magnetic fields with biological materials, the *in vivo* information of physical properties can be obtained. However, some interaction, especially by time-varying magnetic fields, leads to unexpected/unintended phenomena, which could be hazardous to the body under examination. Safety issues of medical noninvasive devices are discussed in this book. There are also many phenomena in biomagnetics that still need further study to promote their utilization. In turn, the authors note the possible future trends of the field, which may encourage young aspiring scholars to work in the field.

Seiji Ogawa, PhD
Professor
Kansei Fukushi University
Aobaku, Sendai, Japan

Preface

Biomagnetics is an interdisciplinary field where magnetics, biology, and medicine overlap. Recent advances in biomagnetics have enabled us not only to detect extremely weak magnetic fields from the human brain but also to control cell orientation and cell growth by extremely high magnetic fields. Pulsed magnetic fields are used for transcranial magnetic stimulation of the human brain, and both high-frequency magnetic fields and magnetic nanoparticles have promising therapeutic applications for treatments of cancers and brain diseases such as Parkinson's and Alzheimer's.

On the imaging front, magnetic resonance imaging (MRI) is now a powerful tool for basic and clinical medicine. New methods of MRI based on the imaging of impedance of the human body, called impedance MRI, and the imaging of neuronal current activities in the human brain, called current MRI, are also being developed.

To understand the most advanced technologies in biomagnetics, basic sciences such as physics, biology, chemistry, magnetics, electromagnetics, physiology, and neurophysiology are important to build a foundation and construct bridges between science and medicine. These basic items are included in this book.

This book, *Biomagnetics: Principles and Applications of Biomagnetic Stimulation and Imaging*, is focused on three important fields: (1) magnetic nerve stimulation and transcranial magnetic stimulation (TMS), (2) biomagnetic measurements and imaging of the human brain by advanced technologies of magnetoencephalography (MEG) and MRI, and (3) biomagnetic approaches to potential treatments of cancers, pain, and other neurological and psychiatric diseases such as Alzheimer's disease and depression. Biomagnetic approaches to advanced medicine such as regenerative medicine and rehabilitation medicine are also included.

The core parts of the book are based on lecture notes that the editors of the book have used for students in graduate courses at the Department of Electronic Engineering, Graduate School of Engineering, and the Department of Biomedical Engineering, Graduate School of Medicine, University of Tokyo for more than 20 years. Dr. Ueno has been continuously revising and improving his notes since 1994 when he started his Laboratory of Biomagnetics and Bioimaging at the Department of Biomedical Engineering, Faculty of Medicine, at the University of Tokyo, Tokyo, Japan. Dr. Ueno retired from the University of Tokyo as professor emeritus. Masaki Sekino is now responsible for a course of biomedical engineering at the Graduate School of Engineering, University of Tokyo.

In this book, the editors asked five colleagues to join chapter contributors to further explore and study the field of biomagnetics and bioimaging.

The book is composed of 10 chapters, and a brief outline of each follows.

Chapter 1. Introduction (Shoogo Ueno)

A history of biomagnetic research is described, introducing Gilbert's book *De Magnete* published in 1600. The chapter also includes brief summaries of studies of biological effects and medical applications of magnetic and electromagnetic fields, biomagnetic stimulation and transcranial magnetic stimulation (TMS) of the human brain, biomagnetic measurements and magnetoencephalography (MEG), biomagnetic imaging and magnetic resonance imaging (MRI), and magnetic approaches to cancer therapy and other treatments.

Chapter 2. Principles of Biomagnetic Stimulation (Shoogo Ueno and Masaki Sekino)

The history and principles of magnetic nerve and muscle stimulation, and TMS of the human brain are introduced. Principles of electromagnetics, fundamentals for capacitor banks, coil circuits, and types of coils to stimulate the brain by a transcranial method, and localization of stimulation areas are introduced. Computational biomagnetic design for biomagnetic stimulation is also discussed.

Chapter 3. Applications of Biomagnetic Stimulation for Medical Treatments and for Brain Research (Shoogo Ueno and Masaki Sekino)

Potential medical applications of TMS and repetitive TMS, called rTMS, are introduced. Reduction of pain sensation by rTMS is described first. Studies of treatments of depression by TMS and rTMS are also reviewed. Potential therapeutic applications of rTMS for neurophysiological and neuropathological diseases such as Parkinson's and Alzheimer's are also briefly described. For the medical usages of rTMS, experimental studies using rat hippocampus neurons are introduced. Both classical and recent studies of cognitive functions using TMS are discussed.

Chapter 4. Biomagnetic Measurements (Sunao Iwaki)

The history of biomagnetic measurements, called biomagnetism, is briefly introduced first. Then, the measurements of extremely weak magnetic fields produced by the brain's electrical activities, called magnetoencephalography (MEG), are discussed. The measurement technique for MEG is described, introducing a superconducting quantum interference device (SQUID) system. Some studies of functional brain activities and cognitive functions revealed by MEG measurements are briefly reviewed.

Chapter 5. Principles of Magnetic Resonance Imaging (Masaki Sekino, Norio Iriguchi, and Shoogo Ueno)

Principles of nuclear magnetic resonance and magnetic resonance imaging are introduced. Hardware and three types of coils are explained. Pulse sequences related to various imaging techniques are described. Basic keywords, such as Ramor frequency, T1 and T2 relaxations, and free induction decay to advanced imaging such as echo planer imaging (EPI), diffusion tensor imaging (DTI), and functional MRI (fMRI) based on blood oxygenation level dependent (BOLD) effects are described.

Chapter 6. Prospects of Magnetic Resonance Imaging of Impedance and Electric Currents (Masaki Sekino, Norio Iriguchi, and Shoogo Ueno)

Imaging of electrical parameters such as imaging of impedance, called impedance MRI, and imaging of neuronal electric currents in the brain, called current MRI, are discussed. First, the history and the importance of impedance and electric current imaging are described. Various types of impedance imaging by magnetic resonance techniques are introduced. Then, recent studies on electric current MRI are introduced and discussed, introducing the images obtained by *in vitro* and *in vivo* experiments. Recent advances in MR neuroimaging are also discussed.

Chapter 7. Magnetic Control of Biological Cell Growth (Shoogo Ueno and Sachiko Yamaguchi-Sekino)

Mechanisms of the biological effects of static and pulsed magnetic fields are reviewed. The importance of the role of diamagnetic properties in biological systems is emphasized when they are exposed to static magnetic fields.

On the basis of the mechanisms of anisotropy of diamagnetic materials, magnetic control of biological cell orientation and cell growth are discussed. Biomagnetic approaches to tissue engineering and regenerative medicine are discussed, introducing new findings related to bone acceleration by strong static magnetic fields, alignment of blood vessel tissues by magnetic fields, and magnetic control of orientation of nerve axons during sprouting processes obtained by *in vitro* and *in vivo* experiments.

Cancer cell destruction by pulsed magnetic fields is also reviewed, introducing, for example, physical destruction of leukemia TCC-S cells combined with magnetizable beads by an antigen–antibody reaction. Pulsed magnetic stimulation effectively damages only the cancer cells targeted by an antigen–antibody reaction. Cancer therapy with repetitive pulsed magnetic stimulation is a promising method in the near future.

Chapter 8. Effects of Radio Frequency Magnetic Fields on Iron Release and Uptake from and into Cage Proteins (Oscar Cespedes and Shoogo Ueno)

An interesting effect of radio frequency magnetic fields on iron ion release and uptake from and into ferritins, iron cage proteins, is introduced. The chapter emphasizes that the effects of radio frequency magnetic fields on iron cage proteins may be used in the medical diagnosis and treatment of brain diseases, for example, for the imaging and dissolution of the β-amyloid aggregates related to Alzheimer's disease.

Chapter 9. Safety Aspects of Magnetic and Electromagnetic Fields (Sachiko Yamaguchi-Sekino, Tsukasa Shigemitsu, and Shoogo Ueno)

Biological effects of magnetic and electromagnetic fields are discussed. Safety aspects of magnetic fields and electromagnetic fields used in MRI, TMS, and other medical equipment are important issues for patients and workers. Safety issues regarding industrial equipment such as welding machines are also discussed.

Recent guidelines of IEC and EU directives are discussed. Mechanisms and possible effects of static, low-frequency, and radio frequency electromagnetic fields are reviewed and discussed.

Chapter 10. New Horizons in Biomagnetics and Bioimaging (Shoogo Ueno and Masaki Sekino)

Recent advances in biomagnetic imaging and biomagnetic stimulation are discussed. Functional imaging of biological systems by the combination of low magnetic field MRI and SQUID (superconducting quantum interference device) systems, molecular bioimaging, and various coil array systems for biomagnetic stimulation are reviewed and discussed.

This book provides an explanation of physical principles of biomagnetic stimulation and imaging, as well as applications of these techniques in neuroscience, clinical medicine, and healthcare. The book aims to educate graduate students, young scientists, and engineers on how these techniques are used in hospitals and why they are so promising. A brief overview of the recent research trend in biomagnetics serves as an informative guide for researchers and engineers to further study this field.

Acknowledgments

We thank the chapter authors for their intellectual contributions. Further, we thank the staff of CRC Press/Taylor & Francis, in particular, Francesca McGowan, editor, physics, for her consistent support and valuable advice regarding the book proposal and all editing processes; Sarfraz Khan, editorial assistant, for his assistance; Amber Donley, project coordinator, for her editorial project work; the staff of the editorial project development team; and John Navas, former senior acquisition editor, who first contacted Shoogo Ueno to discuss a book on biomagnetics in 2012. Without the enthusiastic and consistent support and help of all the people related to this project, this book would not have been published.

Finally, but not least, we greatly appreciate Dr. Seiji Ogawa, the inventor of functional MRI based on the BOLD effect, for kindly agreeing to write the Foreword for this book. We are honored.

Shoogo Ueno
Masaki Sekino

Editors

Shoogo Ueno, PhD, earned his BS, MS, and PhD (Dr Eng) degrees in electronic engineering from Kyushu University in 1966, 1968, and 1972, respectively. Dr. Ueno was an associate professor with the Department of Electronics, Kyushu University, from 1976 to 1986. From 1979 to 1981, he spent his sabbatical with the Department of Biomedical Engineering, Linkoping University, Linkoping, Sweden, as a guest scientist. He served as a professor in the Department of Electronics, Graduate School of Engineering, Kyushu University from 1986 to 1994. He subsequently served as a professor in the Department of Biomedical Engineering, Graduate School of Medicine, University of Tokyo from 1994 to 2006. During this time, he also served as a professor in the Department of Electronic Engineering, Graduate School of Engineering, University of Tokyo. In 2006 he retired from the University of Tokyo as professor emeritus. Since then, he has been a professor with the Department of Applied Quantum Physics, Graduate School of Engineering, Kyushu University, Fukuoka, Japan, and he served as the dean of the faculty of medical technology, Teikyo University, Fukuoka, Japan from 2009 through 2012.

Dr. Ueno is a fellow of the Institute of Electrical and Electronics Engineers, IEEE (2001), and a life fellow (2011) and a fellow of the American Institute for Medical and Biological Engineering, AIMBE (2001). He is a fellow (2006) and secretary (2012) of the Governing Council of the International Academy for Medical and Biological Engineering, IAMBE. He was a president of the Bioelectromagnetics Society, BEMS (2003–2004), chairman of the Commission K on Electromagnetics in Biology and Medicine of the International Union of Radio Science (URSI) (2000–2003). He received the Doctor Honoris Causa from Linkoping University, Linkoping, Sweden (1998). He was a 150th Anniversary Jubilee Visiting Professor at Chalmers University of Technology, Gothenburg, Sweden (2006) and a visiting professor at Simon Frasier University, Burnaby, Canada (1994) and Swinburne

University of Technology, Hawthorn, Australia (2008). Dr. Ueno was awarded the IEEE Magnetics Society Distinguished Lecturer during 2010 and the d'Arsonval Medal from the Bioelectromagnetics Society (2010).

Masaki Sekino, PhD, earned his BS, MS, and PhD (Dr Eng) degrees in mechanical engineering and electronic engineering from the University of Tokyo, Tokyo, Japan, in 2000, 2002, and 2005, respectively. Dr. Sekino was a research associate with the Department of Biomedical Engineering, Graduate School of Medicine, University of Tokyo, from 2005 to 2006. He was subsequently a research associate (2006–2010), a lecturer (2010–2011), and an associate professor (2011–2015) with the Department of Electrical Engineering and Information Systems, Graduate School of Engineering, University of Tokyo, Tokyo, Japan.

Dr. Sekino was awarded the International Union of Radio Science Young Scientist Award (2002), and the International Conference on Complex Medical Engineering Best Conference Paper Award (2012).

Contributors

Oscar Cespedes
School of Physics and Astronomy
University of Leeds
Leeds, United Kingdom

Norio Iriguchi
Center for Multimedia and
  Information Technology
Graduate School of Science and
  Technology
Kumamoto University
Kumamoto, Japan

Sunao Iwaki
Automotive Human Factors
  Research Center
National Institute of Advanced
  Industrial Science and
  Technology (AIST)
Tsukuba, Japan

Masaki Sekino
Department of Electrical
  Engineering and Information
  Systems
Graduate School of Engineering
University of Tokyo
Tokyo, Japan

Tsukasa Shigemitsu
Central Research Institute of
  Electric Power Industry
Tokyo, Japan

Shoogo Ueno
Department of Biomedical
  Engineering
Graduate School of Medicine
University of Tokyo
Tokyo, Japan

and

Department of Applied Quantum
  Physics
Graduate School of Engineering
Kyushu University
Fukuoka, Japan

Sachiko Yamaguchi-Sekino
Division of Health Effects Research
National Institute of Occupational
  Safety and Health (JNIOSH)
Kawasaki, Japan

CHAPTER 1

Introduction

Shoogo Ueno

CONTENTS

1.1 HISTORY OF BIOMAGNETICS

1.1.1 From Gilbert (1600) to d'Arsonval (1896)

Biomagnetics is an interdisciplinary field where magnetics, biology, and medicine overlap. Biomagnetics is also called biomagnetism, which is often used in medicine and biology. In this book, however, biomagnetics and biomagnetism are used interchangeably. Biomagnetics has a long history, beginning in 1600 when William Gilbert (1544–1603) published his groundbreaking book *De Magnete*.[1] In it, he stated that the Earth is itself a huge magnet (Figure 1.1). Gilbert's illustration describes that the origin of Earth's magnetic fields should exist inside the Earth. He also noted the relationship between magnetism and human life:

> Magnetic force is animate or imitates life; and in many things surpasses human life, while this is bound up in the organic body

WILLIAM GILBERT, 1600

William Gilbert, Father of Magnetism
"The Earth is itself a huge magnet"

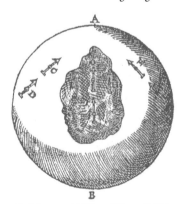

De Magnete, William Gilbert (1600)

FIGURE 1.1   William Gilbert (1544–1603) and the Earth.

The original book was written in Latin, and the above quote is taken from Busby.[2]

Gilbert's hypothesis is very thought-provoking and suggests the study of interactions between magnetism and human life. It was difficult, however, to verify Gilbert's profound hypothesis in 1600 because science and technology were not well developed at that time. Gilbert was a physicist as well as a physician for Queen Elizabeth I (1533–1603) in England.

In France in 1896, Jacques-Arsene d'Arsonval (1851–1940) reported a phenomenon called magnetophosphene, a visual sensation induced in human subjects when exposed to alternating magnetic fields.[3] The magnetophosphene reported by d'Arsonval was among the first reliable phenomena related to the interactions of human life with magnetism (Figure 1.2).

The magnetophosphenes were repeated by other groups; for example, S. P. Thompson in the United Kingdom observed that peripheral areas in the eye see light more clearly than in the center of the eye.[4] C. E. Magnusson and H. C. Stevens in the United States observed frequency characteristics in light sensation, changing the frequency of alternating magnetic fields, and determined that the light is sensitive around 20 Hz.[5] P. Lovsund, P. A. Oberg, and S. E. G. Nilsson in Sweden quantitatively

FIGURE 1.2   Jacques-Arsene d'Arsonval (1851–1940). (From Wikipedia, the free encyclopedia, Jacques-Arsene d'Arsonval.)

studied the magnetophosphenes in a dark room and observed that the threshold for light sensation is 10 mT at 20 Hz.[6] Magnetophosphene is a good model for evaluating biological effects of extremely low frequency (ELF) magnetic fields, and this phenomenon is often used for safety guidelines in international organizations such as the International Commission on Non-Ionizing Radiation Protection (ICNIRP) and the International Electrotechnical Commission (IEC).

1.1.2 Modern Biomagnetics in the 1970s and 1980s

In 1970, a special issue on biomagnetics was published in *IEEE Transactions on Magnetics*.[7] This issue is comprised of selected papers discussed at an international symposium on the application of magnetism in bioengineering that was held in Rehovot, Israel, and organized by E. H. Frei at Weizmann Institute of Science in 1969. It is amazing that most of the important topics in biomagnetics are discussed in this special issue. These topics include measurements of magnetic fields produced from the human heart and brain, theoretical study on magnetocardiography (MCG) and magnetoencephalography (MEG), the possibility of magnetic stimulation of peripheral nerves and heart muscles, contrast medium of barium meal mixed with magnetic materials for enhancing X-ray images, internal magnetic capsules driven and guided by external pulsed magnetic fields, and biological effects of magnetic fields. The fundamental framework of biomagnetics for medical applications was introduced in this issue.

In the 1970s, modern biomagnetics accelerated along with developments of new technologies in quantum physics, superconducting physics, and electronic engineering; for example, superconducting quantum interference device (SQUID) systems and magnetic resonance imaging (MRI) enabled us to accelerate brain research and medical applications.

During the period when SQUID technology was not yet available, G. Baule and R. McFee[8] measured MCG signals by a flux gate magnetometer with a pair of ferrite cores wound by two millions turns of coils in 1963. Subsequently, D. Cohen[9] measured spontaneous MCG activity by a SQUID system in magnetically shielded room at the Massachusetts Institute of Technology in 1970, and his group[10] successfully measured the so-called ST segment shift in MCG direct current signals related to the injury current produced by coronary artery occlusion. D. Cohen[11] also measured a spontaneous MEG activity related to alpha rhythmic activity without averaging the signals in 1972.

After the success of Cohen's MEG study, magnetic fields evoked by visual, somatic, and auditory stimuli were measured by different groups. For example, D. Brenner and his coworkers[12] measured visually evoked magnetic fields in the human brain in 1975, and they also measured somatically evoked magnetic fields[13] in 1978. M. Reite and his coworkers[14] measured auditory evoked magnetic fields in 1978, and G. L. Romani and his coworkers[15] observed the tonotopic organization of the human auditory cortex by MEG study in 1982. These pioneering studies opened a new horizon in functional brain research called neuromagnetism in the 1980s.[16]

Magnetic stimulation of nerves and muscles was studied in the 1970s to identify ways to avoid undesirable problems in electrical stimulation through electrodes such as decaying of stimulatory effects due to the electrochemical reactions at the interface between electrodes and living tissues, by D. D. Irvin and his group,[17] J. A. Maass and M. M. Asa[18] in 1970, and by P. A. Oberg[19] in 1973.

In magnetic nerve stimulation, Maass and Asa[18] proposed a transformer type of stimulation in which a nerve bundle was threaded through a magnetic core as the secondary winding. These authors showed that the flux change in the core could be used to excite nerves. Oberg[19] proposed an air gap type of stimulation in which a bundle of nerves was exposed to alternating magnetic fields. These studies demonstrated experimentally that induced eddy currents in the membrane tissues could be expected to stimulate nerves, though the underlying nerve-excitation process from magnetic stimulation was not understood. S. Ueno and his coworkers[20] pointed out that a capacitive stimulatory effect could contribute to magnetic nerve stimulation in 1978.

To shed light on the basic excitation mechanisms, studies of a single nerve axon were carried out by S. Ueno and his coworkers, focusing on the effects of time-varying magnetic fields on the action potential in lobster giant axon.[21,22] The results obtained from the lobster experiments suggested that nerve excitation by magnetic field influence is mediated via the induction of eddy currents in the tissue surrounding the nerve. The macroscopic eddy currents that flow along the nerve and in the tissues surrounding the nerve are important, as they contribute to the depolarization of the membrane.[21,22] After the study of single nerve axons, a new type of magnetic nerve stimulation without interlinkage between the nerve and magnetic flux was proposed by S. Ueno and his coworkers[23] in 1984.

Basic studies of magnetic nerve stimulation have opened a new area of research for transcranial magnetic stimulation (TMS) of human subjects in the middle of the 1980s. Stimulation of human brain by TMS with a round coil was reported by A. T. Barker and his coworkers[24] in 1985. The success of the human brain stimulation by TMS made a strong impact on the scientific community. By TMS with a round coil, however, it was difficult to stimulate a localized area of the brain. A method of localized brain stimulation by TMS with a figure-eight coil was proposed by S. Ueno and his coworkers[25] in 1988, and stimulation of the human motor cortex within a 5-mm resolution was successfully realized in 1989.[26–28] The functional map of the human motor cortex was obtained in a 5-mm resolution by this method in 1990.[29] TMS with a figure-eight coil is now used worldwide in cognitive brain research and clinical medicine.[30]

It is also remarkable that a magnetotactic bacteria was discovered by R. P. Blakemore in 1975.[31] Blakemore observed that the magnetotactic bacteria swim along Earth's magnetic field. Blakemore and his colleagues identified a chain of crystals of magnetite (Fe_3O_4) that exist inside the body of magnetotactic bacteria.[32,33] Magnetotactic bacteria can sense magnetic torque acting on the angle between the direction of Earth's magnetic field and a long chain of magnetites inside the body so as to minimize the magnetic torque during swimming. The discovery of magnetotactic bacteria triggered the studies of biomineralization and biomagnetic sensing mechanisms in fish, birds, honeybees, and other animals.[34–36] The interactions of biological systems with a weak magnetic field or the Earth's magnetic field with a 30–50 μT order are interesting because all living species on Earth are exposed to the its magnetic fields.

In contrast, magnetic orientation of fibrin was observed when the polymerization process from fibrin monomer to fibrin polymer was exposed to a very strong magnetic field of 11 T by J. Torbet in 1981.[37] Through this study, the importance of diamagnetism was recognized in biological effects of strong magnetic fields.

Effects of magnetic fields on chemical reactions were observed in the 1970s. Magnetic field effects on photochemical reactions were observed in different groups[38–40] in 1976. Photochemical reactions produced by a radical-pair intermediate in a solution can be expected to show magnetic field effects that arise from an electron Zeeman interaction, electron-nuclear hyperfine interaction, or hyperfine interaction mechanism including an electron-exchange interaction in a radical-pair intermediate. That is, a common mechanism of magnetic field effects on chemical processes is that a chemical yield of the cage or escape product shows a magnetic field dependence when a singlet–triplet intersystem crossing is subject to magnetic perturbations.[41,42] This area has been well developed as dynamic spin chemistry in the 1980s and 1990s.[41,42] Possible biomagnetic and chemical effects can be expected when biological systems are exposed to both magnetic fields and other energy such as light and radiation.[43] However, biological effects of magnetic fields based on dynamic spin chemistry with a radical-pair model have not yet been clarified.[44]

Basic studies on biomagnetics and the related fields in the 1970s and 1980s have been explored for further developments in the 1990s and 2000s.

1.2 BIOMAGNETIC PHENOMENA AND MEDICAL APPLICATIONS

1.2.1 Biomagnetic Phenomena at Different Intensities and Their Frequencies

Various biomagnetic phenomena for different intensities and their frequencies are shown in Figure 1.3.[45,46]

Magnetic stimulation of the brain, (i.e., transcranial magnetic stimulation [TMS] of the human brain), requires strong pulsed magnetic fields

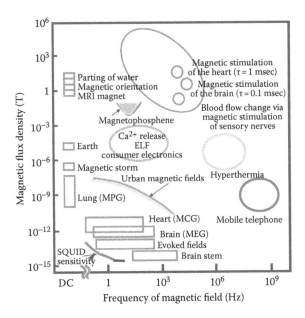

FIGURE 1.3 Various biomagnetic phenomena observed or used at different frequencies (Hz) and magnetic field intensities (T).

of 1–2 T for a short period of time (0.1–0.2 msec) to get sufficient induced electric fields for the stimulation of neurons in the brain. Magnetic stimulation of the heart requires more strong pulsed magnetic fields (1–5 T) with a 1–2 msec pulse width to excite the cardiac muscles. In Figure 1.3, the frequencies of the pulsed magnetic fields are converted from pulse width.

Rapid changes in blood flow in the hands of human subjects are observed when the hand is exposed to alternating magnetic fields of 32 mT at 3.8 kHz.[47] The explanation for the blood flow changes following magnetic field exposure is that nerve system information from the skin sensory receptors is sent via the spinal cord to the central nervous system, which modulates the efferent nerve activity and affects the blood flow of the skin.

The magnetophosphene shows a V-shaped curve for light sensation with a minimum threshold of 10 mT at 20 Hz.[6]

Effects of magnetic fields on biological systems are observed mostly in the range of magnetic fields higher than the Earth's magnetic field of 30–50 µT.

Health effects of extremely low frequency (ELF) electromagnetic fields are discussed, introducing possible mechanisms including a model of calcium release from living cells exposed to ELF fields. Amplitudes and frequencies of ELF electromagnetic fields related to consumer electronics and power lines are in the oval area marked in Figure 1.1.

Magnetic fields used in clinical MRI systems today are 0.3 T for permanent magnets and 1.5–3.0 T for superconducting magnet systems. For research purposes, 4–7 T MRI systems are used. Magnetic orientation of biological materials is observed in 4–8 T magnetic fields. "Parting of water" or the "Moses effect" is observed in 8-T magnetic fields.

In contrast, biomagnetic fields or magnetic fields produced from the living systems are extremely weak, 10^{-15} to 10^{-11} T or 1 fT to 10 pT order of magnetic fields, buried with urban noise. These biomagnetic fields are usually measured by superconducting quantum interference device (SQUID) systems in a magnetically shielded room.

1.2.2 Biomagnetic Measurements of Magnetic Fields Produced by Living Systems

Magnetoencephalography (MEG) is a most promising tool for studying functional organization of the brain noninvasively with a high spatial (millimeter) and high temporal (millisecond) resolution. The magnetic signals are 10^{-14} to 10^{-12} T or 10 fT to 1 pT order of magnetic fields for evoked and spontaneous brain electrical activities. The magnetic fields produced from the human brain stem evoked by auditory stimuli, called auditory evoked brain stem magnetic fields, are extremely weak, 5×10^{-15} T or 5 fT magnetic fields, and these magnetic fields are detected by SQUID systems by averaging several thousand times above signal-to-noise ratio. Spontaneous brain magnetic activities such as alpha MEG activities of 10^{-12} T or pT order of magnetic fields are detected by SQUID systems without averaging the signals.

Magnetocardiography (MCG) is a noninvasive modality for detecting electrical activities of the heart, and the biomagnetic signals are 10^{-12} to 10^{-10} T or 1–100 pT order of magnetic fields.

Magnetopneumography (MPG) is a technique to measure magnetic fields from the lung.[48,49] The MPG measures remnant magnetic fields produced by magnetic contaminants in the lung or deposition of inhaled magnetic nanoparticle dusts in the lung. The lung is magnetized by external magnetic fields and a relaxation process from a peak of the magnetized vector of MPG signals to a low level of signals due to randomization of the magnetized vector. The relaxation process is reflected by the ability of phagocytosis of alveolar macrophages.

Origins or sources of magnetic fields of MEG and MCG are electrical currents generated by electrical activities of neurons and muscles, whereas the magnetic fields of MPG come from magnetic dust in and around alveolar macrophages in the lungs.

In biomagnetic measurements, spatiotemporal distributions of magnetic fields over the surface of the body are obtained by multichannel SQUID systems, and the sources of biomagnetic fields are estimated by solving inverse problems. From a mathematical point of view, there are no unique solutions in inverse problems, and localizations of sources are estimated by adding some assumptions and boundary or constraint conditions.

1.2.3 Biomagnetic Stimulation and Magnetic Treatments

Magnetic stimulation of the brain requires around 1–2 T of pulsed magnetic fields with 0.1–0.2 msec pulse width. Magnetic stimulation of the heart requires around 2–4 T of pulsed magnetic fields with 1–2 msec duration.[50,51] The pulse width of stimulation is determined by the chronaxie in the strength-duration curve, which is roughly equal to the time constant of the excitation of electrical characteristics of membrane tissues. The optimal pulse duration for membrane excitation is around 0.1–0.2 msec for nerve tissues and is around 1–2 msec for cardiac muscles. The rapid changes in these pulsed magnetic fields induce electric fields in the living body, and the nervous and muscular systems are stimulated by the magnetically induced electric fields.

For therapeutic applications of magnetic brain stimulation, rapid-rate transcranial magnetic stimulation or repetitive TMS (rTMS) with a series of repetitive pulses at several or several tens of hertz is used, for example, as an alternative approach to induce seizure for the treatment of depression in place of electroconvulsive therapy (ECT).[52,53] This approach could allow selective stimulation of brain sites that are involved in depression, thus reducing side effects (e.g., memory deficits) due to electrical disruption of function in unrelated sites.[52,53] The technique of rTMS has promising potential applications not only for the treatment of depression but also for Parkinson's disease, dementia, and other brain diseases, as well as for reduction of intractable pain and rehabilitation or recovery process of an injured brain after a stroke.

From the viewpoint of a therapeutic application of rTMS for neurological disorders, a number of animal studies testing the basic mechanisms of rTMS-induced alternations of neurotrophic factors, gene expression, and changes in plasticity have been conducted. The experimental results obtained by M. Fujiki and O. Stewart[54] and by other groups[55–58] suggest that there is strong evidence that the expression of certain genes such as the immediate early gene, astrocyte-specific glial fibrillary acidic protein (GFAP) messenger ribonucleic acid (mRNA)[54] and brain-derived neurotropic factor (BDNF)[55] are altered in response to rTMS. This indicates that the measurable effects of rTMS reach the cellular and molecular levels.[54,58] The effects of rTMS on neuronal electric activities in the rat hippocampus were studied, focusing on long-term potentiation (LTP) by M. Ogiue-Ikeda and her colleagues.[56,57]

Risk and safety aspects of rTMS with pulse trains of high frequencies need to be investigated.[59,60]

1.2.4 Magnetic Resonance Imaging of the Living System

Magnetic resonance imaging (MRI) is an indispensable tool in medicine. The Nobel Prize in Physiology and Medicine in 2003 was awarded to Paul Christian Lauterbur (1929–2007) of the United States and Peter Mansfield (1933–) of the United Kingdom, for the discovery of principles of nuclear magnetic resonance imaging. P. C. Lauterbur proposed a principle of imaging by induced local interactions in magnetic resonant frequencies with linear gradient magnetic fields in addition to the main static magnetic field in 1973.[61] P. Mansfield developed pulse sequences for imaging such as echo planar imaging in 1977.[62] MRI is a noninvasive imaging tool, and because of its variety of potential modality, MRI has evolved and is still evolving.

Seiji Ogawa (1934–)[63] of Japan proposed a functional MRI (fMRI) based on blood oxygenation level dependent (BOLD) effects to visualize functional organizations in the brain in 1990. This BOLD-fMRI is a powerful tool for functional brain research and clinical medicine because it requires no injection of contract agents into human subjects.

A method of diffusion weighted imaging (DWI) was studied in the 1990s by different groups including D. Le Bihan and coworkers,[64] and through a theoretical study of estimation of the effective self-diffusion tensor from NMR spin echo by P. J. Basser and his coworkers,[65] the DWI has developed a new imaging tool called the diffusion tensor imaging (DTI) to visualize structural networks of neural fibers in the brain in the 2000s.

In 1998, S. Ueno and N. Iriguchi[66] proposed a method to visualize electrical impedance distributions in the living body called impedance MRI. H. Kamei and his coworkers[67] proposed a method to image neuronal currents in the brain called current MRI in 1999. The imaging of electrical currents and impedance distributions in the head based on new principles of MRI will be very important for studying brain functions of humans.[68] The signal-to-noise ratio is essentially low in current MRI, and the issues of fundamental factors such as a sensitivity limit, RF inhomogeneity, the dielectric resonance effect, and so on need to be investigated. M. Sekino and his coworkers estimated a theoretical limit of sensitivity of 10^{-8} to 10^{-9} T for current MRI and obtained a transient decrease in signal intensity in the rat brain using a 4.7-T MRI system.[69]

Recent advances in superconducting technology have enabled us to use high-field MRI systems. The 1.5-T and 3-T MRI systems are used in clinical diagnoses, and 4- and 7-T MRI systems are used for human research.

Ultrahigh 11-T MRI systems for human subjects are also being developed. Safety aspects of potential health effects of these ultrahigh field MRI systems need to be clarified.

1.2.5 Magnetic Control of Cell Manipulation, Cell Orientation, and Cell Growth

A phenomenon, called parting of water or Moses effect, where water is divided into two parts by magnetic fields of up to 8 T, was observed by S. Ueno and M. Iwasaka[70,71] in 1994. When a horizontally positioned half-filled water chamber is exposed to 8-T magnetic field with a field gradient of 50 T/m, water is parted in two making two water walls, and the bottom of the water chamber appears in between water walls.[70,71] This effect will be useful in manipulating biological cells magnetically.

Magnetic alignment or magnetic orientation of biological materials is observed when the materials are exposed to strong static magnetic fields of a 5–10 T order.[37] Fibrin,[37] collagen, and other adherent cells such as osteoblasts, vascular endothelial cells, smooth muscle cells, and Schwann cells orient either parallel or perpendicular to magnetic fields, depending on the anisotropy of magnetic susceptibility of biological materials. Using this effect, for example, bone growth acceleration is observed when samples of mixture of bone morphogenetic protein 2 (BMP-2) and collagen are exposed to 8-T magnetic fields for 60 hours in the beginning period in bone formation.[72] The magnetic orientation of biological materials has promising potential applications for tissue engineering and regenerative medicine where the cells or tissues should be aligned in preferred directions.[73,74]

1.2.6 Biological Interactions of ELF, Pulsed, and RF Magnetic Fields

There are many studies on the interactions of extremely low frequency (ELF) magnetic and electromagnetic fields with biological systems. The low-energy level ELF magnetic and electromagnetic fields are produced by consumer electronics and power lines.

Hyperthermia is a medical method for cancer therapy where cancer cells are heated by electromagnetic induction. Magnetic nanoparticles are often used for hyperthermia to increase heat effectively around the targeted lesions. New biomagnetic approaches to cancer therapy by pulsed magnetic fields, without cell heating, are also studied.[75,76]

Studies of possible health effects of radio frequency (RF) electromagnetic fields have become important because of the rapid increase in mobile telephones and the rapid increase in high-field MRIs. Different approaches to the relationship between human health and mobile telephony are needed, from biological cellular level study (i.e., *in vitro* study and animal study, *in vivo* study and human study, or epidemiological study). In parallel to these animal and human studies, physical and numerical studies or dosimetry studies are essential to determine the level and distributions of energy absorbed in the living body exposed to the RF fields.

1.3 FUNDAMENTAL BASES FOR UNDERSTANDING BIOMAGNETIC PHENOMENA

1.3.1 Electromagnetic Induction in Living Tissues

Biological effects of magnetic and electromagnetic fields are classified into three categories:[43] (1) effects of time-varying magnetic fields, (2) effects of static magnetic fields, and (3) effects of multiplications of both static magnetic fields and other energy such as light and radiation.

When we discuss the effects of time-varying magnetic fields, the relationship between magnetism and electricity is important. Hans C. Oersted (1777–1851) discovered a phenomenon that electric currents produce magnetic fields in 1820 in Denmark. Michael Faraday (1791–1867) discovered a phenomenon called electromagnetic induction in 1831 in England. Electric currents are induced in the secondary coil when the currents in the primary coil change rapidly. In other words, electric fields are induced by time-varying magnetic fields. James C. Maxwell (1831–1879) derived the fundamental four equations, called Maxwell's equations, to describe the relationship between magnetism and electricity mathematically in 1864. That is, electromagnetic phenomena consist of electric and magnetic fields that change with time and space. Maxwell's equations are laws that define the relationship between temporally and spatially averaged electric and magnetic fields.

Because the living body is composed of conductive tissues and materials, electric fields are induced in it when it is exposed to time-varying magnetic fields. In the case of low-frequency magnetic fields lower than several 100 kHz or pulsed magnetic fields, stimulatory effects of excitable membranes such as nervous system and muscular system are dominant. Transcranial magnetic brain stimulation is realized by this principle.

In contrast, in the case of high-frequency magnetic and electromagnetic fields higher than several 100 kHz and in the range of RF fields, thermal effects of living systems are dominant. Health effects of RF electromagnetic fields used in mobile telephony and MRI systems are discussed mainly on the basis of the thermal effects. The specific absorption rate (SAR) (W/kg) is used for the evaluation of biological effects of RF fields.

When an electric field E (V/m) is induced in a tissue exposed to RF electromagnetic fields, SAR (W/kg) in the tissue is determined by

$$SAR = \sigma E^2 / \rho_m$$

where σ is electrical conductivity (S/m) and ρ_m is the mass density (kg/m^3) of tissue. The induced electric field E (V/m) is dependent on external RF fields and electromagnetic properties of tissues represented by permittivity (dielectric constant and electrical conductivity).

Mechanisms of biological effects of magnetic and electromagnetic fields are summarized in Table 1.1.[46]

TABLE 1.1 Mechanisms of Biological Effects of Electromagnetic Fields

Type of Field	Formulae	Examples
1. Time-varying magnetic field		
a. Low frequencies Eddy currents	$J = -\sigma \dfrac{\partial B}{\partial t}$	Nerve stimulation
b. High frequencies Heat	$\mathrm{SAR} = \sigma \dfrac{E^2}{\rho}$	Thermal effects
2. Static magnetic field		
a. Homogeneous magnetic field Magnetic torque	$T = -\dfrac{1}{2\mu_0} B^2 \Delta\chi \sin 2\theta$	Magnetic orientation of biological cells
b. Inhomogeneous magnetic field Magnetic force	$F = \dfrac{\chi}{\mu_0}(\mathrm{grad}B)B$	Parting of water by magnetic fields (Moses effect)
3. Multiplication of magnetic fields and other energy		
a. Photochemical reactions with radical pairs		Yield effect of cage product and escape product
b. Singlet–triplet intersystem crossing		

1.3.2 Magnetism of Living Systems and Materials

When we discuss biological effects of static magnetic fields, magnetic properties of living systems and materials are important to understand the biomagnetic effects. From the magnetism viewpoint, all materials are classified into three categories: diamagnetic, paramagnetic, and ferromagnetic materials. The origins of magnetic properties of these materials are explained by behaviors of magnetic moments and spins of nuclei and electrons where interactions between neighboring spins are characterized by an exchange interaction in quantum physics. In this section, however, we discuss the magnetic properties by phenomenological classification as follows.

Diamagnetic materials are magnetized in the opposite direction of applied magnetic fields when the materials are exposed to the external magnetic fields. The magnetization is extremely weak, and the magnetization disappears reversibly when the external magnetic fields disappear. Diamagnetic materials are almost nonmagnetic materials and "transparent" for magnetic fields in usual daily life. Most of the human body is composed of diamagnetic materials. Water is a typical example of diamagnetism. Oxyhemoglobin is also a diamagnetic material. Diamagnetic materials are only slightly subject to the opposed direction of magnetic force. However, if magnetic force is very strong with a strong magnetic field of Tesla order and a high gradient magnetic field of several 10 T/m order, we can observe a visible phenomenon such as parting of water or the Moses effect.[70,71] The magnetic force acting on materials is proportional to the product value of the magnetic susceptibility of the materials and magnetic field and the spatial gradient of magnetic field. Although the magnetic susceptibility of water is very small, magnetic force acting on water is strong enough to push back water when the product values of the magnetic field and the spatial gradient of magnetic field are very strong.

One of the important characteristics of diamagnetic materials is anisotropy of magnetic susceptibility of materials. When diamagnetic materials are exposed to external magnetic fields, the materials act to rotate to a preferred direction so as to minimize magnetic torque induced by the anisotropy of magnetic susceptibility of the materials. Magnetic orientation of biological materials is observed by this principle.

Paramagnetic materials are magnetized in the same direction of applied magnetic fields when the materials are exposed to the external magnetic fields. The magnetization is not strong, and the magnetization disappears reversibly when the external magnetic fields disappear. Paramagnetic materials are attracted to magnetic force, but the attractive force is not strong compared with the force in the case of ferromagnetic materials. Oxygen is a paramagnetic molecule. Deoxyhemoglobin is a paramagnetic material. Free radicals that are produced transiently and dynamically on the processes of chemical reactions are paramagnetic species.

Using the paramagnetic property of oxygen, extinguishing a candle flame by magnetic fields was demonstrated by S. Ueno in 1989.[77] Candle flames are extinguished by a closed curtain of paramagnetic oxygen gas called the "magnetic curtain." When candle flames are captured by the magnetic curtain, they produce magnetic fields of 1.0–1.5 T with a high gradient magnetic field of 100–300 T/m.

The BOLD-fMRI by S. Ogawa[63] visualizes the changes in MRI signals based on the changes in magnetic property of paramagnetic deoxyhemoglobin and diamagnetic oxyhemoglobin in blood in capillaries and veins in the brain.

Ferromagnetic materials are magnetized strongly in the same direction of applied magnetic fields when the materials are exposed to the external magnetic fields. The magnetization is strong, and magnetization does not disappear irreversibly when the external magnetic fields disappear. Ferromagnetic materials are attracted by magnetic force strongly.

Magnets and iron nails are ferromagnetic materials. Iron, nickel, and cobalt are typical ferromagnetic metallic elements.

Ferrimagnetic materials are in the category of ferromagnetic materials and are strongly attracted by magnetic force. Magnetites (Fe_3O_4), ferrites, and many other iron oxides are ferrimagnetic materials.

When the magnetic materials are nanostructured below the single domain size, the materials behave as superparamagnetic particles. The superparamagnetic particles have a large magnetic susceptibility well above that of conventional paramagnetic materials. The superparamagnetic materials act as important roles in magnetic nanobioscience and biomagnetics. The effects of RF magnetic fields on iron ion release and uptake from and into iron cage proteins are investigated, focusing on the role of superparamagnetic ferrihydrate nanoparticles inside ferritins.[78]

1.4 SUMMARY

In this biomagnetics overview, we study biomagnetics, an interdisciplinary field. We focus on basic principles and medical applications of biomagnetic stimulation and imaging in the following chapters. We hope the readers enjoy the essence of recent advances in biomagnetics in this book. We recommend that readers refer to other bibliographies related to biomagnetics, bioelectromagnetics, brain science, neuroimaging, and regenerative medicine, as well as basic sciences to expand their exposure to and knowledge of the world of biomagnetics. We provide some other review articles, book chapters, and books.[79–89]

REFERENCES

1. Gilbert, W. 1600. *De Magnete*. Translation to English by Mottelay (1958). Dover, New York.
2. Busby, D. E. 1968. Space biomagnetics. *Space Life Science* 1: 23–63.
3. D'Arsonval, J. A. 1896. Disrositifs pour la mesure des courants alternatives des toutes frequencies. *Comptes Rendus l'Académie des Sciences* 48: 450–451.
4. Thompson, S. P. 1910. A physiological effect of an alternating magnetic field. *Proceedings of the Royal Society, Series B* 82: 396–398.
5. Magnusson, C. E., and Stevens, H. C. 1911. Visual sensations caused by changes in the strength of a magnetic field. *American Journal of Physiology* 29: 124–136.
6. Lovsund, P., Oberg, P. A., and Nilsson, S. E. G. 1980. Magnetophosphenes: A quantitative analysis of thresholds. *Medical and Biological Engineering and Computing* 28: 326–334.

7. Frei, E. H. (Ed.) 1970. Introduction to the symposium on application of magnetism in bioengineering. *IEEE Transactions on Magnetics* **MAG-6**: 307–375.

8. Baule, G., and McFee, R. 1963. Detection of the magnetic field of the heart. *American Heart Journal* **66**: 95–97.

9. Cohen, D. 1970. Review of measurements of magnetic fields produced by natural ion currents in humans. *IEEE Transactions on Magnetics* **MAG-6**: 344–345.

10. Cohen, D., and Kaufman, L. A. 1975. Magnetic detection of the relationship between the S-T segment shift and the injury current produced by coronary artery occlusion. *Circulation Research* **36**: 414–424.

11. Cohen, D. 1972. Magnetoencephalography: Detection of the brain's electrical activity with a superconducting magnetometer. *Science* **175**: 664–666.

12. Brenner, D., Williamson, S. J., and Kaufman, L. 1975. Visually evoked magnetic fields of the human brain. *Science* **190**: 480.

13. Brenner, D., Lipton, J., Kaufman, L., and Williamson, S. J. 1878. Somatically evoked magnetic fields of the human brain. *Science* **199**: 81–83.

14. Reite, M., Edrich, J., Zimmerman, J. T., and Zimmerman, J. E. 1978. Human magnetic auditory evoked fields. *Electroencephalography and Clinical Neurophysiology* **40**: 59–66.

15. Romani, G. L., Williamson, S. J., and Kaufman, L. 1982. Tonotopic organization of the human auditory cortex. *Science* **216**: 1339–1340.

16. Williamson, S. J., and Kaufman, L. 1983. Application of SQUID sensors to the investigation of neural activity in the human brain. *IEEE Transactions on Magnetics* **MAG-19**: 835–844.

17. Irwin, D. D., Rush, S., Evering, R., Lepeschkin, E., Montgomery, D. B., and Weggel, R. J. 1970. Stimulation of cardiac muscle by a time-varying magnetic field. *IEEE Transactions on Magnetics* **MAG-6**: 321–322.

18. Maass, J. A., and Asa, M. M. 1970. Contactless nerve stimulation and signal detection by inductive transducer. *IEEE Transactions on Magnetics* **MAG-6**: 322–326.

19. Oberg, P. A. 1973. Magnetic stimulation of nerve tissue. *Medical and Biological Engineering* **11**: 55–64.

20. Ueno, S., Matsumoto, S., Harada, K., and Oomura, Y. 1978. Capacitive stimulatory effect in magnetic nerve stimulation of nerve tissue. *IEEE Transactions on Magnetics* **MAG-14**: 958–960.

21. Ueno, S., Lovsund, P., and Oberg, P. A. 1981. On the effects of alternating magnetic fields on action potential in lobster giant axon. In *Proceedings of the 5th Nordic Meeting on Medical and Biological Engineering*, Linkoping, Sweden and the 25th Anniversary, Swedish Society for Medical Physics and Medical Engineering, Umea, Sweden: 262–264.

22. Ueno, S., Lovsund, P., and Oberg, P. A. 1986. Effect of time-varying magnetic fields on the action potential in lobster giant axon. *Medical and Biological Engineering and Computing* **24**: 521–526.

23. Ueno, S., Harada, K., Ji, C., and Oomura, Y. 1984. Magnetic nerve stimulation without interlinkage between nerve and magnetic flux. *IEEE Transactions on Magnetics* **MAG-20**: 1660–1662.
24. Barker, A. T., Jalinous, R., and Freeston, I. L. 1985. Non-invasive magnetic stimulation of human motor cortex. *The Lancet* **i**: 1106–1107.
25. Ueno, S, Tashiro, T., and Harada, K. 1988. Localized stimulation of neural tissues in the brain by means of a paired configuration of time-varying magnetic fields. *Journal of Applied Physics* **64**: 5862–5864.
26. Ueno, S., Matsuda, T., Fujiki, M., and Hori, S. 1989. Localized stimulation of the human motor cortex by means of pair of opposing magnetic fields. *Digest of International Magnetics Conference, Washington, D.C.*: GD-10.
27. Ueno, S., Matsuda, T., and Fujiki, M. 1989. Localized stimulation of the human cortex by opposing magnetic fields. In *Advances in Biomagnetism*, Williamson, S. J., Hoke, M., Stroink, G., and Kotani, M., eds. Springer, New York: 529–532.
28. Ueno, S., Matsuda, T., and Fujiki, M. 1989. Localized stimulation of the human brain by a pair of opposing pulsed magnetic fields. *Memoirs of the Faculty of Engineering, Kyushu University* **49**: 161–173.
29. Ueno, S., Matsuda, T., and Fujiki, M. 1990. Functional mapping of the human motor cortex obtained by focal and vectorial magnetic stimulation of the brain. *IEEE Transactions on Magnetics* **26**: 1539–1544.
30. Ueno, S. (Ed.) 1994. *Biomagnetic Stimulation*. Plenum Press, New York: 1–136.
31. Blakemore, R. P. 1975. Magnetotactic bacteria. *Science* **190**: 377–379.
32. Kalmijn, A. J., and Blakemore, R. P. 1977. Geomagnetic orientation in marine mud bacteria. *Proceedings of International Union of Physiological Society* **13**: 364.
33. Frankel, R. B., Blakemore, R. P., and Wolf, R. S. 1979. Magnetite in freshwater magnetotactic bacteria. *Science* **203**: 1355–1356.
34. Kirshvink, J. L., Jones, D. S., and MacFadden, B. J. (Eds.) 1985. *Magnetite Biomineralization and Magnetoreception in Organisms: A New Biomagnetism*. Plenum Press, New York: 1–682.
35. Kirshvink, J. L., Kobayashi-Kirshvink, A., and Woodford, B. J. 1992. Magnetite biomineralization in the human brain. *Proc. Natl. Acad. Sci., USA (Biophysica)* **89**: 7683–7687.
36. Matsunaga, T., Nakamura, C., Burgess, J. G., and Sode, K. 1992. Gene transfer in magnetic bacteria; transposon mutagenesis and cloning of genomic DNA fragments for magnetosome synthesis. *Journal of Bacteriology* **174**: 2748–2753.
37. Torbet, J, Freyssinet, M., and Hudy-Clergeon, G. 1981. Oriented fibrin gels formed by polymerization in strong magnetic fields. *Nature* **289**: 91–93.
38. Hata, N. 1976. The effect of external magnetic field on the photochemical reaction of isoquinoline N-oxide. *Chemical Letters* **5**: 547–550.
39. Shulten, K., Staek, H., Welter, A., Wernar, H. J., and Nickel, B. 1976. Magnetic field dependence of the geminate recombination of radical ion pairs in polar solvents. *Zeitschrift für Physikalische Chemie* **101**: 371–390.

40. Tanimoto, Y., Hayashi, H., Nagakura, S., Sakaguchi, H., and Tokumaru, K. 1976. The external magnetic field effect on the singlet sensitized photolysis of dibenzoyl peroxide. *Chemical and Physical Letters* **41**: 267–269.

41. Nagakura, S., and Molin, Y. (Eds.) 1992. Magnetic field effects upon photophysical and photochemical phenomena. *Chemical Physics, Special Issue* **162**: 1–234.

42. Nagakura, S., Hayashi, H., and Azumi, T. (Eds.) 1998. *Dynamic Spin Chemistry: Magnetic Controls and Spin Dynamics of Chemical Reactions.* John Wiley, New York: 1–297.

43. Ueno, S., and Harada, K. 1986. Experimental difficulties in observing the effects of magnetic fields on biological and chemical processes. *IEEE Transactions on Magnetics* **MAG-22**: 868–873.

44. Ueno, S. (Ed.) 1996. *Biological Effects of Magnetic and Electromagnetic Fields.* Plenum Press, New York: 1–243.

45. Ueno, S., and Iwasaka, M. 1996. Magnetic nerve stimulation and effects of magnetic fields on biological, physical and chemical processes. In *Biological Effects of Magnetic and Electromagnetic Fields*, Ueno, S., ed. Plenum Press, New York: 1–21.

46. Ueno, S., and Shigemitsu, T. 2006. Biological effects of static magnetic fields. In *Handbook of Biological Effects of Electromagnetic Fields*, 3rd ed. *Bioengineering and Biophysical Aspects of Electromagnetic Fields*, Barnes, F. S., and Greenebaum, B., ed. CRC Press, Boca Raton, Fla.: 203–259.

47. Ueno, S., Lovsund, P, and Oberg, P. A. 1986. Effects of alternating magnetic fields and low-frequency electrical currents on human skin blood flow. *Medical and Biological Engineering and Computing* **24**: 57–61.

48. Cohen, D. 1973. Ferromagnetic contamination in the lungs and other organs of the human body. *Science* **180**: 745–748.

49. Kalliomaki, P-L., Kalliomaki, K, Aittoniemi, K., Korhonen, O, Pasanen, J., and Moilanen, M. 1983. In *Biomagnetism: An Interdisciplinary Approach*, Williamson, S. J., Romani, G-L., Kaufman, L., and Modena, I., eds. Plenum Press, New York: 533–568.

50. Bourland, J. D., Mouchawar, G. A., Nyenhuis, J. A., Geddes, L., A., Foster, K. S., Jones, J. T., and Graber, G. P. 1990. Transchest magnetic (eddy-current) stimulation of the dog heart. *Medical and Biological Engineering and Computing* **28**: 196–198.

51. Bourland, J. D., Mounchawar, G. A., Nyenhuis, J. A., Geddes, L. A., Foster, K. S., Jones, J. T., and Graber, G. P. 1992. Closed-chest cardiac stimulation with a pulsed magnetic field. *Medical and Biological Engineering and Computing* **30**: 162–168.

52. Georg, M. S., and Wassermann, E. M. 1994. Rapid-rate transcranial magnetic stimulation and ECT. *Convulsive Therapy* **10**: 251–254.

53. Chandos, B., Khan, A., Lai, H., and Lin, J. C. 1996. The application of electromagnetic energy to the treatment of neurological and psychiatric diseases. In *Biological Effects of Magnetic and Electromagnetic Fields*, Ueno, S., ed. Plenum Press, New York: 161–169.

54. Fujiki, M., and Stewart, O. 1997. High frequency transcranial magnetic stimulation for protection against delayed neuronal death induced by transient ischemia. *Journal of Neurosurgery* **99**: 1063–1069.

55. Muller, M. B., Toschi, N., Kresse, A. E., Post, A., and Keck, M. E. 2000. Long-term repetitive transcranial magnetic stimulation increases the expression of brain-derived neurotrophic factor and cholecystokinin mRNA, but not neuropeptide tyrosine mRNA in specific areas of rat brain. *Neuropsychopharmacology* **23**: 205–215.

56. Ogiue-Ikeda, M., Kawato, S., and Ueno, S. 2003. The effect of repetitive transcranial magnetic stimulation on long-term potentiation in rat hippocampus depends on stimulus intensity. *Brain Research* **993**: 222–226.

57. Ogiue-Ikeda, M., Kawato, S, and Ueno, S. 2005. Acquisition of ischemic tolerance by repetitive transcranial magnetic stimulation in the rat hippocampus. *Brain Research* **1037**: 7–11.

58. Ueno, S., and Fujiki, M. 2007. Magnetic stimulation. In *Magnetism in Medicine A Handbook Second Completely Revised and Enlarged Edition*, Andra, W., and Nowak H., ed. Wiley-VHC, Weinheim: 511–528.

59. Wassermann, E. M. 1998. Risk and safety of repetitive transcranial magnetic stimulation: Report and suggested guidelines from the International Workshop on the Safety of Repetitive Transcranial Magnetic Stimulation, June 5-7, 1996. *Electroenchephalography and Clinical Neurophysiology* **108**: 1–16.

60. Pascual-Leone, A., Davey, N., Rothwell, J., Wassermann, E. M., and Puri, B. K. 2002. *Handbook of Transcranial Magnetic Stimulation*. Hodder Arnold, London.

61. Lauterbur, P. C. 1973. Image formation by induced local interactions: Examples employing nuclear magnetic resonance. *Nature* **242**: 190–191.

62. Mansfield, P. 1977. Multi-planar image formation using NMR spin echoes. *Journal of Physics C: Solid State Physics* **10**: L55–L58.

63. Ogawa, S., Lee, T. M., Kay, A. R., and Tank, D. W. 1990. Brain magnetic resonance imaging with contrast dependent on blood oxygenation. *Proceedings of the National Academy of Sciences USA* **87**: 9868–9872.

64. Le Bihan, D., Turner, R., Douek, P., and Patronas, N. 1992. Diffusion MR imaging: Clinical applications. *American Journal of Roentgenology* **159**: 591–599.

65. Basser, P. J., Mattiello, J., and Le Bihan, D. 1994. Estimation of the effective self-diffusion tensor from the NMR spin echo. *Journal of Magnetic Resonance B* **103**: 247–254.

66. Ueno, S., and Iriguchi, N. 1998. Impedance magnetic resonance imaging. *Journal of Applied Physics* **83**: 6450–6452.

67. Kamei, H., Iramina, K., Yoshikawa, K., and Ueno, S. 1999. Neuronal current distribution imaging using magnetic resonance. *IEEE Transactions on Magnetics* **35**: 4109–4111.

68. Ueno, S. 1999. Biomagnetic approaches to studying the brain. *IEEE Engineering in Medicine and Biology* **18**: 108–120.

69. Sekino, M., Ohsaki, H., Yamaguchi-Sekino, S., and Ueno, S. 2009. Toward detection of transient changes in magnetic-resonance signal intensity arising from neuronal electrical activities. *IEEE Transactions on Magnetics* **45**: 4841–4844.

70. Ueno, S., and Iwasaka, M. 1994. Properties of diamagnetic fluid in high gradient magnetic fields. *Journal of Applied Physics* **75**: 7177–7179.

71. Ueno, S., and Iwasaka, M. 1994. Parting of water by magnetic fields. *IEEE Transactions on Magnetics* **30**: 4698–4700.

72. Kotani, H., Kawaguchi, H., Shimoaka, T., Iwasaka, M., Ueno, S., Ozawa, H., Nakamura, K., and Hoshi, K. 2002. Strong static magnetic field stimulates bone formation to a definite orientation in vivo and in vitro. *Journal of Bone and Mineral Research* **17**: 1814–1821.

73. Ueno, S., Ando, J., Fujita, H., Sugawara, T., Jimbo, Y., Itaka, K., Kataoka, K., and Ushida, T. 2006. The state of the art of nanobioscience in Japan. *IEEE Transactions on Nanobioscience* **5**: 54–65.

74. Ueno, S., and Sekino, M. 2006. Biomagnetics and bioimaging for medical applications. *Journal of Magnetism and Magnetic Materials* **304**: 122–127.

75. Ogiue-Ikeda, M., Sato, Y., and Ueno, S. 2003. A new method to destruct targeted cells using magnetizable beads and pulsed magnetic force. *IEEE Transactions on Nanobioscience* **2**: 262–265.

76. Yamaguchi, S., Ogiue-Ikeda, M., Sekino, M., and Ueno, S. 2006. The effect of repetitive magnetic stimulation on tumor and immune functions in mice. *Bioelectromagnetics* **27**: 64–72.

77. Ueno, S. 1989. Quenching of flames by magnetic fields. *Journal of Applied Physics* **65**: 1243–1245.

78. Cespedes, O., and Ueno, S. 2009. Effects of radio frequency magnetic fields on iron release from cage proteins. *Bioelectromagnetics* **30**: 336–342.

79. Ueno, S. 2012. Studies on magnetism and bioelectromagnetics for 45 years: From magnetic analog memory to human brain stimulation and imaging. *Bioelectromagnetics* **33**: 3–22.

80. Ueno, S. 2008. New horizon in biomagnetics and bioimaging. *Electromagnetic Field, Health and Environment. Proceedings of EHE'07*, Krawczyk, A., Kubacki, R., Wiak, S., and Antunes, C. L., eds. IOS Press, Amsterdam: 8–17.

81. Andra, W., and Nowak, H. (Eds.). 1998. *Magnetism in Medicine A Handbook*, Wiley-VCH, Berlin.

82. Andra, W., and Nowak, H. (Eds.). 2007. *Magnetism in Medicine A Handbook*, 2nd ed. Wiley-VCH, Weinheim.

83. Barnes, F. S., and Greenebaum, B. (Eds.). 2007. *Handbook of Biological Effects of Electromagnetic Fields*, 3rd ed., Bioengineering and Biophysical Aspects of Electromagnetic Fields (one part) and Biological and Medical Aspects of Electromagnetic Fields (one part). CRC Press, Boca Raton, Fla.

84. Lin, J. C. (Ed.). 2000. *Advances in Electromagnetic Fields in Living Systems*, vol. 3. Kluwer/Plenum, New York.

85. Lin, J. C. (Ed.). 2012. *Electromagnetic Fields in Biological Systems*. CRC Press, Boca Raton, Fla.

86. Williamson, S. J., Romani, G. L., Kaufman, L., and Modena, I. (Eds.). 1983. *Biomagnetism: An Interdisciplinary Approach. Series A: Life Sciences* **66**. Plenum Press, New York.

87. Reilly, J. P. 1998. *Applied Bioelectricity: From Electrical Stimulation to Electropathology.* Springer-Verlag, New York.

88. McLean, M. J., Engstrom, S., and Holcomb, R. R. (Eds.) 2003. *Magnetotherapy: Potential Therapeutic Benefits and Adverse Effects.* TFG Press, New York.

89. Rosch, P. J., and Markov, M. S. (Eds.). 2004. *Bioelectromagnetic Medicine.* Marcel Dekker, New York.

CHAPTER 2

Principles of Biomagnetic Stimulation

Shoogo Ueno and Masaki Sekino

CONTENTS

2.1 INTRODUCTION

2.1.1 Basic Principles

Electromagnetic induction, which provides the fundamental principle of biomagnetic stimulation, was discovered by Michael Faraday in 1831. When a pair of coils is situated close to one another and the current flowing in one coil varies, an electromotive force arises in the other coil. The current flowing in the primary coil generates a magnetic field, and a rapid switching of the current causes a steep variation in the magnetic flux linkage. The magnitude of electromotive force in the circuit is proportional to the time rate of change in this flux linkage. The electromotive force is directed so that the magnetic field arising from the secondary coil current inhibits the variation of flux linkage. Because biological tissues and fluids are electrically conductive, the electromagnetic induction occurs also in the body to induce electric fields and resulting currents. When a rapidly varying current is applied to a coil located on the surface of the body, electric fields are induced in nearby tissues owing to the varying magnetic fields. The magnetic stimulation is understood as an electric stimulation mediated by magnetic fields because the resulting biological effects originate from the induced electric fields. Considering that the magnetic permeabilities of biological tissues are nearly equal to that of free space, electric properties of biological tissues play important roles in biological effects of alternating and pulsed magnetic fields. At low frequencies below tens of kilohertz, depolarization of excitable membrane is the dominant mechanism of biological effects, and thermal effects are negligible in many cases.

In 1896, d'Arsonval reported magnetophosphene in which an exposure of the head to an alternating magnetic field evokes a visual perception.[1] This phenomenon has been investigated in many studies afterward and treated as a valuable experimental model for assessing the effects of magnetic fields on the human body. The magnetophosphene exhibits a significant dependence on the frequency of magnetic field, and the most clearly perceptible magnetophosphene occurs in the vicinity of 20 Hz.[2] At this frequency, magnetic fields above 10 mT applied to the eyes evoke magnetophosphene. Several mechanisms have been assumed for the magnetophosphene. It is generally believed that induced electric fields stimulate the retinas and optic nerves.

Subsequently, biomagnetic stimulation of nerves and muscles by means of applying pulsed magnetic fields has attracted attention. Several pioneering studies on the methodology of magnetic stimulation have been carried out since 1965.[3-5] These studies did not reach clinical applications because of technical problems like difficulty measuring evoked electromyograms. Barker et al.[6] succeeded in magnetically stimulating the brain in 1985 by applying magnetic fields transcranially to the brain. They used a circular coil producing pulsed magnetic fields with the intensity of around 1 T. Ueno et al.[7,8] proposed a method for locally stimulating the brain using a figure-eight coil and succeeded in stimulating the human motor cortex with a 5-mm resolution. These achievements led to the subsequent progress of methodological and clinical studies of transcranial magnetic stimulation (TMS).[9-12] Inducing electric fields specifically in the target region of the brain enables us to activate neurons in this region. Figure 2.1a shows a TMS of the primary motor cortex. Stimulation causes a contraction of the corresponding muscles. Figure 2.1b and c show spatial patterns of electric fields induced in the brain by stimulations using a circular coil and a figure-eight coil. The TMS is now widely used for diagnostics, therapies, and basic studies of brain function.

Magnetic field is generated as a pulse with durations usually shorter than 1 msec. Figure 2.2a and b show the waveforms of monophasic and biphasic pulses. Intensity of the magnetic field generated from the coil is proportional to the coil current.

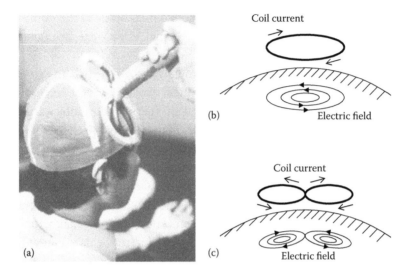

FIGURE 2.1   (a) Transcranial magnetic stimulation. (b) Electric fields induced by a circular coil. (c) Electric fields induced by a figure-eight coil.

In addition to the brain, the spinal root, diaphragm, and heart can also be magnetically stimulated. Applications of magnetic stimulations to artificial respiration and defibrillation have been explored.

2.1.2 Advantages of Magnetic Stimulation

Because limited techniques are available for directly stimulating the human brain, TMS is expected to be developed by continuously taking advantage of its unique feature. The brain is surrounded by the skull, which has a high electric resistivity. When electric stimulations are delivered through electrodes attached to the surface of the scalp, electric fields hardly penetrate into the brain. Though the brain stimulations through the electrode are possible using a quite strong electric field, the resulting intense pain has put restrictions on clinical applications of such stimulations. Biological tissues such as the brain, skull, and scalp are nonmagnetic. Magnetic fields penetrate into the brain without any hindrance by intermediate tissues and efficiently induce electric fields in the brain. Unlike the electric stimulation, TMS does not give strong stimulations to pain receptors in the scalp. Owing to this major advantage of reduced pain, TMS obtained a wide field of applications.

By comparing these two methods for delivering electric stimulations to the body, applying electric field through the electrodes, and inducing

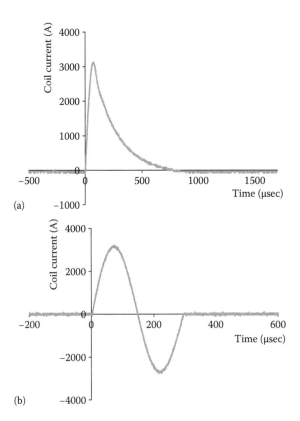

FIGURE 2.2   (a) Monophasic and (b) biphasic waveforms of current flowing in the coil. Magnetic field intensity is proportional to the current.

the electric field by pulsed magnetic field, one can find a difference in the resulting electric field distributions. In the case of using electrodes, the streamline of electric field flows out from the positive electrode, diffuses in the body, and flows into the negative electrode. In the case of TMS, the induced fields flow along closed-loop pathways in the body, as shown in Figure 2.1b and c.

Recent clinical studies have shown that TMS improves various symptoms of neurological and psychiatric diseases. It should be noted that the TMS exhibits therapeutic effects even for a percentage of drug-resistant patients. TMS is advantageous also in terms of rarely causing side effects, which are sometimes problematic in drug treatment. As discussed primarily in Chapter 3, TMS is a promising technique for a variety of fields such as brain research, diagnosis of the nervous system, treatment of diseases, and recovery of brain function.

2.2 MAGNETIC STIMULATOR

2.2.1 Stimulator Coils

2.2.1.1 Circular Coil

Circular coils have been used since TMS was invented.[6] Owing to the simplest structure among various types of coils, the circular coils are still widely used today. As shown in Figure 2.1b, the coil conductors are wound up circularly. Figure 2.3a shows a circular coil for basic studies. The generated magnetic fields are perpendicular to the surface of the head below the coil center and are tangential to the surface below the coil conductor. The electric fields generated in the brain form circular pathways along the surface of the head and in the direction opposite the coil current. The magnitude of electric field is zero below the center of the coil and is highest below the coil conductors. As a result, the electric fields distribute in an area whose dimension is comparable to the coil diameter. Because the magnitude of magnetic field decreases with the distance from the coil, the electric fields are highest at the surface of the head and attenuate at deep regions in the brain. Coils with smaller diameters

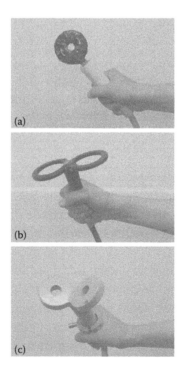

FIGURE 2.3 (a) Circular coil, (b) figure-eight coil, and (c) double-cone coil. These coils are widely used for magnetic stimulation.

exhibit steeper attenuations of fields. In order for electric fields to reach deeper regions in the brain, using a coil with large diameter is desirable.

The spatial distribution of magnetic field generated from a circular coil can be analytically formulated. For simplicity, a circular coil with a radius of a lies in the x–y plane with its center coinciding with the origin. A current I is applied to the coil. The following formulas give the magnetic field vector generated from the circular coil in a spherical coordinate:

$$B_r = \frac{1}{r\sin\theta}\frac{\partial}{\partial\theta}(A_\varphi \sin\theta) \tag{2.1}$$

$$B_\theta = -\frac{1}{r}\frac{\partial}{\partial r}(rA_\varphi) \tag{2.2}$$

$$B_\varphi = 0 \tag{2.3}$$

where $\mu_0 = 4\pi \times 10^{-7}$ H/m is the permeability of free space. The transform between the Cartesian coordinate and the spherical coordinate is given by $x = r\sin\theta\cos\varphi$, $y = r\sin\theta\sin\varphi$, and $z = r\cos\theta$. A_φ is the magnetic vector potential:

$$
\begin{aligned}
A_\varphi &= \frac{\mu_0 I a}{4\pi}\int_0^{2\pi}\frac{\cos\varphi'd\varphi'}{(a^2+r^2-2ar\sin\theta\cos\varphi')^{1/2}} \\
&= \frac{\mu_0}{4\pi}\frac{4Ia}{\sqrt{a^2+r^2+2ar\sin\theta}}\left[\frac{(2-k^2)K(k)-2E(k)}{k^2}\right]
\end{aligned} \tag{2.4}
$$

$$k^2 = \frac{4ar\sin\theta}{a^2+r^2+2ar\sin\theta} \tag{2.5}$$

where $K(k)$ and $E(k)$ are elliptic integrals of the first and second kinds, respectively.

$$K(k) = \int_0^{\pi/2}\frac{1}{\sqrt{1-k^2\sin^2\theta}}d\theta \tag{2.6}$$

$$E(k) = \int_0^{\pi/2}\sqrt{1-k^2\sin^2\theta}\,d\theta \tag{2.7}$$

Various coils introduced below can be modeled as a set of circular coils for analyzing the magnetic field distributions. The above equations can be applied to such analyses.

The current flowing in the coil is subject to Lorentz force owing to the magnetic field generated from the coil itself. The generation of pulsed force causes a vibration of coil conductors. Acoustic noise produced from the coil at the moment of delivering stimulation originates from the Lorentz force. In a circular coil, force is generated outwardly so that the coil gets slightly distended.

2.2.1.2 Figure-Eight Coil

The figure-eight coil was originally developed by Ueno et al.[7,8] This coil is now in widespread clinical use along with the single circular coil. The figure-eight coil consists of a pair of circular coils adjacent to one another, and the current is applied in the opposite directions to these circular coils. Magnetic fields are produced from the two circular coils in the opposite directions, and the magnetic fields induce electric fields in the brain. As shown in Figure 2.1c, the electric field flows in the brain in the reversed direction, forming two loop pathways. The electric fields converge in the target region in the brain below the intersection of the figure-eight coil, resulting in an electric field intensity approximately 3× higher than those around the target.[11] In the single circular coil, however, the electric fields form a circular pathway. The resulting fields get the maximum intensity below the coil conductors, and the intensity becomes zero below the center of the circle.

The figure-eight coil enables one to stimulate a target area in the human cerebral cortex with a spatial resolution as high as 5 mm.[7] Thus, this coil is extremely effective when stimuli should be delivered to a specific target area. Another advantage of this coil is the ability to provide a vectorial stimulation because the electric fields in the target area are induced in a specific direction along the surface of the head. The attitude of the coil defines the direction of electric field.

Figure 2.3b and c show two major figure-eight coils: planar coil and double-cone coil. In the double-cone coil, the outer conductors come close to the surface of the head, which intensify the electric field at the target area. The electric fields in the brain generally attenuate with the distance from the coil conductors. An increase in the diameter of the coil leads to stronger electric fields in deep parts of the brain.

2.2.1.3 Coil Array

Researchers have investigated potential benefits of composing an array of small coils distributed over the surface of the head for TMS. A variety of stimulations can be delivered according to spatial and temporal patterns of currents applied to the coil elements. The magnetic field arising from the coil array is the superposition of magnetic fields from each coil element. In principle, the circular and figure-eight coils can be imitated with correspondingly designed current patterns. Various coils including these two could be replaced with one coil array. Mathematical methods such as lead field and minimum-norm estimation can be used to obtain the pattern of coil currents for a given distribution of stimulating electric fields in the brain. A unique advantage of the coil array is the ability to deliver a train of stimulating pulses for different target areas with a short pulse interval.

Advantages of a coil array have been evaluated based on numerical simulations.[13,14] Figure 2.4 shows coil elements distributed over the surface of a numerical human head model. This simulation showed that the focality and depth of stimulation can be improved by properly selecting the coil current direction and current phases in the neighboring coils.

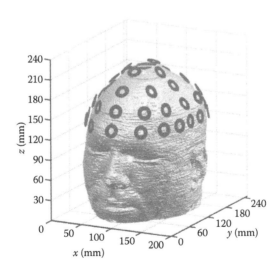

FIGURE 2.4   Coil array designed on a numerical model of human head. The array consisted of 40 elements. (From Lu, M. et al., *IEEE Transactions on Magnetics* **45**: 1662–1665, 2009.)

While these advantages have been reported, development of a proto-type coil is not technically easy. Thickness of the coil conductor is a few millimeters in many cases, allowing pulsed currents of kiloamperes to flow through the conductor. Because the conductor should be wound mul-tiple turns, miniaturization of the element coil is limited. When driving currents are applied to multiple coils for imitating a large coil, adjacent coils may have currents in the opposite directions. Such currents do not contribute to the generation of magnetic fields but cause ohmic loss and decreased efficiency. In addition, the coil array requires much larger driv-ing circuits and cables compared with the single coils. The research and development of coil array will be carried out considering the balance of its unique advantages and technical challenges.

2.2.1.4 Emerging Coil Designs

Various coil designs have been explored for achieving high performances such as higher focality, deeper penetration, and improved efficiency of stimulations. By using four or more coil elements over the head, high flexibility in the design of the spatiotemporal pattern of magnetic field pulses becomes possible.[15] Model simulations showed that a good control of the excited area was achieved by proper positioning of the coil. The use of three-dimensional configurations of multiple coil elements (such as a slinky coil or a differential coil) can produce more focused stimulation than is possible with planar coils.[16,17] In the H-coils, coil conductors were distributed along the surface of the brain to deliver electric fields to deeper parts of the brain.[18] Deeper distribution of the electric fields would lead to improved efficacy of TMS especially in therapeutic applications.

Sekino et al.[19-21] proposed an eccentric figure-eight coil that induces strong electric fields in the target brain tissue with a given coil current. This coil is a modification of the figure-eight coil. The center of the outer circumference in the ordinary figure-eight coil coincides with the cen-ter of the inner circumference, as shown in Figure 2.5a. In the eccentric figure-eight coil, the inner circumferences are shifted closer together, as shown in Figure 2.5b. The dense conductors at the middle of the coil lead to intensified stimulating currents. Numerical analyses were carried out to obtain magnetic field distribution of the eccentric figure-eight coil and electric fields in the brain. The eccentric coil required a driving current intensity of approximately 18 percent less than that required by the con-centric coil to cause comparable electric field intensities within the brain. The electric field localization of the eccentric coil was slightly higher than

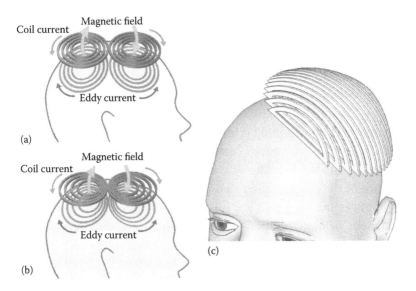

FIGURE 2.5 (a) Eddy currents induced by an ordinary figure-eight coil. (b) Eddy currents induced by an eccentric figure-eight coil. (c) Schematic of a bowl-shaped coil.

that of the concentric coil. A prototype eccentric coil was designed and fabricated. Instead of winding a wire around a bobbin, eccentric spiral slits were formed on the insulator cases, and a wire was woven through the slits. The coils were used to deliver TMS to healthy subjects. The current slew rate corresponding to motor threshold values for the concentric and eccentric coils were 86 and 78 A/μsec, respectively. Results indicated that the eccentric coil consistently requires a lower driving current to reach the motor threshold than the concentric coil does.

As discussed in Section 3.4.4, the installation of a TMS system at a patient's home has been proposed for providing the continuous therapeutic effects of daily sessions. Because the figure-eight coil induces a highly localized electric field, it is challenging to achieve accurate coil positioning above the targeted brain area without help by a medical expert. A bowl-shaped coil for stimulating a localized but wider area of the brain was proposed.[22,23] The bowl-shaped coil consists of the conductors aligned tangentially to the surface of the head and the counter conductors running over the tangential conductors, as shown in Figure 2.5c. Electric fields are induced along the aligned tangential conductors. The length and width of the arrayed tangential conductors determine the localization of stimulation. Thus, the bowl-shaped coil exhibits a moderate and tunable localization of electric field distribution. In order to design a compact stimulator coil, the counter conductor elements were arrayed above the field-inducing conductor. The coil's electromagnetic characteristics were analyzed using numerical simulations, and the analysis showed that the bowl-shaped coil induced electric fields in a wider area of the brain model compared with the figure-eight coil. The expanded distribution of the electric field led to greater robustness to the positioning error.

Efficacy of introducing iron cores into stimulator coils has been evaluated in many studies.[24,25] The iron core leads to a reduction of magnetic reluctance of the coil, enabling us to drive the coil with smaller currents. This produces the benefit of a miniaturized driving circuit. In addition, the magnetic field distribution of the coil can be modified by placing iron cores that have a high magnetic permeability close to the coil. Technical challenges associated with the use of iron core are increased weight of the coil and heat generation due to iron loss.

When a metal plate is introduced between the coil and the head, eddy currents induced in the metal plate provide a shielding effect on the magnetic field.[26] Focality of electric field in the brain can be improved using this effect.

2.2.1.5 Cooling Systems

Electric currents in the coil cause heat generation due to the resistance of the coil.[27,28] The amount of heat is proportional to the square of coil current and to the repetition rate of pulses. Heat generation is prominent when stimulating with high currents and high repetition rates. This issue is particularly important in therapeutic applications of TMS because stimulations are delivered at a repetition rate higher than five pulses per second. Guidelines are provided in each country that define the maximum temperature on the coil surface attached to the human body. Coils should be designed and tested following those guidelines.

One approach for addressing heat generation is to introduce a heat insulator between the coil conductor and the human head. The heat insulator suppresses the rising temperature on the side attached to the head, while radiating heat from the opposite side.

Coils should be actively cooled when heat generation is relatively high. Two major methods are air cooling and liquid cooling. Air-cooling is provided from a fan installed above the coil or from an external blower through a tube. Air cooling can be implemented with simple machinery. For liquid cooling, the coolant is circulated between the coil and an external heat exchanger. It has higher capacity: however, the circulation system with sealed fluid channel and pump complicates the design.

2.2.1.6 Measurement of Magnetic Field and Coil Current

Magnetic fields for stimulation are generated typically with intensities of approximately 1 T and pulse widths of hundreds of microseconds. A search coil is suitable for measuring such magnetic fields. The search coil is fabricated by winding a lead wire several times with a diameter of a few millimeters. When the magnetic field at the search location coil varies with time, a voltage is induced in the search coil because of electromagnetic induction. Instantaneous induced voltage is proportional to the time rate of change in magnetic field intensity. The induced voltage of search coil reflects the magnetic field component parallel to the winding axis of search coil. Figure 2.6a shows a measurement of magnetic field using a search coil. This search coil consists of five 6.3-mm diameter inner loop turns and four 7.1-mm diameter outer loop turns. The search coil is attached

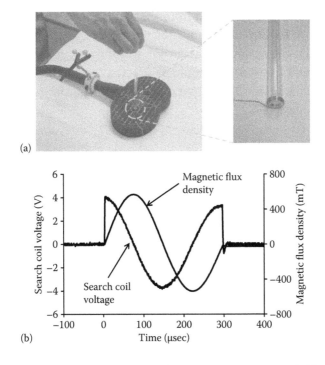

FIGURE 2.6 (a) Homemade search coil for measuring magnetic fields generated from the stimulating coil. (b) Measured search coil voltage and estimation of magnetic field intensity.

to the stimulator coil with its axis perpendicular to the winding plane of the stimulator coil, and the search coil measures the magnetic field component of this axis. Induced voltage is recorded using an oscilloscope. The following formula gives the induced voltage V:

$$V = \pi a^2 N \frac{dB}{dt} \tag{2.8}$$

where B is the generated magnetic field and a and N are the respective radius and number of turns of the search coil. The integral of voltage with respect to time gives the estimation of magnetic field intensity.

$$B = \frac{1}{\pi a^2 N} \int_0^t V(\tau) d\tau \tag{2.9}$$

Figure 2.6b shows the waveforms of measured voltage and estimated magnetic field intensity.

The magnetic field generated from the coil is proportional to the current flowing in the stimulator coil. The current of the stimulator coil can be measured using a commercial current transformer, as shown in Figure 2.2. The current transformer should be selected considering current magnitudes of kiloamperes. The output of the current transformer is recorded using the oscilloscope.

2.2.2 Driving Circuit

2.2.2.1 Hardware

The driving circuit for magnetic stimulation supplies pulsed electric currents to the stimulator coil. Figure 2.7 shows a simplified circuit diagram for generating biphasic pulses.[29] The circuit is composed of a power supply, a high voltage generator, and a semiconductor switch. The power supply circuit generates DC voltages from the AC input. The high voltage generator outputs voltages of kilovolts for charging the capacitor. Resistance limits the current for charging the capacitor. The mechanism of storing electric charges in the capacitor and then discharging for providing currents to the coil enables the generation of strong current from relatively small equipment. When the thyristor is turned on after the positive charging of the capacitor, electric current begins to flow in the coil. Here, the circuit is equivalent to the coil L and capacitor C connected serially. Sinusoidal current flows in these elements with the pulse width T given by

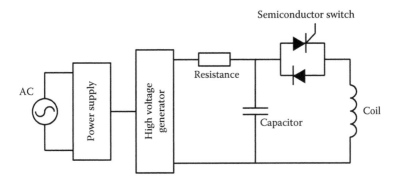

FIGURE 2.7 Diagram of driving circuit for magnetic stimulator. The circuit provides pulsed currents to the coil. (From Sekino, M. et al., *Proceedings of the ICME International Conference on Complex Medical Engineering* 728–733, 2012.)

$$T = 2\pi\sqrt{LC} \tag{2.10}$$

After turning on the thyristor, the current gradually increases and reaches the maximum at the moment of the charge in capacitor reducing to zero. The current subsequently begins to decrease, and the capacitor is negatively charged. When the current reduces to zero, the negative capacitor voltage reaches the minimum. Then, the current takes the same time course with the reversed polarity. The diode that is connected parallel to the thyristor in the opposite direction allows the reverse current to flow. When the coil current reduces to zero again, the capacitor is positively charged. The circuit is deactivated at that moment because the thyristor is turned off. The capacitor voltage ideally recovers to the level it had before the pulse generation. In actual circuits, however, voltage drop occurs because of resistance in the coil and capacitor. The capacitor is charged again before subsequent pulse generation so that the voltage recovers to the initial value. The repetition rate of pulse generation is controlled by the internal unit, allowing the circuit to generate single pulse and repeated pulses with rates ranging typically from 1 to 20 Hz.

Circuit elements such as the capacitor and thyristor have maximum rated powers. The circuit can be substantially downsized if one uses single pulse only.[30,31] Pulse generations with high repetition rates require relatively large elements. Some of the repetitive pulse generator is equipped with multiple capacitors. As described in the next section, a generation of monophasic pulse leads to significant discharge of the capacitor. In this case, preparation for the subsequent pulse generation takes a longer time in comparison with the case of biphasic pulse. Generation of the monophase pulse train usually requires multiple capacitors.

2.2.2.2 Analysis of Driving Circuit

This section shows an analysis of pulse waveform for the case of monophasic pulse. As shown in Figure 2.8, the major elements in the circuit consists of the inductance L of the stimulator coil, DC voltage source V_0, capacitor C, resistance R, diode D, and switches S_1 and S_2. First, the capacitor C is charged by opening the switch S_2 and closing the switch S_1. No current flows in the diode D and resistance R at this moment because the diode allows the current to flow in only one direction downward. After sufficient charging of capacitor C, switch S_1 is opened and switch S_2 was closed. Current begins to flow in the coil L, and the capacitor C begins to be discharged. The voltage v of capacitor is equal to that of coil. The following equations give the relations among coil current i_L, capacitor current i_C, and voltage v:

$$v = L\frac{di_L}{dt} \tag{2.11}$$

$$i_C = C\frac{dv}{dt} \tag{2.12}$$

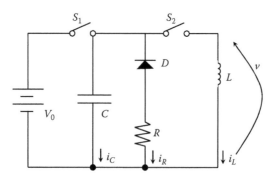

FIGURE 2.8 Simplified circuit for analyzing a monophasic pulse current. (From Mano, Y., and Tsuji, S., *Magnetic Stimulation—Basic Principles and Applications.* Ishiyaku Publishers, Tokyo, 2005.)

The relation $i_L + i_C = 0$ holds when no current flows in the capacitor R. The following equation holds for the time course of i_L:

$$L\frac{d^2 i_L}{dt^2} + \frac{i_L}{C} = 0 \qquad (2.13)$$

where the moment of closing switch S_2 is defined as $t = 0$. The above equation gives a well-known simple harmonic oscillation. The following is one solution satisfying the initial condition of $i_L = 0$ at $t = 0$:

$$i_L = V_0 \sqrt{\frac{C}{L}} \sin\frac{t}{\sqrt{LC}} \qquad (2.14)$$

The coil current initially exhibits a sinusoidal increase. Coil voltage is given by

$$v = L\frac{di_L}{dt} = V_0 \cos\frac{t}{\sqrt{LC}} \qquad (2.15)$$

After the closing of switch S_2, voltage v decreases and becomes zero at the time $t_1 = (\pi/2)(LC)^{1/2}$. The voltage v then turns to be negative, and the diode allows the current to flow also in the resistance R. The Ohm's low holds between the voltage v and current i_R in the resistance.

$$v = Ri_R \qquad (2.16)$$

The continuity of currents $i_R + i_L + i_C = 0$ leads to

$$C\frac{d^2 v}{dt^2} + \frac{1}{R}\frac{dv}{dt} + \frac{v}{L} = 0 \qquad (2.17)$$

When the relation $L > 4R^2C$ holds, among the solution of the above equation the following one is connected to Equation 2.15 at $t_1 = (\pi/2)(LC)^{1/2}$:

$$v = -2V_0 R\sqrt{\frac{C}{L - 4R^2C}}\exp\left(-\frac{t - t_1}{2RC}\right)\sinh\left(\sqrt{\frac{1}{(2RC)^2} - \frac{1}{LC}}\cdot(t - t_1)\right) \qquad (2.18)$$

$$i_L = -C\frac{dv}{dt} - \frac{v}{R} = V_0 \sqrt{\frac{C}{L - 4R^2C}} \exp\left(-\frac{t - t_1}{2RC}\right)$$

$$\times \left[\sqrt{1 - \frac{4R^2C}{L}} \cosh\left(\sqrt{\frac{1}{(2RC)^2} - \frac{1}{LC}} \cdot (t - t_1)\right)\right.$$

$$\left. + \sinh\left(\sqrt{\frac{1}{(2RC)^2} - \frac{1}{LC}} \cdot (t - t_1)\right)\right] \tag{2.19}$$

Both voltage v and current i_L attenuate toward zero over time. The combination of Equations 2.14 and 2.15 gives the monophasic waveform shown in Figure 2.2. The magnetic field generated from the coil is proportional to the current i_L, and the electric field induced in the body is proportional to the time derivative of current di_L/dt.

2.2.3 Coil Navigation System

Figure 2.9 shows a typical coil navigation system for transcranial magnetic stimulation. A dedicated marker is equipped with three reflective spheres. Markers are fixed on the stimulator coil and on the bed for patient. An infrared binocular camera observes the three reflective spheres, which enable the system to estimate the position and orientation of the markers. The position of the center of the coil relative to the marker is measured beforehand using a dedicated probe. In addition, the positions of several landmarks

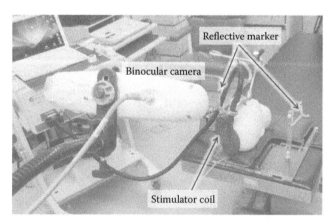

FIGURE 2.9 Coil navigation system using a near-infrared binocular camera and reflective markers.

on the head, such as the nasion and preauricular points, are measured. Placing the patient supine leads to stabilized positioning between the head and marker. The system enables us to see the position of the stimulator coil relative to the head in real time. The system reads the patient's magnetic resonance images for graphically showing the location of the coil and the target position on the brain. The binocular camera achieves submillimeter accuracy in estimating the position of markers. Introducing such a navigation system enables us to precisely and quickly place the stimulator coil above the target position. This is particularly valuable for highly localized stimulations using figure-eight coils.

2.2.4 Multimodal Imaging Using Magnetic Stimulation

A combination of magnetic stimulation and other modalities such as electroencephalography (EEG) and magnetic resonance imaging (MRI) is effective for measuring the evoked brain activities. The combinations give rise to technical challenges for preventing influences of magnetic field pulses on measuring methods. The EEG is sensitive to external noise due to the requirement of recording small evoked potentials. Magnetic field of MRI causes strong electromagnetic forces on the stimulator coil.

Iramina et al.[32] reported the measurement of evoked potentials induced by magnetic stimulation for observing the neuronal connectivity in the brain. Figure 2.10a shows an EEG measurement system to eliminate the electromagnetic interaction caused by the stimulation. Using this artifact free amplifier, TMS-evoked EEG responses were successfully measured, as shown in Figure 2.10b. Several components of the evoked potential appeared at 9, 20, and 50 msec after a cerebellar stimulation. During motor area stimulation, there was no clear peak of the waveforms within 10 msec latency. Occipital stimulation caused more evoked responses to spread to the center of the brain than at other areas of stimulation. The evoked signal by TMS was possibly conducted posteriorly to anteriorly along the pathways of the neuronal fiber exiting the cerebellum into the cerebral cortex.

Reaction maps of the stimulation at the occipital area were then obtained using the 9-msec evoked potential component.[33] From the reaction map, it was possible to consider that the fast component of evoked signal by stimulation to the cerebellum was conducted posteriorly to anteriorly along the pathways of the neuronal fiber exiting the cerebellum to the cerebral cortex. The response at the right brain was large when the left

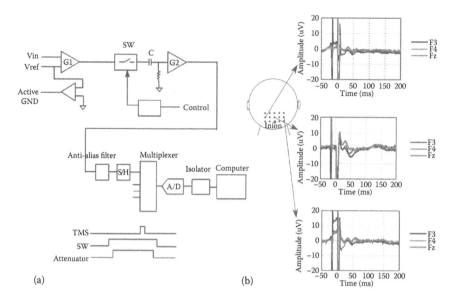

FIGURE 2.10 (a) Block diagram of TMS-compatible EEG amplifier and timing chart of the circuit. (b) Measurement points of EEG and stimulated point on the head. The stimulus point, which is located on the cerebellum, was 20 mm superior to the inion. (From Iramina, K. et al., *Journal of Applied Physics* **93**: 6718–6720, 2003.)

cerebellum was stimulated. The response at the left brain was large when the right cerebellum was stimulated. The response map gave information of the neuronal connectivity. Magnetic stimulation will become a useful tool for the study of cortical reactivity and neuronal connectivity of the brain.

An MRI is useful for observing spatial distribution of brain activities. For demonstrating a magnetic stimulation in the MRI system, the head and coil are placed in a static magnetic field. Applying a pulsed electric current to the stimulator coil under the magnetic field causes an impulsive Lorentz force. The coil and supporting structure should be durable against the force. In addition, the magnetic field generated in the MRI system affects acquisition of images. The timings of pulse generation and MRI acquisition should be staggered with each other. The cable between the coil and driving circuit should be long enough so that the driving circuit is placed distant from the MRI magnet. A lot of studies have been carried out using simultaneous operation of magnetic stimulation and MRI.[34,35]

2.3 MECHANISMS OF MAGNETIC STIMULATION

2.3.1 Fundamentals of Electrophysiology

2.3.1.1 Resting Membrane Potential

Intracellular space has a lower electric potential compared with that of extracellular space. The difference in potential is caused by cations accumulating on the outer surface of the cell membrane and anions on the inner surface. The potential of intracellular space normally ranges from −40 to −80 mV with respect to the extracellular space, which is called the resting membrane potential. Many biological macromolecules exist as organic anions that have negative net charges. When large organic anions remain in intracellular space because of the impermeability through the membrane, the membrane-transportable cations and anions are affected by these organic anions and distribute with different concentrations in both sides of the membrane. This phenomenon is called Donnan's effect. Intracellular inorganic ions are composed mainly of potassium (K^+) and small amounts of sodium (Na^+) and chloride (Cl^-). On the other hand, Na^+ and Cl^- compose the greater part of extracellular inorganic ions.

When the difference in concentration occurred between the internal and external side of the membrane only owning to Donnan's effect, the resting membrane potential is expressed by Nernst's formula as follows:

$$E = \frac{RT}{zF} \ln \frac{C_o}{C_i} \qquad (2.20)$$

where E is an intracellular electric voltage on the basis of the extracellular voltage, C_o is an extracellular ion concentration, C_i is an intracellular ion concentration, R is gas constant, T is a temperature, F is Faraday constant, and z is a valence of ion. Because the potassium channels are opening while the sodium ones are closing at the resting state, the resting potential mostly corresponds with the difference in potassium concentration between the intra-extra cellular spaces. Additionally, the cellular membrane has an active ion-exchanging mechanism (Na^+–K^+ exchange pump) that selectively intake K^+ to intracellular space and outflows intracellular Na^+ to extracellular space. This active pump system is driven by proteins on cellular membrane selectively transporting ions. Some of the ion-channel proteins have a voltage-dependent gating system, and it contributes to generating the action potential (details are described at the next section). Ion concentration differences between the intra-extracellular cellular spaces represent a complex result of this active pump system and Donnan's effect.

2.3.1.2 Action Potential

When a local depolarization occurred at the cellular membrane by the excitation stimuli, a voltage-dependent Na⁺ channel on the membrane opens. Subsequently, Na⁺ flows into intracellular space due to the gradient of concentration and electrical potential between intra-extra cellular spaces. Corresponding to the decrease in net negative potential of the membrane owing to the Na⁺ inflow, large numbers of the Na⁺ channels open and the greater Na⁺ influx occurs. According to the Na⁺ gate opening, an electric current caused by Na⁺ inflow exceeds those of K⁺ from K⁺ leak channels, and finally the membrane potential shows a positive charge at the intracellular side, which means the resting potential is reversed. This phenomenon is known as action potential. The depolarization and subsequent Na⁺ channels opening spread to surrounding area, and the propagation of action potential occurs.

The basic properties of the action potential propagation are analyzable using the electric circuit model of the cellular membrane. As shown in Figure 2.11a, a cellular membrane has electric properties r_m and c_m. The r_m represents penetration resistance per unit length at an ion transition, and c_m is an electric capacity of the membrane. Additionally, a nerve cell also

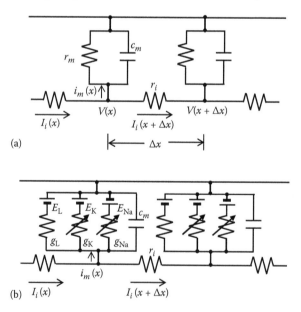

FIGURE 2.11 (a) A passive cable model for analyzing propagation of action potential. (b) An equivalent circuit of nerve fiber with the Hodgkin-Huxley model of voltage-gated channels.

has a resistance per unit length r_i along to the direction of nerve fibers. Extracellular cellular resistance is negligible compared with r_m; thus, it is considered as zero in this case. On the basis of these conditions, a membrane electric current i_m per unit length is represented as

$$i_m = c_m \frac{\partial V}{\partial t} + \frac{V}{r_m} \tag{2.21}$$

where V is a membrane potential in this formula. An intracellular electric current, along with the direction of the nerve fiber, follows Ohm's law.

$$r_i I_i = -\frac{\partial V}{\partial x} \tag{2.22}$$

where x represents the position at the direction of the fiber. A membrane electric current i_m satisfy a current continuity equation:

$$i_m = -\frac{\partial I_i}{\partial x} \tag{2.23}$$

With the combination of these three formulae, the cable properties of the cellular membrane are represented as

$$\lambda^2 \frac{\partial^2 V}{\partial x^2} - V = \tau \frac{\partial V}{\partial t}$$

where the constants λ and τ are given by

$$\lambda = \sqrt{\frac{r_m}{r_i}} \tag{2.24}$$

$$\tau = c_m r_m \tag{2.25}$$

This passive cable model is quite useful to examine the propagation characteristic of the action potential.

Actual membrane resistance depends on the voltage due to the actions of voltage-dependent ion channels. To express this characteristic, r_m for each type of ion is treated as separately and voltage dependent. Moreover, Nernst's potential of each ion is put into the model as a resource of electric voltage. This model is named the Hodgkin-Huxley model, and the electric property of an action potential propagation is well described by Hodgkin-Huxley equations.

2.3.2 Transcranial Magnetic Stimulation

2.3.2.1 Resting Motor Threshold

Cortial motor threshold (CMT) is an index of the stimulation intensity for TMS. The CMT is defined as the intensity of motor cortex stimulation that induces the motor evoked potential (MEP) higher than 50 μV with the probability greater than 50 percent. The CMT is indicated as the relative intensity to the maximum (approximately 2–3 T, defined as 100 percent). The CMT measured at the resting state is called the resting motor threshold (RMT). The RMT is distinguished from the active motor threshold (AMT) whose value is often lower than the RMT. These definitions are based on the guideline released from the International Federation of Clinical Neurophysiology (IFCN). Generally, the CMT is determined by stimulating the motor cortex approximately 10 times by changing the intensity and the stimulus intensity, which induces MEPs larger than 50 μV, where a probability of greater than 50 percent is assessed. In the formal procedure defined by IFCN, the initial intensity is set at 35 percent and the operator tries to look for the coil strength that shows MEP 5 times or more out of 10 stimulations with increasing intensity by a 5 percent step. Then, the same procedure is repeated by lowering the coil intensity by a 1 percent step to find the index, which indicates MEP less than 5 times in 10 trials. Finally, CMT is determined by adding 1 percent to this index.[36] On the other hand, the simplified manner has reported that the observation of muscle twitches more than 50 percent instead of CMT, and evaluated the interindividual variability.[37] RMT is known to have a large variation among subjects and shows approximately 20 percent change of the value owing to the coil position even in the subject. The numerical calculation by finite element method reported that the eddy current density in the brain was 6 A/m^2 when the human head model was stimulated by the pulsed magnetic field strength, which is equal to RMT.[38]

2.3.2.2 Stimulation Site

Neurons are sensitive to externally applied electric fields generated parallel to dendrites and axons. In transcranial electric stimulation, as shown in

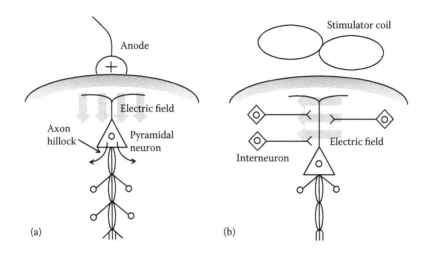

FIGURE 2.12 (a) Electric stimulation of neurons in the brain. (b) Magnetic stimulation of neurons in the brain. Interneurons are susceptible to the induced currents.

Figure 2.12, excitation occurs at the axon hillock where the axon runs out from the cell body of pyramidal neurons oriented perpendicular to the surface of the brain. In TMS, on the other hand, the electric field parallel to the brain surface mainly excites interneurons running in this direction. This leads to an excitation of connecting pyramidal neurons, and muscle contraction finally occurs. A stimulation of the primary motor cortex evokes a twitch. A shift of coil position from the maximally sensitive position causes a delay of 0.7–0.8 msec in the latency of motor evoked potentials. This delay corresponds to the time passing one synapse.[39] The distribution of induced electric field strongly depends on the structure of gyrus and sulcus.[40]

2.3.3 Peripheral Nerve Stimulation

2.3.3.1 Activating Function

In the peripheral nervous system, nerve fibers form bundles. Assuming a straight running nerve fiber, the mechanism of nerve excitation by magnetic stimulation can be investigated. Roth and Basser[41] proposed a physical model of stimulating a peripheral nerve fiber and introduced a term of electric field induced by time-varying magnetic field into the cable equation of nerve fiber. On the basis of this model, the cable equation is revised as

$$\lambda^2 \frac{\partial^2 V}{\partial x^2} - V = \tau \frac{\partial V}{\partial t} + \lambda^2 \frac{\partial \varepsilon_x}{\partial x} \tag{2.26}$$

where the second term in the right side expresses the influence of magnetic stimulation. This equation shows an interesting point that the influence of magnetic stimulation is not proportional to the electric field strength itself but proportional to the spatial gradient of electric field $\partial e_x/\partial x$. This term is referred to as the activating function. When an inhomogeneous magnetic field is induced depending on the geometry of stimulator coil, peripheral nerve stimulation is likely to occur at the location exhibiting high activating function.

2.3.3.2 Stimulation Site

The conductivity of biological tissue differs between tissue types. At the boundary between bone and muscle, for example, there is a discontinuous change in conductivity. Liu and Ueno[42] calculated the activating function to study the influence of the interface between tissues with different conductivities on nerve excitation. A Neumann-type boundary condition is derived for applying the finite element method to calculate the induced electric field. The spatial distributions of electric fields and activating functions in homogeneous and inhomogeneous volume conductors are calculated for comparison. The results show that the interface between conductors of different conductivities exhibits a large magnitude of activating function. Such locations are susceptible to magnetic nerve excitations. In addition, the point where a nerve fiber bends is susceptible to magnetic stimulation because of a high activating function.

2.4 NUMERICAL ANALYSES OF MAGNETIC FIELD AND EDDY CURRENT

2.4.1 Numerical Methods

2.4.1.1 Finite Element Method

The finite element method (FEM) is a numerical framework for solving partial differential equations. The method originally evolved in the field of structural mechanics. The basic principle is based on the variation method that has been known through the ages. First, a functional is derived so that a first-order minute error in the field distribution gives a resulting influence in only second order. Then, the volume for analysis is divided into meshes. Tetrahedral or hexahedral elements are usually used for a simple generation of meshes and flexible modeling of curving surfaces. The use of tetrahedral elements enables extremely flexible generation of meshes according to the target of calculation. Because of this characteristic, the mesh density can be increased locally at locations where the field steeply varies or the target object has complex geometry. Various algorithms for automatically generating meshes have been investigated. After the generation of meshes, the nodes are assigned on the sides or on the vertexes of each element, and an expansion of field equation is derived in terms of unknown parameters on the nodes. Substitution of this equation into the functional gives an expansion of functional with the unknown parameters. Because of the variation principle, this expression takes an extremal value with respect to a minimal variation of the unknown parameters. The differential of the expression by the unknown parameters is equal to zero. Doing this calculation for the all unknown parameters gives a simultaneous linear equation for the unknown parameters. Numerical analyses lead to the solution for the unknown parameters.

Because the FEM allows the use of the tetrahedral element, which is applicable to complicated geometries, this method is appropriate for the analyses of objects with complicated geometries. However, many numerical models of the human body consist of cubic voxels originating from voxels of cross-sectional images. In this case, the boundaries between tissues should be reconstructed to fit the tetrahedral meshes. In the case of the brain, methods for segmenting tissues from image pixels are developed. Software for generating tetrahedral mesh is available.

2.4.1.2 Scalar Potential Finite Difference Method

The scalar potential finite difference (SPFD) method is widely used for calculating induced electric fields in biological bodies.[43] The target body is modeled with voxels, and the equation of the electromagnetic field is solved with electric scalar potentials at the nodes contained as unknown parameters. For using this method, the skin depth should be much larger than the dimension of the calculated body, and the secondary magnetic field arising from induced current should be much smaller than the primary current. These conditions are satisfied in typical electromagnetic fields in TMS. The following equation is obtained by combining Faraday's law, current continuity equation, and Ohm's law and by discretizing the equation about the node 0:

$$\sum_{n=1}^{6} s_n \phi_n - \left(\sum_{n=1}^{6} s_n \right) \phi_0 = j\omega \sum_{n=1}^{6} (-1)^n s_n l_n A_{0n} \qquad (2.27)$$

where ϕ_n is the electric scalar potential at the node n, A_{0n} is the external magnetic vector potential parallel to the side n of the voxel connecting the nodes 0 and n, l_n is the length of side n, and s_n is the conductance of side n given by

$$s_n = \bar{\sigma}_n a_n / l_n \tag{2.28}$$

Here, a_n is the area of face perpendicular to the side n and σ_n is the mean conductivity of four voxels sharing the node n.

For calculating the electric field in the voxel, first, a system of equations is constructed from Equation 2.27 with the electric scalar potentials ϕ_n included as unknown parameters and for all nodes having finite conductivities. The electric scalar potential is obtained by solving the system of equations using sparse matrix methods such as successive overrelaxation. Then, the electric fields on the sides of voxel are calculated from magnetic vector potentials and electric scalar potentials. The electric field of voxel is obtained for x, y, and z components as the average of electric fields on the four sides.

2.4.2 Numerical Model of the Brain

Analyses of induced electric field distributions have been carried out using simplified models consisting of concentric spheres or standard human head models.[44] The use of a simplified spherical model is effective because of a lower amount of calculation, especially for theoretically evaluating the coil performance depending on coil design. The standard human head model providing precise anatomy of the brain, on the other hand, is valuable for estimating the electric field that is not able to be measured experimentally. Examples of analyses using the standard model will be introduced below. The results of these analyses reveal the influence of complex brain geometries such as gyrus and sulcus on the induced electric fields.[45]

In TMS, the resting motor threshold and therapeutic effect have individual variations. Differences in the electric field according to the shape of the brain may underlie this individual variation. The geometry of the standard head model is different from that of individual subjects, and it has been difficult to investigate the individual variation using the standard model. To understand the origin of individual variation, the brain model should be generated individually and perform analyses of electric fields.

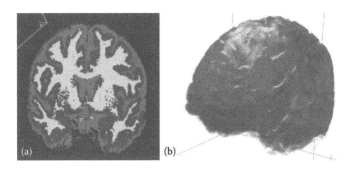

FIGURE 2.13   (a) A numerical model of the brain consisting of gray matter, white matter, and cerebrospinal fluid. (b) Numerical simulation of induced current in the brain using the scalar-potential finite difference method.

In some recent studies, numerical models based on individual brain shapes have been generated and used for analyses.[46] The aim was to show the individual variation in electric fields and to investigate the correlation with motor evoked potentials. The individual numerical models are generated from the MRI of subjects. Figure 2.13 shows an example of an individual head model consisting of gray matter, white matter, and cerebrospinal fluid and the result of electric field analysis using the SPFD method.

2.4.3 Analyses of Transcranial Magnetic Stimulation

2.4.3.1 Comparison of Electroconvulsive Therapy and Magnetic Stimulation

Electroconvulsive therapy (ECT) is a psychiatric treatment in which electric currents applied to the brain generate a cerebral seizure. The efficacy of ECT has been reported for the treatment of some symptoms of mental illnesses and neurological disorders that are not medicinally curable, including depression, schizophrenia, and Parkinson's disease. Safer methods of ECT administration have been introduced including the use of anesthetics. However, ECT is still an invasive treatment because the applied currents flow not only in the target region but throughout the brain as well. There are many side effects associated with ECT such as severe headaches and partial loss of memory after treatment. Because the localized stimulation reduces the risk of side effects, TMS is a more desirable method of treatment than ECT. To induce a comparable neural activation to ECT, it is important to evaluate current distributions generated inside the brain in TMS.

Sekino and Ueno[47] compared current density distributions in ECT and TMS by numerical simulations, as shown in Figure 2.14.[48] In the ECT model, electric currents were applied through electrodes with a voltage of 100 V. In the TMS model, a figure-eight coil was placed on the vertex. TMS generated comparable current densities to ECT. While the skull significantly affected current distributions in ECT, TMS efficiently induced eddy currents in the brain. In addition, the dependences of the current distribution in TMS on stimulus conditions were investigated.[49] Magnetic stimulations were performed using figure-eight coils with diameters of 50, 75, and 100 mm, and coil positions varied from the vertex to the forehead. The difference in current distributions in ECT and TMS decreased with

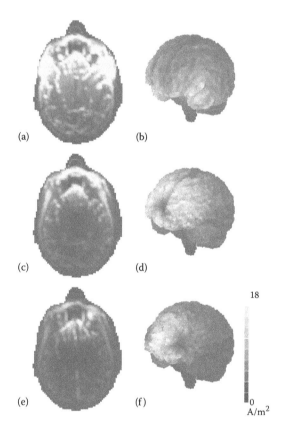

FIGURE 2.14   (a, b) Current distributions in electroconvulsive therapy (ECT). (c, d) Current distributions in magnetic stimulation using a 100-mm circular coil. (e, f) Current distributions in magnetic stimulation using a 75-mm figure-eight coil. (From Sekino, M., and Ueno, S., *Neurology and Clinical Neurophysiology* **88**: 1–5, 2004.)

the coil position approaching the forehead. The difference decreased with an increase in the coil diameter.

2.4.3.2 Magnetic Stimulation to the Cerebellum

The cerebellum is responsible for maintaining the body's posture and balance, as the antigravity muscles of the lower limbs are under the direction of the cerebellum to maintain standing posture. In addition, the cerebellum controls timings of muscular contractions for smooth motions. The cerebellum is connected to the motor cortex via neuronal fiber tracts, and activities in these areas are influenced by each other. TMS to the cerebellum enhances or depresses activities in the motor cortex, which reveals functional relations between these areas. Contractions of skeletal muscles evoked by cerebellar TMS reflect neuronal pathways from the cerebellum to peripheral nerves.

While the distance between the cerebrum and the stimulating coil is smaller than 2 cm, the distance between the cerebellum and the stimulator coil is 3 or 4 cm. It is important to clarify whether the increase in the distance affects the localization of eddy current. Sekino et al.[50] performed numerical simulations of the eddy current induced by TMS to the cerebellum.

Solutions were obtained on a three-dimensional human head model with inhomogeneous conductivity, as shown in Figure 2.15. The figure-eight coil had a diameter of 110 mm. Electric current applied to the coil had an intensity of 44.2 kA turns, which resulted in a magnetic field intensity of 0.56 T at the center of the coil element. The maximum current density in the cerebellum was 2.9 A/m². Distribution of the eddy current in the cerebellum was limited to approximately 1 cm beneath the surface of the cerebellum. The eddy current had a localized distribution in the cerebellum, while the magnetic field had a broad distribution.

2.4.4 Use of Numerical Methods in Coil Design

The electromagnetic characteristics of the coil depend on the design parameters such as shape, diameter, and number of turns. As described above, electric field distributions can be estimated for a given coil design using numerical methods. In addition, spatial integration of the magnetic fields generated around the coil gives the estimation of coil inductance as follows:

$$L = \frac{1}{\mu_0 I^2} \int B^2 \, dV \qquad (2.29)$$

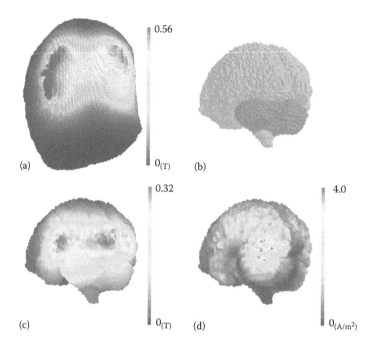

0.56

0(T)

(a)   (b)

0.32   4.0

(c)   0(T)   (d)   0(A/m²)

FIGURE 2.15   (a) Distribution of magnetic field on the surface of the head model in magnetic stimulation to cerebellum. (b) Three-dimensional view of the brain. (c) Distribution of magnetic field on the surface of the brain. (d) Distribution of eddy current on the surface of the brain. (From Sekino, M. et al., *IEEE Transactions on Magnetics* 42: 3575–3577, 2006.)

Figure 2.16 shows the electromagnetic characteristics of eccentric figure-eight coils introduced in Section 2.2.1.4. Here, three parameters of inner diameter, outer diameter, and number of turns were varied. The resulting coil inductance, pulse width, maximum eddy current density, working voltage, and coil resistance were estimated.

The increases in inner diameter, outer diameter, and number of turns caused increases in the eddy current density, inductance, and working voltage. The working voltage should not be very high for ensuring safety. The pulse width of around 200 μsec is close to the values of clinically used stimulators. Resistance increases with the winding length. The heat generation increases in proportion to the resistance. When eddy current increases given a constant coil current, stimulation can be performed with smaller coil currents to obtain the required current density in the brain. Taking this point into consideration, enlarging the diameters or increasing the number of turns enables us to reduce the heat generation.

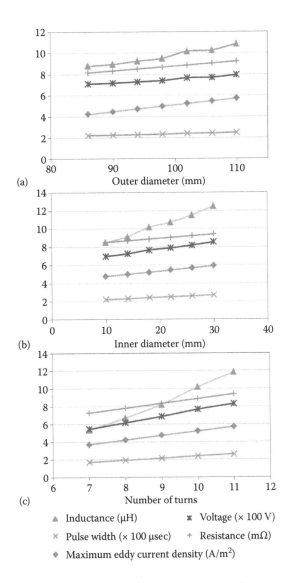

(a)

(b)

(c)

▲ Inductance (μH)   ✻ Voltage (× 100 V)

✕ Pulse width (× 100 μsec)   ✛ Resistance (mΩ)

◆ Maximum eddy current density (A/m²)

FIGURE 2.16   Influences of (a) outer diameter, (b) inner diameter, and (c) number of turns on the characteristics of eccentric figure-eight coil. (From Kato, T. et al., *Journal of Applied Physics* **111**: 07B322, 2012.)

2.5 SUMMARY AND FUTURE PROSPECTS

As discussed in this chapter, the unique feature of TMS is its noninvasiveness for stimulating the brain. Using a figure-eight coil, the brain can be stimulated with a resolution as high as 5 mm. Design of stimulator coil has a high degree of freedom, and focality and efficiency of stimulation strongly depend on the design parameters. Novel coil designs are still explored in many recent studies. Research and development of the driving circuit is actively progressing for achieving high output power, downsized equipment, and high efficiency.

One of the technical issues in TMS is stimulation of a deep part of the brain. Coil designs for deeper penetration of induced fields have been investigated. The ultimate goal is to achieve a three-dimensional localization of induced fields at a deep part of the brain. Though no method has been proposed for achieving such stimulation so far, further research is needed on this challenging but interesting physical problem. For preventing unexpected side effects, brain areas outside the target should not be strongly stimulated. Development of a systematic method for designing the coil windings from the given electric field distribution in the brain would be quite valuable.

Methods for making numerical brain models of individual subjects are now becoming available. This technique has medical significance if the effect of TMS can be predicted for each patient beforehand. More TMS therapeutic applications are be performed. The prediction of individual TMS therapeutic effect will enable one to provide more effective and more efficient therapy.

REFERENCES

1. d'Arsonval, A. 1896. Dispositifs pour la mesure des courants alternatifs de toutes frequencies. *Comptes Rendus de Societe Biologique* **2**: 450–451.
2. Magnusson, C. E., and Stevens, H. C. 1911. Visual sensations caused by changes in the strength of a magnetic field. *American Journal of Physiology* **29**: 124–136.
3. Irwin, D. D. Rush, S. Evering, R. Lepeschkin, E. Montgomery, D. B., and Weggel, R. J. 1970. Stimulation of cardiac muscle by a time-varying magnetic field. *IEEE Transactions on Magnetics* **6**: 321–322.
4. Maass, J. A., and Asa, M. M. 1970. Contactless nerve stimulation and signal detection by inductive transducer. *IEEE Transactions on Magnetics* **6**: 322–326.
5. Ueno, S., Harada, K., Ji, C., and Oomura, Y. 1984. Magnetic nerve stimulation without interlinkage between nerve and magnetic flux. *IEEE Transactions on Magnetics* **20**: 1660–1662.
6. Barker, A. T. Jalinous, R., and Freeston, I. L. 1985. Non-invasive magnetic stimulation of human motor cortex. *Lancet* **1**: 1106–1107.
7. Ueno, S., Tashiro, T., and Harada, K. 1988. Localized stimulation of neural tissues in the brain by means of a paired configuration of time-varying magnetic fields. *Journal of Applied Physics* **64**: 5862–5864.

8. Ueno, S., Matsuda, T., and Fujiki, M. 1990. Functional mapping of the human motor cortex obtained by focal and vectorial magnetic stimulation of the brain. *IEEE Transactions on Magnetics* **26**: 1539–1544.

9. Ueno, S. 1994. *Biomagnetic Stimulation*. New York: Plenum Press.

10. Hallett, M. 2000. Transcranial magnetic stimulation and the human brain. *Nature* **406**: 147–150.

11. Mano, Y., and Tsuji, S. 2005. *Magnetic Stimulation—Basic Principles and Applications*. Ishiyaku Publishers, Tokyo.

12. Ueno, S., and Fujiki, M. 2007. Magnetic stimulation. In *Magnetism in Medicine*, Andra, W., and Nowak, H. (Eds.). Wiley-VCH, Weinheim: 511–528.

13. Ruohonen, J., and Ilmoniemi, R. J. 1998. Focusing and targeting of magnetic brain stimulation using multiple coils. *Medical & Biological Engineering & Computing* **36**: 297–301.

14. Lu, M., Ueno, S., Thorlin, T., and Persson, M. 2009. Calculating the current density and electric field in human head by multichannel transcranial magnetic stimulation. *IEEE Transactions on Magnetics* **45**: 1662–1665.

15. Grandori, F., and Ravazzani, P. 1991. Magnetic stimulation of the motor cortex: Theoretical considerations. *IEEE Transactions Biomedical Engineering* **38**: 180–191.

16. Lin, V. W., Hsiao, I. N., and Dhaka, V. 2000. Magnetic coil design considerations for functional magnetic stimulation. *IEEE Transactions on Biomedical Engineering* **47**: 600–610.

17. Hsu, K. H., and Durand, D. M. 2001. A 3-D differential coil design for localized magnetic stimulation. *IEEE Transactions on Biomedical Engineering* **48**: 1162–1168.

18. Zangen, A., Roth, Y., Voller, B., and Hallett, M. 2005. Transcranial magnetic stimulation of deep brain regions: Evidence for efficacy of the H-coil. *Clinical Neurophysiology* **116**: 775–779.

19. Kato, T., Sekino, M., Matsuzaki, T., Nishikawa, A., Saitoh, Y., and Ohsaki, H. 2011. Fabrication of a prototype magnetic stimulator equipped with eccentric spiral coils. *Proceedings of the Annual International Conference of the IEEE Engineering in Medicine and Biology Society*, 1985–1988.

20. Kato, T., Sekino, M., Matsuzaki, T., Nishikawa, A., Saitoh, Y., and Ohsaki, H. 2012. Electromagnetic characteristics of eccentric figure-eight coils for transcranial magnetic stimulation: A numerical study. *Journal of Applied Physics* **111**: 07B322.

21. Sekino, M., Ohsaki, H., Takiyama, Y., Yamamoto, K., Matsuzaki, T., Yasumuro, Y., Nishikawa, A., Maruo, T., Hosomi, K., and Saitoh, Y. 2015. Eccentric figure-eight coils for transcranial magnetic stimulation. *Bioelectromagnetics* **36**: 55–65.

22. Sekino, M., and Suyama, M. 2014. A magnetic stimulator coil with high robustness to positioning error. *URSI General Assembly and Scientific Symposium*, Beijing, China.

23. Yamamoto, K. Suyama, M. Takiyama, Y. Kim, D. Saitoh, Y. Sekino, M. Characteristics of bowl-shaped coils for transcranial magnetic stimulation. *Journal of Applied Physics* **117**: 17A318.

24. Epstein, C. M., and Davey, K. R. 2002. Iron-core coils for transcranial magnetic stimulation. *Journal of Clinical Neurophysiology* 19: 376–381.
25. Han, B. H., Lee, S. Y., Kim, J. H., and Yi, J. H. 2003. Some technical aspects of magnetic stimulation coil design with the ferromagnetic effect. *Medical & Biological Engineering & Computing* 41: 516–518.
26. Lu, M., and Ueno, S. 2009. Calculating the electric field in real human head by transcranial magnetic stimulation with shield plate. *Journal of Applied Physics* 105: 07B322.
27. Bischoff, C., Machetanz, J., Meyer, B. U., and Conrad, B. 1994. Repetitive magnetic nerve stimulation: Technical considerations and clinical use in the assessment of neuromuscular transmission. *Electroencephalography and Clinical Neurophysiology* 93: 15–20.
28. Weyh, T., Wendicke, K., Mentschel, C., Zantow, H., and Siebner, H. R. 2005. Marked differences in the thermal characteristics of figure-of-eight shaped coils used for repetitive transcranial magnetic stimulation. *Clinical Neurophysiology* 116: 1477–1486.
29. Sekino, M., Kato, T., Ohsaki, H., Saitoh, Y., Matsuzaki, T., and Nishikawa, A. 2012. Eccentric figure-eight magnetic stimulator coils. *Proceedings of the ICME International Conference on Complex Medical Engineering*, 728–733.
30. Epstein, C. M. 2008. A six-pound battery-powered portable transcranial magnetic stimulator. *Brain Stimulation* 1: 128–130.
31. De Sauvage, R. C., Beuter, A., Lagroye, I., and Veyret, B. 2010. Design and construction of a portable transcranial magnetic stimulation (TMS) apparatus for migraine treatment. *Journal of Medical Devices* 4: 015002-1-6.
32. Iramina, K., Maeno, T., Nonaka, Y., and Ueno, S. 2003. Measurement of evoked electroencephalography induced by transcranial magnetic stimulation. *Journal of Applied Physics* 93: 6718–6720.
33. Iramina, K., Maeno, T., and Ueno, S. 2004. Topography of EEG responses evoked by transcranial magnetic stimulation to the cerebellum. *IEEE Transactions on Magnetics* 40: 2982–2984.
34. Shitara, H., Shinozaki, T., Takagishi, K., Honda, M., and Hanakawa, T. 2011. Time course and spatial distribution of fMRI signal changes during single-pulse transcranial magnetic stimulation to the primary motor cortex. *Neuroimage* 56: 1469–1479.
35. Leitão, J., Thielscher, A., Werner, S., Pohmann, R., and Noppeney, U. 2013. Effects of parietal TMS on visual and auditory processing at the primary cortical level—A concurrent TMS-fMRI study. *Cerebral Cortex* 23: 873–884.
36. Groppa, S., Oliviero, A., Eisen, A., Quartarone, A., Cohen, L. G., Mall, V., Kaelin-Lang, A., Mimah, T., Rossi, S., Thickbroom, G. W., Rossini, P. M., Ziemann, U., Valls-Solém, J., and Siebner, H. R. 2012. A practical guide to diagnostic transcranial magnetic stimulation: Report of an IFCN committee. *Clinical Neurophysiology* 123: 858–882.
37. Balslev, D., Braet, W., McAllister, C., and Miall, R. C. 2007. Inter-individual variability in optimal current direction for transcranial magnetic stimulation of the motor cortex. *Neuroscience Methods* 162: 309–313.

38. Kowalski, T., Silny, J., and Buchner, H. 2002. Current density threshold for the stimulation of neurons in the motor cortex area. *Bioelectromagnetics* 23: 421–428.
39. Mano, Y., Morita, Y., Tamura, R., Morimoto, S., Takayanagi, T., and Mayer, R. F. 1993. The site of action of magnetic stimulation of human motor cortex in a patient with motor neuron disease. *Journal of Electromyography & Kinesiology* 3: 245–250.
40. Lu, M., Ueno, S., Thorlin, T., and Persson, M. 2008. Calculating the activating function in the human brain by transcranial magnetic stimulation. *IEEE Transactions on Magnetics* 44: 1438–1441.
41. Roth, B. J., and Basser, P. J. 1990. A model of the stimulation of a nerve fiber by electromagnetic induction. *IEEE Transactions on Biomedical Engineering* 37: 588–597.
42. Liu, R., and Ueno, S. 2000. Calculating the activating function of nerve excitation in inhomogeneous volume conductor during magnetic stimulation using the finite element method. *IEEE Transactions on Magnetics* 36: 1796–1799.
43. Dawson, T. W., and Stuchly, M. A. 1996. Analytic validation of a three-dimensional scalar-potential finite-difference code for low-frequency magnetic induction. *Applied Computational Electromagnetics Society Journal* 11: 63–71.
44. Nagaoka, T., Watanabe, S., Sakurai, K., Kunieda, E., Watanabe, S., Taki, M., and Yamanaka, Y. 2004. Development of realistic high-resolution whole-body voxel models of Japanese adult males and females of average height and weight, and application of models to radio-frequency electromagnetic-field dosimetry. *Physics in Medicine and Biology* 49: 1–15.
45. Laakso, I., Hirata, A., and Ugawa, Y. 2014. Effects of coil orientation on the electric field induced by TMS over the hand motor area. *Physics in Medicine and Biology* 59: 203–218.
46. Opitz, A., Legon, W., Rowlands, A., Bickel, W. K., Paulus, W., and Tyler, W. J. 2013. Physiological observations validate finite element models for estimating subject-specific electric field distributions induced by transcranial magnetic stimulation of the human motor cortex. *Neuroimage* 81: 253–264.
47. Sekino, M., and Ueno, S. 2002. Comparison of current distributions in electroconvulsive therapy and transcranial magnetic stimulation. *Journal of Applied Physics* 91: 8730–8732.
48. Sekino, M., and Ueno, S. 2004. Numerical calculation of eddy currents in transcranial magnetic stimulation for psychiatric treatment. *Neurology and Clinical Neurophysiology* 88: 1–5.
49. Sekino, M., and Ueno, S. 2004. FEM based determination of optimum current distribution in transcranial magnetic stimulation as an alternative to electroconvulsive therapy. *IEEE Transactions on Magnetics* 40: 2167–2169.
50. Sekino, M., Hirata, M., Sakihara, K., Yorifuji, S., and Ueno, S. 2006. Intensity and localization of eddy currents in transcranial magnetic stimulation to the cerebellum. *IEEE Transactions on Magnetics* 42: 3575–3577.

Applications of Biomagnetic Stimulation for Medical Treatments and for Brain Research

Shoogo Ueno and Masaki Sekino

CONTENTS

3.1 INTRODUCTION

3.1.1 Applications in Brain Research

It is well known that each part of the brain has a different function. In the early days of neuroscience, knowledge about localization of brain function was accumulated through observations of patients suffering brain disorders. Cerebrovascular disorders or injuries occurring in a specific part of the brain often cause a particular brain function abnormality. This reveals that a specific part of the brain is necessary for fulfilling the function. Broca and Welnicke identified the speech center based on such an approach. Observations of patients with brain disorders played important roles in understanding the localization of brain function. However, a methodological limitation existed owing to the uncontrollability of the location of the disorders. Penfield performed direct electric stimulations to a patient's brain during a surgical treatment of epilepsy and recorded the evoked responses. Various mental activities arose according to the positions of stimulation, which provided a brief map of the localized brain function. Because transcranial electric stimulations of the brain cause pain, mapping of brain function using electric stimulation could not be performed practically except for patients under craniotomy.

As discussed in Chapter 2, transcranial magnetic stimulation (TMS) enables noninvasive and localized stimulation of the brain. Such localized stimulation, resembling Penfield's work, makes it possible to depict a functional map of the brain even for normal subjects. Moreover, magnetic stimulation exerted on a region where spontaneous brain activities occur transiently interferes with these activities. This TMS ability represents the feasibility of producing a virtual lesion that imitates the above disorders. Such interference of brain function, once controlled, owns superiority in terms of reproducibility and safety, and this approach affirmatively facilitates systematic investigation of functional localization of the brain.

3.1.2 Applications in Medicine

TMS is used for clinical examinations of neuronal pathways that connect the primary motor cortex and peripheral muscles. To be specific, TMS applied to the primary motor cortex evokes contraction of muscles, for example, in a finger, and the motor evoked potential (MEP) measured through electrodes provides a criterion. It has been well studied that amplitudes and latencies of MEP are affected by neurological diseases such as cerebral infarction and spinal cord injury. These examinations using TMS are now widely used in clinical practice.

Therapeutic applications of TMS for neurological diseases and mental illnesses have also been significantly developed in recent years. Repetitive TMS (rTMS) that leads to functional changes in the central nervous system is reported to be effective for relieving symptoms of a variety of diseases, including depression, Parkinson's disease, and neuropathic pain. TMS treatment is effective even for some patients who failed in drug treatment, and TMS hardly causes side effects. The Food and Drug Administration (FDA) gave clearance for TMS treatment on depression. In addition, animal experiments showed that preconditioning of the hippocampus using rTMS induced ischemic tolerance of the hippocampus. With the successive clinical results of TMS treatment on other diseases, therapeutic applications of rTMS will continue to progress in the future.

In addition to the brain, magnetic stimulation is applied to nerves in the pelvic floor to treat urinary incontinence and to relieve muscle fatigue. The rTMS is a novel treatment complementary to drugs, and it will is expected to improve through further studies and clinical trials.

3.2 FUNCTIONAL MAPPING OF THE BRAIN

3.2.1 Principles

3.2.1.1 Characteristics of Figure-Eight Coils

TMS is capable of stimulating targeted areas in the brain. Among various coils that have been developed, the figure-eight coil is one of the most widely used.[1,2] The figure-eight coils are advantageous in stimulating localized

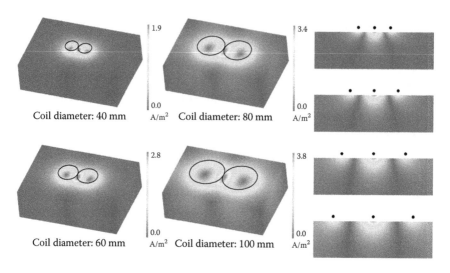

FIGURE 3.1   Distributions of eddy currents in a conductive cuboid induced by figure-eight coils with different diameters. Large diameter leads to broad and deep distributions of eddy currents.

targets compared with other types of coils. When an electric current flows through the figure-eight coil, the current generates time-varying magnetic fields from the two circular coils in the opposite directions. Eddy currents are induced in the target in the direction opposite the coils' currents so that the magnetic fields arising from the eddy currents attenuate the primary magnetic fields generated from the figure-eight coil. Eddy currents in the two loops flow in the same direction below the middle of the coil. This convergence of eddy currents leads to the locally increased current density. Furthermore, the figure-eight coil induces eddy current in a specific direction at this location. Figure 3.1 shows that the intensity of the eddy current density in the target increases as the diameter of figure-eight coils becomes larger. The maximum eddy current density induced by the 100-mm coil was twice as high as that of the 40-mm coil. The maximum eddy current density was observed on the surface of the target for all the coils. The magnetic fields attenuate with the distance from the coil. The above mentioned two characteristics of figure-eight coils, i.e., the intensity of eddy current density and the penetration depth, can be adjusted by varying the diameters.

3.2.1.2 Spatial Resolution of Brain Mapping
The brain is divided into small areas according to their specific functions. When the motor cortex in the right hemisphere relevant to the thumb is

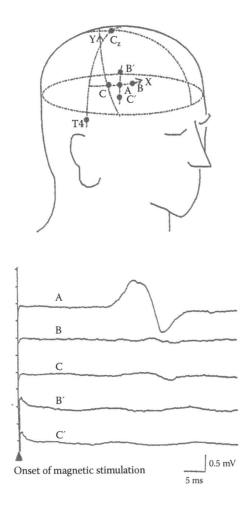

FIGURE 3.2   EMG responses to TMS. A: Thenar point stimulation. B: Stimulation at a point 5 mm posterior to the thenar point. C: Stimulation at a point 5 mm anterior to the thenar point. B′: Stimulation at a point 5 mm above the thenar point. C′: Stimulation at a point 5 mm lower from the thenar point.[2]

stimulated, the left thumb twitches. Figure 3.2 shows the recorded MEPs with varied locations of stimulation. A clear MEP was observed when the motor cortex corresponding to the left thumb, point A, was stimulated. No specific waveform was acquired at points B, B′, C, and C′, whose distances from A were all 5 mm. This result indicates that the localized stimulation of the human motor cortex was achieved with a spatial resolution of 5 mm. Figure 3.3 shows the functional mapping of the regions relevant to the hand and foot. The arrows show the directions of eddy currents

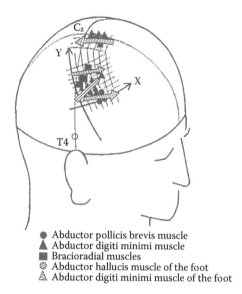

● Abductor pollicis brevis muscle
▲ Abductor digiti minimi muscle
■ Bracioradial muscles
⊘ Abductor hallucis muscle of the foot
⚠ Abductor digiti minimi muscle of the foot

FIGURE 3.3 Functional distribution of the human motor cortex related to the hand and foot areas. The arrows show current directions for neural excitation. The distance between grid points is 5 mm.[2]

exhibiting the maximum excitability. For example, when the stimulation was applied at point A with the opposite direction, the thumb did not respond. The regional and directional dependences of excitability reflect the structures of gyri and sulci of the brain.

3.2.1.3 Electroencephalography Responses Evoked by Magnetic Stimulation

The combination of TMS with functional brain imaging such as functional MRI (fMRI), positron emission tomography (PET), and electroencephalography (EEG) is an effective approach for investigating the functional connectivity in the brain. EEG is a useful tool for functional mapping because of its high temporal resolution. The combination of TMS with simultaneous EEG provides us the possibility to noninvasively probe the brain's excitability, time-resolved activity, and functional connectivity.[3] The experimental setup for recording EEG evoked by TMS is shown in Figure 3.4. The electrical brain activities are recorded through the electrodes on the surface of the head. The example of recorded EEG is shown in Figure 3.5.

However, combining TMS and EEG is not simple work owing to the heat and artifact induced by the interference between TMS and electrodes. Dhuna et al.[4] reported the possibility of burn injury by heating the

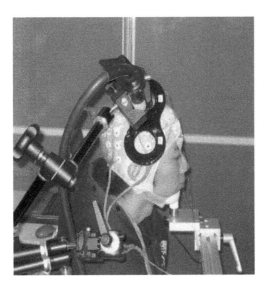

FIGURE 3.4 Experimental setup for recording electroencephalograms evoked by TMS.

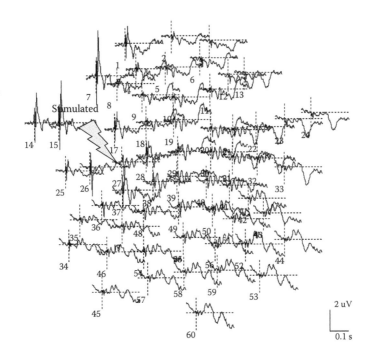

FIGURE 3.5 Recorded electroencephalogram. Because of the dedicated amplifier and electrode for combination with TMS, the electroencephalograms were not affected by the artifact of TMS.

electrode (Ag-AgCl). Eddy current induced by time-varying magnetic field generates the heat. In order to reduce the eddy current in the electrode, an electrode with slits was developed. With this electrode, heating was reduced by 32 percent.[4] In order to suppress the artifact, TMS-compatible EEG systems also have been developed. EEG activities evoked by TMS were measured successfully with a sample-and-hold circuit controlled by the trigger signal from the magnetic stimulator. In addition to these techniques, the independent component analysis was also introduced to remove the stimulus artifact in off-line analysis.[5]

3.2.2 Medical Applications

Functional mapping of the brain is applied also for neurosurgery.[6] Identifying motor and language areas before surgery is important for preventing damage to these areas. Identification using TMS is shown in Figure 3.6.

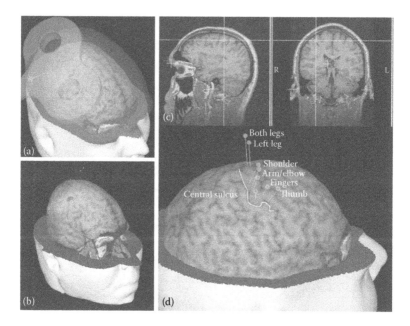

FIGURE 3.6 Representative demonstration of navigated brain stimulation for human brain mapping.[6] (a, b) Stimulation area is reached by utilizing on-line, interactive interface that guides the stimulation coil to the target. The maximum electric field induced in the brain is visualized, and the strength of the field in the target is determined. (c) The system allows precise determination of a stimulation target from other imaging modalities or according to anatomical landmarks. (d) Reliable map of motor areas can be reached by recording the muscle responses to stimulation.

The functional map that Penfield obtained using electric stimulations can now be noninvasively reproduced using TMS with a figure-eight coil.

There are several techniques available for functional mapping of the brain. For example, motor area identification can be conducted using fMRI, and language area identification can be conducted with the Wada test. However, the accuracy of the identification using fMRI has not yet been established, and the Wada test has a problem of invasiveness. Identification using TMS is totally noninvasive, and TMS can be performed even in an operating room. For these reasons, TMS is useful for surgical planning.

3.3 APPLICATIONS IN BRAIN RESEARCH

3.3.1 Virtual Lesion

TMS enables spatiotemporal controls of brain activities, with both excitation and disturbance, with varied pulse intensities and frequencies. By virtue of this feature, neuroscientists use the virtual lesion approach in which TMS causes a transient disturbance in the brain function. Brain function is generally realized through interactions in multiple regions of the brain. For instance, in visual perception processes such as visual search, letter recognition, and face recognition, giving TMS to a specific part of the brain causes a reduction in the performance of visual perception in some cases. These findings indicate that the stimulated area is necessary for the realization of visual perception. The influences of TMS on visual perception have been investigated with precisely controlled locations and timing for administering virtual lesions. This methodology provides a powerful tool for investigating the mechanism of spatiotemporal information processing in the brain related to visual perception. Virtual lesion facilitates the accumulation of knowledge on the localization of brain function.

3.3.2 Applications in Cognitive Neuroscience

3.3.2.1 Episodic Memory

Episodic memory is a sequential memory related to memorable things that contain time, place, and emotion of the memories. It is a kind of long-term memory and is sometimes related to personal experiences or social occurrences. It is generally recognized that the prefrontal area has a role in generation of episodic memories, but there are still different hypotheses about the localization of the detailed function in the prefrontal area.

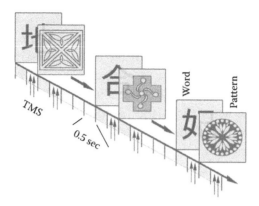

FIGURE 3.7   Timeline (left to right) for presentation of Kanji words and abstract patterns in one trial of 9-sec duration.[7] Pulses of TMS are indicated by paired vertical arrows. Shaded areas beneath the baseline indicate when cognitive stimuli are visible.

Epstein et al.[7] carried out a study to reveal this localization. As shown in Figure 3.7, ten Japanese subjects underwent sequential visual stimuli, which contained 18 sets of Kanji words and abstract patterns, and TMS pulses were delivered between the visual stimulations. The TMS coil was placed at the left dorsolateral prefrontal cortex (DLPFC), the right DLPFC, the cranial vertex, and off the head. After the set of stimuli, subjects took a test to check the memory correctness of the pair of Kanji and abstract patterns. As shown in Figure 3.8, the subjects had fewer correct responses with the right DLPFC stimulation when compared with the stimulations on other areas. This result indicates that the right DLPFC has a role in generating the episodic memory. As in this example, TMS can elucidate the mechanisms of higher brain function.

3.3.2.2 Influence on P300 Evoked Potentials

P300 is a kind of EEG response that appears when subjects undergo some stimulation or task and pay attention or memorize something. The memorable stimulations have a lot of variety. When the subjects notice something, P300 is observed after 300 msec of the stimulation.

The study by Iwahashi et al.[8] using TMS revealed the detailed brain function related to P300 generation. According to this study, when a TMS was applied on the left supramarginal gyrus after 200 and 250 msec of the auditory stimulation, the latency of P300 was delayed about 50 msec. However, in the case of applying TMS after 100 and 150 msec of the

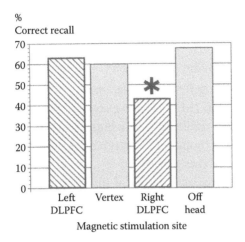

FIGURE 3.8 Overall percent correct for subsequent recall of paired associates according to site of TMS.[7] The comparison of interest is between the left and right DLPFC, with an active TMS control at the vertex. In a second control, the magnetic coil is off the head. The asterisk indicates $P < 0.05$ for right versus left DLPFC.

auditory stimulation, the delay was not observed. Because the TMS inhibits the function of the stimulated area, these results suggest that the cortex around the left supramarginal gyrus contributes to the generation of P300 at around 200 msec after the auditory stimulation.

3.3.3 Paired-Pulse Stimulation

Paired-pulse stimulation is a protocol that has a short time interval between the first and second stimulation. The first stimulation is applied with a low intensity that does not evoke MEP, and the second stimulation is applied with a high intensity that evokes MEP. The influence of the first stimulation on MEP is evaluated, and using this methodology, one can reveal the synaptic function in the motor cortex or discuss the related physiological mechanisms.

In the study by Kobayashi et al.[9] the stimulation intensity was fixed at 70–80 percent for the first one, and 120–140 percent for the second one, and the interval was changed from 1 to 15 msec to investigate the effects on MEP. Observed MEP had two peaks, N1 and N2, and the increase in the interval resulted in the higher N2 amplitude, while the N1 amplitude was not affected. The authors concluded that the N1 was the direct wave (D wave) resulting from the directly stimulated pyramidal neurons, and the N2 was the indirect wave (I wave) resulting from the pyramidal neurons indirectly stimulated through interneurons. This study is important as evidence that there are two types of responses evoked by TMS.

3.3.4 Repetitive Magnetic Stimulation

3.3.4.1 Changes in Long-Term Potentiation

Long-term potentiation (LTP) is a long-lasting increase in the efficacy of synaptic transmission resulting from a high-frequency stimulation.[10] LTP in the hippocampus is thought to be a typical model of synaptic plasticity related to learning and memory.

The most commonly used protocol for LTP induction is an electric stimulation to hippocampus slices. Electric stimulation to the hippocampus slices, typically to the Schaffer collaterals, generates excitatory postsynaptic potentials (EPSPs) in the postsynaptic CA1 cells. If the Schaffer collaterals are stimulated only two or three times per minute, the magnitude of the evoked EPSP in the CA1 neurons remains constant. However, a brief, high-frequency train of stimuli to the same neurons causes LTP, which is evident as a long-lasting increase in EPSP magnitude (Figure 3.9). This high-frequency electric stimulation of a neuron is called tetanus stimulation.

Recently, rTMS has become an increasingly important tool for the potential treatment of neurological and psychiatric disorders. Depression and Parkinson's disease are major targets of the rTMS treatment.[11,12] Many animal studies have reported rTMS-induced changes in cellular functions.[13] Gene expressions, such as c-fos, glial fibrillary acidic protein (GFAP), and brain-derived neurotrophic factor (BDNF), were enhanced in the rat brain by rTMS.[14-17] And rTMS-related effects in the hippocampus

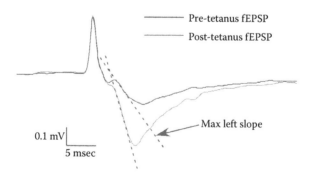

FIGURE 3.9 Basic mechanisms of LTP.[13,20] The plots show the changes in field EPSP (fEPSP) after LTP induction stimuli (tetanus stimulation).

have been investigated on, e.g., monoamine release, neurogenesis, and memory function.[17–19]

Ogiue-Ikeda et al.[13,20] investigated the effects of rTMS on rat hippocampus, focusing on changes in LTP after rTMS. The authors firstly investigated the influence of rTMS on field excitatory postsynaptic potentials (fEPSP) after applying tetanus stimulation to the rat hippocampus.[20] The rats were magnetically stimulated by a round coil positioned over the head. The stimulator delivered biphasic cosine current pulses 238 μsec in duration. The peak magnetic fields were set to 0.50 T (lower than motor threshold) and 1.25 T (above motor threshold) at the center of the coil. The motor threshold was defined as the intensity with which the hind limbs moved as a result of the magnetic stimulation. Rats received 10 sec trains of 25 pulses/sec with a 1-sec intertrain interval four times per day for 7 days. After magnetic stimulation of 7 days, hippocampus slices were prepared and fEPSPs were recorded from the dendrites of CA1 pyramidal cells by stimulating Schaffer collaterals. There was no significant difference between the LTP of the 0.50-T stimulated and sham control groups. The LTP of the 1.25-T stimulated group, however, was inhibited compared with the LTP of the sham control group, suggesting that the synaptic plasticity in the hippocampus was impaired by strong TMS, as shown in Figure 3.10b. The calculation of eddy current distribution showed that the estimated eddy currents around the hippocampus were approximately 4 A/m^2 (0.5 T) and 10 A/m^2 (1.25 T), respectively.

The authors also demonstrated the same rTMS experiment to examine the dependency on the field intensity.[13] Rats were magnetically stimulated for 7 days with the condition described above, but the peak magnetic fields at the center of the coil were modified as 0.75 and 1.00 T. LTP enhancement was observed only in the 0.75-T rTMS group, as shown in Figure 3.10a, while no change was observed in the 1.00-T rTMS group. These results suggest that the effect of rTMS on LTP depends on the stimulus intensity.

3.3.4.2 Ischemic Tolerance of Hippocampus

A mild exposure to cerebral ischemia confers transient tolerance to a subsequent ischemic challenge in the brain. This phenomenon of ischemic tolerance has been confirmed in various animal models of forebrain ischemia and focal cerebral ischemia. On the contrary, CA1 pyramidal neurons in the hippocampus are highly vulnerable to cerebral ischemia.[21] Brief periods (minutes) of severe ischemia cause neuronal degeneration

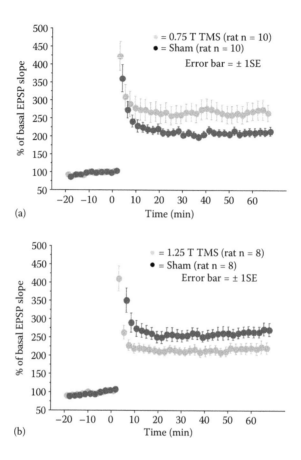

FIGURE 3.10   Changes in LTP at rat hippocampus slices after rTMS[13,20]: (a) LTPs of 0.75-T stimulated and sham control groups and (b) LTPs of 1.25-T stimulated and sham control groups.

in the CA1 region 3 to 7 days after the ischemia by apoptosis.[22] Recent studies have focused on applying electric stimulation to the brain via electrodes to potentially reduce ischemic symptoms.[23] However, the invasiveness of electric stimulation (e.g., surgery and anesthetization) reduces its feasibility as a therapeutic treatment.

Instead of invasive electric stimulation, Ogiue-Ikeda et al.[21] investigated the acquisition of ischemic tolerance in the rat hippocampus using rTMS. On the basis of the authors' previous studies about rTMS effects on hippocampus LTP,[13,20] the stimulus intensity was set at 0.75 T (the estimated maximum eddy current was approximately 9 A/m² in the brain). The rats received 0.75-T rTMS of 1000 pulses/day for 7 days. The fEPSPs were measured in the hippocampal CA1. After rTMS treatment, the hippocampus

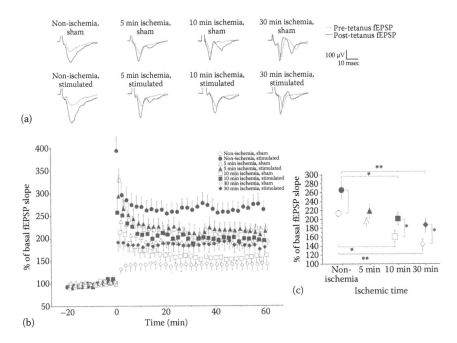

(a)

(b)

(c)

FIGURE 3.11 Effects of rTMS on ischemic tolerance of hippocampus.[21] (a) Examples of pre-tetanus and post-tetanus fEPSPs after each ischemic condition. (b) LTP in the stimulated and sham control groups in each ischemic condition. (c) Average value of the maintenance phases of LTP (ranging from 10 min after tetanus stimulation to 60 min). Open circle: LTPs in the nonischemia group. Open circle, square, triangle, and diamond represent LTPs in the nonischemia group, postischemia 5, 10, and 30 min after ischemia in the sham control group, respectively. Closed circle, square, triangle, and diamond represent LTPs in the nonischemia group, postischemia 5, 10, and 30 min after ischemia in the sham control group, respectively.

slices were exposed to ischemic conditions to examine the influence of rTMS on ischemic tolerance. LTP was induced by the same protocol as described in the previous section. The LTP of the stimulated group was enhanced compared with the LTP of the sham control group in each ischemic condition (Figure 3.11), suggesting that rTMS has the potential to protect hippocampal function from ischemia.

3.3.4.3 Neuronal Plasticity

Studies on neuronal plasticity attract attention in association with learning, memory, and recovery of brain function through rehabilitation. Recent neuroimaging studies using functional MRI and magnetoencephalography have revealed that cerebral infarction and limb amputation cause partial changes in the localization of brain function.

The primary somatosensory cortex and primary motor cortex are located posteriorly and anteriorly, respectively, in the central sulcus. Assume, for example, that one has had the right-hand amputated. The brain is intact, and the neurons in the motor and somatosensory areas corresponding to the right hand suddenly lose their roles. In the following days and weeks, the localization of brain function changes so that adjacent brain areas, corresponding to the lip, for example, expand their areas to irrupt into the right-hand area. Similar changes occur also in the motor cortex. TMS may be used to interpose these processes for the patient to obtain better quality of life afterward. TMS may enable such rehabilitation after a cerebral infarction by stimulating the primary motor area, supplementary motor area, and prefrontal area with prospectively programmed patterns for appropriate stimulations.

3.4 APPLICATIONS IN MEDICINE

3.4.1 Diagnosis of Nervous System

The most popular method of the magnetic stimulation to examine the nervous system is the measurement of central motor conduction time (CMCT).[24] The CMCT is the time taken for neural impulses to travel through the central nervous system on their way to the target muscles. In this method, the primary motor cortex (M1) is stimulated by TMS and MEP is recorded. Subsequently, the nerve root on the neck or lumbar is stimulated with recording MEP. Finally, the CMCTs with these MEP latencies (cortex latency–spinal cord latency) are calculated. The delay of CMCT reflects the disturbance of pyramidal tract.

A great number of TMS (or magnetic stimulation) studies for the inspection of nerve disorders have been reported.[25-32] Guillain–Barré syndrome (GBS) and chronic inflammatory demyelinating polyneuropathy (CIDP) are the representative inflammatory demyelinating peripheral nerve diseases. In several reports, TMS has been applied to these diseases to inspect a conduction disturbance.[25-27] Oshima et al.[28] reported the use of magnetic simulation as a tool to determine the involvement of the corticospinal tract in GBS patients with hyperreflexia. The authors examined the central motor conduction, and the results revealed axonal motor neuropathy and normal F-wave conduction in these patients. In this case, the inspection of nerve conduction by the magnetic stimulation indicated the functional corticospinal tract involvement in patients with a GBS variant. Charcot-Marie-Tooth (CMT) disease is the representative inherited disorders of the peripheral nervous system, and CMT has four variants. Clays et al.[29] reported the differences in CMCT in CMT variants using magnetic stimulation. The CMCT in CMT1 showed a remarkable delay while that of CMT2 delayed slightly. Diabetes leads to peripheral neuropathy. Tchen[30] reported that TMS for diabetes showed a delay of CMCT. Maetzu et al.[31] used TMS for diabetes and measured the CMCT based on the calculation using F-wave latency. The authors showed that motor transmissions were not damaged in diabetic patients. Öge reported that CMCT extended in n-hexane polyneuropathy,[32] and Chokroverty stimulated the phrenic nerve at the neck and root and showed that magnetic stimulation is potentially useful for clinical application.[33]

3.4.2 Treatment of Neurological Diseases

3.4.2.1 Parkinson's Disease

Parkinson's disease (PD) is a neurodegenerative brain disorder that occurs when neurons in the substantia nigra die or become impaired. Normally, these cells produce dopamine, which allows smooth, coordinated function of the body's muscles; therefore, less dopamine leads the patient with less ability to regulate their movements and emotions.

In recent years, some clinical trials of single or rTMS for PD have been performed. A pioneering study has been published by Pascual-Leone et al.[34] In the report, effects of 5-Hz rTMS on several neurological parameters such as choice reaction time (cRT), movement time (MT), and error rate (ER) in a serial reaction-time task were investigated in six medicated patients with PD and 10 age-matched normal controls. In normal subjects, subthreshold 5-Hz rTMS did not significantly change cRT, slightly shortened MT, but increased ER. In the patients, rTMS significantly shortened cRT and MT without affecting ER. The authors mentioned that the repetitive, subthreshold motor cortex stimulation can improve performance in patients with PD and could be useful therapeutically.

In animal experiments, Funamizu et al.[35] demonstrated that rTMS showed protecting effects on the lesioned rats by administering a neurotoxin MPTP (l-methyl-4-phenyl-l,2,3,6-tetrahydropyridine). Forty-eight hours before rTMS treatment, MPTP was injected into the rats to induce specific damages in dopaminergic neurons. Subsequently, the rats received 1.25-T rTMS (10 trains of 25 pulses/sec for 8 sec) from the round coil that is described in Section 3.3.4.1. The rTMS effects were examined by histological studies at substantia nigra, which contains a high population of dopaminergic neurons and checking the behavioral parameters (functional observational battery-hunched posture score). The functional observational battery-hunched posture score for the MPTP-rTMS group was significantly lower and the number of rearing events was higher compared with the MPTP-sham group. These behavioral parameters revert to control levels (Figure 3.12a). The double-labeling immunofluoresence experiments using tyrosine hydroxylase as the dopaminergic marker and NeuN as the specific neuronal marker showed that the number of surviving dopamine neurons in the MPTP-sham rats were significantly reduced compared with that of the substantia nigra pars compacta in the undamaged rats, while the number of those neurons in the MPTP-rTMS group was significantly larger than the MPTP-sham group (Figure 3.12b). These

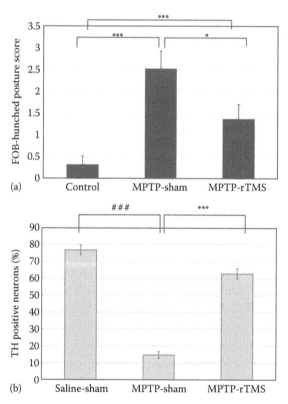

FIGURE 3.12 Effects of rTMS on MPTP-treated rats.[35] (a) Functional observational battery (FOB) of hunched posture score during a 3-min observation period in the home cage 3 days after administration of MPTP. (b) The percentage of TH positive neurons.

results suggested that rTMS treatment reactivates the dopaminergic system in lesion rats.

3.4.2.2 Neuropathic Pain

Neuropathic pain is caused by a lesion or dysfunction in the nerve systems. The nerves affected by some abnormalities (e.g., stroke, cancer, diabetes, and amputation) send wrong pain signals to the brain even without adequate sensory input. Induced pain like burning, aching, stabbing, shooting, or shocking electrically is often described.[36]

As a treatment for the patients suffering neuropathic pain, the efficacy of brain stimulation has been established. Tsubokawa et al.[37] first applied brain stimulation to patients with thalamic pain with implanting electrodes into the cerebral cortex. In this clinical trial, the direct stimulation of the motor cortex was effective for pain relief. Saitoh et al.[38] reported that six out of eight patients experienced pain reduction as a result of motor cortex stimulation, and this kind of stimulation might be effective for treating peripheral as well as central deafferentation pain. The direct stimulation of the motor cortex was effective for relieving neuropathic pain, and the therapeutic effect was considered strongly related to the stimulated area. The acquired pain relief supposed that transmission of the pain signal was modulated by the stimulation of the ascending nerve tracts.

In order to understand the mechanism of pain perception and the therapeutic effect of brain stimulation, studies using PET and fMRI have been performed in both animals and humans.[39,40] The results of these studies showed that brain stimulation of the motor cortex induced brain activations at the areas of thalamus, cognitive cingulate, and insula. Pain-relieving effect by brain stimulation is supposed to have a relationship with brain activations of these areas.

As a noninvasive treatment method of brain stimulation for relieving neuropathic pain, rTMS has been introduced. Recent clinical studies showed that stimulation of the motor cortex by rTMS is also effective in relieving neuropathic pain. Saitoh et al.[41,42] reported that the rTMS of the precentral gyrus at frequencies of 10 Hz reduced pain significantly, compared with lower stimulation of 1 or 5 Hz. The effect of rTMS at higher frequency was greater in patients with a noncerebral lesion than those with a cerebral lesion. These results indicate that patients with a noncerebral lesion are more suitable candidates for high-frequency rTMS of the precentral gyrus.

The therapeutic effect of rTMS for pain relief also depends on the stimulated area. Hirayama et al.[41] examined several cortical areas of motor cortex (M1), the postcentral gyrus (S1), premotor area (preM), and supplementary motor area (SMA) as stimulation targets using a navigation-guided rTMS. The authors reported 10 of the 20 patients (50 percent) indicated that stimulation of M1, but not other areas, provided significant and beneficial pain relief. That group's result indicated that a significant effect was lasting for 3 hours after the stimulation of M1 and stimulation of other targets was not effective. Interestingly, this indicates the M1 is the sole target for treating intractable pain with rTMS, whereas M1, S1, preM, and SMA are located adjacently.

In the future, the motor cortex stimulation using rTMS will be recognized as a treatment complementary to drug therapy.

3.4.3 Treatment of Psychiatric Diseases

A variety of clinical trials toward treatment of psychiatric diseases using TMS have been performed in this decade. So far, treatments by TMS for depression, schizophrenia, obsessive-compulsive disorder, and attention-deficit hyperactivity disorder have been reported. Clinical use of TMS for depression is the most widely investigated. Depression is a disorder of the brain and is caused by a combination of genetic, biological, environmental, and psychological factors. Imbalance of neurotransmitters, in particular, lack of serotonin, leads to depressive illness. Administration of depression medications (antidepressants, e.g., selective serotonin reuptake inhibitors) is the prime choice for depression treatment. However, treatment-resistant depression still remains in approximately 20 percent of the patients, and some of them show severe clinical conditions. Electroconvulsive therapy (ECT) is a technique used to treat patients with depression by injecting currents into the brain through electrodes positioned on the surface of the head.[43] To avoid side effects such as the loss of memory function, TMS is considered as a promising method for the treatment of depression in place of ECT or as an alternative to ECT. In early times of the clinical trial, application of single-pulse TMS to depression has been investigated. And then, George et al.[11] demonstrated rTMS treatment for six medication-resistant depressed inpatients and two patients showed remarkable improvement of clinical status. A number of studies have been performed for clinical trials for depression treatments afterward. Important parameters for treatment are stimulation sites and conditions. A prefrontal cortex (left dorsolateral prefrontal cortex) is well known to exhibit therapeutic effects by TMS in depression treatment.[11] Fast (5–20 Hz) or slow (1 Hz) rTMS with the intensity of 80–110 percent of the motor threshold are commonly used for the treatment. However, fixed parameters have not been reported and the prevention of seizures by rTMS is the most concerning problem.

3.4.4 Development of Magnetic Stimulator for Home Treatment

Current magnetic stimulators are only operated by doctors and are available only at large hospitals. Therefore, only a limited number of patients are able to be treated by rTMS. Moreover, the effect of rTMS is temporary, and the brain must be treated at fixed intervals—everyday, if possible. The development of inexpensive, compact, and energy efficient magnetic stimulators that can be installed at a patient's home is necessary to generalize magnetic stimulation treatment.

Researchers have sought to find a method to increase the efficiency of electromagnetic coils to induce magnetic stimulators operated at low powers. Though a variety of coils have been proposed previously, these coil designs have mainly aimed at controlling the region and depth of stimulation rather than the application of magnetic fields at low driving currents. To attain this objective, several papers have been published on how to increase efficiency of the coil by changing its shape. Sekino et al.[44-46] made the center of the inner peripheral circle of each coil shifted to the middle of the figure-eight coil to enhance the concentration of electric field directly under the center of the figure-eight coil, as shown in Figure 3.13. This coil design had an increased efficiency of 22 percent and decreased heat generation in proportion of the square of the current.

On the other hand, researchers have aimed to compensate the loss of robustness to locational error arising from the high locality of the induced electric field by adequately increasing the range where the electric field deploys. In one approach, Yamamoto et al.[47] proposed a bowl-shaped coil. They increased the uniformity of the induced electric field in the brain by arranging the conductor along the surface of head. Additionally, the intensity and range of induced electric coil is adjustable by changing the density and coverage of the conductor.

To make magnetic stimulators inexpensive, researchers have focused on improving the coil navigation system. In the conventional method, the stimulator coil is placed above the target point using an infrared binocular

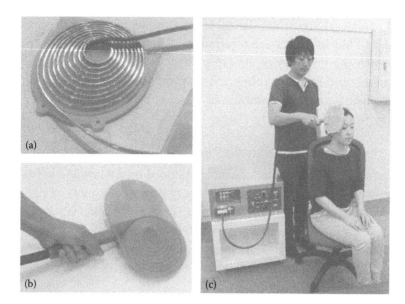

FIGURE 3.13 (a) Formation of an eccentric spiral coil on a plastic insulator case.[47] (b) Finished eccentric coil equipped with a grip and a lead cable. Coil conductors are observed through a 1-mm-thick insulator. (c) A prototype magnetic stimulator system for home treatment.

camera. Unfortunately, this system is not suited to home treatment because of its large size and cost. Accordingly, a compact and inexpensive positioning system has been developed.[48] In this new system, glass that is attached to the magnetic sensor has been developed. This system made it possible not only to decide the position of the coil with higher accuracy by 5 mm but also to decrease the size of the whole positioning system.

3.4.5 Safety of Magnetic Stimulation

Though it is generally recognized that TMS hardly causes adverse side effects, there are several potential side effects to be avoided.[49] The stimulator coil produces acoustic noise that may exceed 140 dB. This is louder than the maximum of the recommended safety level for the auditory system. The patient and operator should use ear protectors. The rTMS may induce a seizure; this is the most severe side effect. To be prepared in case of accidental side effects, the operator should record the patient's physiological signals during stimulations. If the patient has metallic implants near the stimulated region or has taken drugs that lower the threshold of seizure, the rTMS should be avoided.

3.5 SUMMARY AND FUTURE PROSPECTS

The focal and vectorial TMS is suitable for noninvasive and precise investigation of brain function, and this technique has opened a new field in neuroscience. TMS is a unique technique enabling external stimulations to the brain of healthy subjects. The concept of a virtual lesion, combined with a variety of tasks, is utilized as a powerful tool for elucidating functional connectivity in the brain. If coils enabling stimulations to deeper parts of the brain such as the hippocampus are developed, further understanding of the brain function will be realized.

Clinical studies on therapeutic applications of rTMS are actively carried out in the world. Trials for applying rTMS for neurological and psychiatric diseases are carried out as a treatment complementary to drug administrations. Compact and low cost magnetic stimulators are in demand because the treatments using rTMS require daily continuous stimulations in many cases. Further evolution may occur if the control of gene expression using rTMS becomes possible, with facilitation or inhibition of synapse formation or with expression of neuronal cell function around the lesion. Treatment of osteoporosis and other orthopedic diseases will also be explored. In parallel to the studies on the therapeutic effects of TMS through accumulating results of clinical studies and animal experiments, it is important to understand the mechanism of therapeutic effects using EEG and MRI. Deep understandings of the mechanism will lead to a systematic investigation of stimulus conditions for achieving a high efficacy and will realize a prior prediction of the efficacies for individual patients. TMS involves various parameters such as coil geometry, target region, repetition rate, intensity, and number of pulses. The optimal parameters should be established for each disease. Safety of TMS needs to be discussed further, incorporating up-to-date results of basic and clinical studies.

REFERENCES

1. Ueno, S., Tashiro, T., and Harada, K. 1988. Localized stimulation of neural tissues in the brain by means of a paired configuration of time-varying magnetic fields. *Journal of Applied Physics* **64**: 5862–5864.
2. Ueno, S., Matsuda, T., and Fujiki, M. 1990. Functional mapping of the human motor cortex obtained by focal and vectorial magnetic stimulation of the brain. *IEEE Transactions on Magnetics* **26**: 1539–1544.
3. Iramina, K., Maeno, T., and Ueno, S. 2004. Topography of EEG responses evoked by transcranial magnetic stimulation to the cerebellum. *IEEE Transactions on Magnetics* **40**: 2982–2984.

4. Dhuna, A., Gates, J., and Pascual-Leone, A. 1991. Transcranial magnetic stimulation in patients with epilepsy. *Neurology* **41**: 1067–1071.

5. Jung, T. P., Makeig, S., McKeown, M. J., Bell, A. J., Lee, T., and Sejinowski, T. J. 2001. Imaging brain dynamics using independent component analysis. *Proceeding of the IEEE* **89**: 1107–1122.

6. Ueno, S., and Fujiki, M. 2007. Magnetic stimulation. In *Magnetism in Medicine*, Andra, W. Nowak, H. eds. Wiley-VCH. Weinheim: 511–528.

7. Epstein, C. M., Sekino, M., Yamaguchi, K., Kamiya, S., and Ueno, S. 2002. Asymmetries of prefrontal cortex in human episodic memory: Effects of transcranial magnetic stimulation on learning abstract patterns. *Neuroscience Letters* **320**: 5–8.

8. Iwahashi, M., Katayama, Y., Ueno, S., and Iramina, K. 2009. Effect of transcranial magnetic stimulation on P300 of event-related potential. *Proceedings of the Annual International Conference of the IEEE Engineering in Medicine and Biology Society*, 1359–1362.

9. Kobayashi, M., Hyodo, A., Kurokawa, T., and Ueno, S. 1998. The first component of polyphasic motor evoked potentials is resistant to suppression by paired transcranial magnetic stimulation in humans. *Journal of the Neurological Sciences* **159**: 166–169.

10. Bliss, T. V. P., and Lomo, T. 1973. Long-lasting potentiation of synaptic transmission in the dentate area of the anaesthetized rabbit following stimulation of the perforant path. *Journal of Physiology* **232**: 357–374.

11. George, M. S., Wassermann, E. M., Williams, W. A., Callahan, A., Ketter, T. A., Basser, P., Hallett, M., and Post, R. M. 1995. Daily repetitive transcranial magnetic stimulation (rTMS) improves mood in depression. *Neuroreport* **6**: 1853–1856.

12. Mally, J., and Stone, T. W. 1999. Improvement in Parkinsonian symptoms after repetitive transcranial magnetic stimulation. *Journal of the Neurological Sciences* **162**: 179–184.

13. Ogiue-Ikeda, M., Kawato, S., and Ueno, S. 2003. The effect of repetitive transcranial magnetic stimulation on long-term potentiation in rat hippocampus depends on stimulus intensity. *Brain Research* **993**: 222–226.

14. Fujiki, M., and Steward, O. 1997. High frequency transcranial magnetic stimulation mimics the effects of ECS in upregulating astroglial gene expression in the murine CNS. *Molecular Brain Research* **44**: 301–308.

15. Hausmann, A., Weis, C., Marksteiner, J., Hinterhuber, H., and Humpel, C. 2000. Chronic repetitive transcranial magnetic stimulation enhances c-fos in the parietal cortex and hippocampus. *Molecular Brain Research* **76**: 355–362.

16. Muller, M. B., Toschi, N., Kresse, A. E., Post, A., and Keck, M. E. 2000. Long-term repetitive transcranial magnetic stimulation increases the expression of brain-derived neurotrophic factor and cholecystokinin mRNA, but not neuropeptide tyrosine mRNA in specific areas of rat brain. *Neuropsychopharmacology* **23**: 205–215.

17. Czeh, B., Welt, T., Fischer, A. K., Erhardt, A., Schmitt, W., Muller, M. B., Toschi, N., Fuchs, E., and Keck, M. E. 2002. Chronic psychosocial stress and concomitant repetitive transcranial magnetic stimulation: Effects on stress hormone levels and adult hippocampal neurogenesis. *Biological Psychiatry* **52**: 1057–1065.

18. Keck, M. E., Welt, T., Post, A., Muller, M. B., Toschi, N., Wigger, A., Landgraf, R., Holsboer, F., and Engelmann, M. 2001. Neuroendocrine and behavioral effects of repetitive transcranial magnetic stimulation in a psychopathological animal model are suggestive of antidepressant-like effects, *Neuropsychopharmacology* **24**: 337–349.

19. Post, A., Muller, M. B., Engelmann, M., and Keck, M. E. 1999. Repetitive trans-cranial magnetic stimulation in rats: Evidence for a neuroprotective effect in vitro and in vivo. *European Journal of Neuroscience* **11**: 3247–3254.

20. Ogiue-Ikeda, M., Kawato, S., and Ueno, S. 2003. The effect of transcranial magnetic stimulation on long-term potentiation in rat hippocampus. *IEEE Transactions on Magnetics* **39**: 3390–3392.

21. Ogiue-Ikeda, M., Kawato, S., and Ueno, S. 2005. Acquisition of ischemic tolerance by repetitive transcranial magnetic stimulation in the rat hippocampus. *Brain Research* **1037**: 7–11.

22. Kirino, T. 1982. Delayed neuronal death in the gerbil hippocampus following ischemia. *Brain Research* **239**: 57–69.

23. Glickstein, S. B., Ilch, C. P., Reis, D. J., and Golanov, E. V. 2001. Stimulation of the subthalamic vasodilator area and fastigial nucleus independently protects the brain against focal ischemia. *Brain Research* **912**: 47–59.

24. Matsumoto, H., Hanajima, R., Shirota, Y., Hamada, M., Terao, Y., Ohminami, S., Furubayashi, T., Nakatani-Enomoto, S., and Ugawa, Y. 2010. Cortico-conus motor conduction time (CCCT) for leg muscles. *Clinical Neurophysiology* **121**: 1930–1933.

25. Britton, T. C., Meyer, B. U., Herdmann, J., and Benecke, R. 1990. Clinical use of the magnetic stimulator in the investigation of peripheral conduction time. *Muscle & Nerve* **13**: 396–406.

26. Takada, H., and Ravnborg, M. 1997. A comparative study of magnetically evoked motor potentials (MEP) in polyneuropathy: Demyelinating vs. axonal. *Electroencephalography and Clinical Neurophysiology* **103**: 113.

27. Benecke, R. 1996. Magnetic stimulation in the assessment of peripheral nerve disorders. *Bailliere's Clinical Neurology* **5**: 115–128.

28. Oshima, Y., Mitsui, T., Endo, I., Umaki, Y., and Matsumoto, T. 2000. Corticospinal tract involvement in a variant of Guillain-Barre syndrome. *European Neurology* **46**: 39–42.

29. Clays, D., Waddy, H. M., Harding, A. E., Murray, N. M. F., and Thomas, P. K. 1990. Hereditary motor and sensory neuropathies and hereditary spastic paraplegia: A magnetic stimulation study. *Annals of Neurology* **28**: 43–49.

30. Tchen, P. H., Fu, C. C., and Chiu, H. C. 1992. Motor-evoked potentials in diabetes mellitus. *Journal of the Formosan Medical Association* **91**: 20–23.

31. Maetzu, C., Villoslada, C., Cruz Martínez, A. 1995. Somatosensory evoked potentials and central motor pathways conduction after magnetic stimulation of the brain in diabetes. *Electromyography and Clinical Neurophysiology* **35**: 443–448.

32. Öge, A. M., Yazici, J., Boyaciyan, A., Eryildiz, D., Ornek, I., Konyalioğlu, R., Cengiz, S., Okşak, O. Z., Asar, S., and Baslo, A. 1994. Peripheral and central conduction in n-hexane polyneuropathy. *Muscle & Nerve* **17**: 1416–1430.

33. Chokroverty, S., Shah, S., Chokroverty, M., Deutsch, A., and Belsh, J. 1995. Percutaneous magnetic coil stimulation of the phrenic nerve roots and trunk. *Electroencephalography and Clinical Neurophysiology* **97**: 369–374.

34. Pascual-Leone, A., Valls-Sole, J., Brasil-Neto, J. P., Cammarota, A., Grafman, J., and Hallett, M. 1994. Akinesia in Parkinson's disease: II. Effects of subthreshold repetitive transcranial motor cortex stimulation. *Neurology* **44**: 892–898.

35. Funamizu, H., Ogiue-Ikeda, M., Mukai, H., Kawato, S., and Ueno, S. 2005. Acute repetitive transcranial magnetic stimulation reactivates dopaminergic system in lesion rats. *Neuroscience Letters* **383**: 77–81.

36. Siddall, P. J., Taylor, D. A., McClelland, J. M., Rutkowski, S. B., and Cousins, M. J. 1999. Pain report and the relationship of pain to physical factors in the first 6 months following spinal cord injury. *Pain* **81**: 187–197.

37. Tsubokawa, T., Katayama, Y., Yamamoto, T., Hirayama, T., and Koyama, S. 1991. Treatment of thalamic pain by chronic motor cortex stimulation. *Pacing and Clinical Electrophysiology* **14**: 131–144.

38. Saitoh, Y., Shibata, M., Hirano, S., Hirata, M., Mashimo, T., and Yoshimine, T. 2000. Motor cortex stimulation for central and peripheral deafferentation pain: Report of eight cases. *Journal of Neurosurgery* **92**: 150–155.

39. García-Larrea, L., Peyron, R., Mertens, P., Gregoire, M. C., Lavenne, F., Bars, D., Le Convers, P., Mauguière, F., Sindou, M., and Laurent, B. 1999. Electrical stimulation of motor cortex for pain control: A combined PET-scan and electrophysiological study, *Pain* **83**: 259–273.

40. Kim, D., Chin, Y., Reuveny, A., Sekitani, T., Someya, T., and Sekino, M. 2014. An MRI-compatible, ultra-thin, flexible stimulator array for functional neuroimaging by direct stimulation of the rat brain. *Proceedings of the Annual International Conference of the IEEE Engineering in Medicine and Biology Society* 6702–6705.

41. Hirayama, A., Saitoh, Y., Kishima, H., Shimokawa, T., Oshino, S., Hirata, M., Kato, A., and Yoshimine, T. 2006. Reduction of intractable deafferentation pain by navigation-guided repetitive transcranial magnetic stimulation of the primary motor cortex. *Pain* **122**: 22–27.

42. Saitoh, Y., Hirayama, A., Kishima, H., Shimokawa, T., Oshino, S., Hirata, M., Tani, N., Kato, A., and Yoshimine, T. 2007. Reduction of intractable deafferentation pain due to spinal cord or peripheral lesion by high-frequency repetitive transcranial magnetic stimulation of the primary motor cortex. *Journal of Neurosurgery* **107**: 555–559.

43. Ueno, S. 2012. Studies on magnetism and bioelectromagnetics for 45 years: From magnetic analog memory to human brain stimulation and imaging. *Bioelectromagnetics* **33**: 3–22.

44. Kato, T., Sekino, M., Matsuzaki, T., Nishikawa, A., Saitoh, Y., and Ohsaki, H. 2012. Electromagnetic characteristics of eccentric figure-eight coils for transcranial magnetic stimulation: A numerical study. *Journal of Applied Physics* **111**: 07B322.

45. Sekino, M., Kato, T., Ohsaki, H., Saitoh, Y., Matsuzaki, T., and Nishikawa, A. 2012. Eccentric figure-eight magnetic stimulator coils. *Proceedings of the ICME International Conference on Complex Medical Engineering* 728–733.

46. Sekino, M., Ohsaki, H., Takiyama, Y., Yamamoto, K., Matsuzaki, T., Yasumuro, Y., Nishikawa, A., Maruo, T., Hosomi, K., and Saitoh, Y. 2015. Eccentric figure-eight coils for transcranial magnetic stimulation. *Bioelectromagnetics* **36**: 55–65.

47. Yamamoto, K., Suyama, M., Takiyama, Y., Kim, D., Saitoh, Y., and Sekino, M. 2015. Characteristics of bowl-shaped coils for transcranial magnetic stimulation. *Journal of Applied Physics* **117**, 17A318 (2015); http://dx.doi.org /10.1063/1.4914876.

48. Okada, A., Nishikawa, A., Fukushima, T., Taniguchi, K., Miyazaki, F., Sekino, M., Yasumuro, Y., Matsuzaki, T., Hosomi, K., and Saitoh, Y. 2012. Magnetic navigation system for home use of repetitive transcranial magnetic stimulation (rTMS). *Proceedings of the ICME International Conference on Complex Medical Engineering*, 112–118.

49. Rossi, S., Hallett, M., Rossini, P. M., and Pascual-Leone, A. 2009. Safety of TMS consensus group, safety, ethical considerations, and application guidelines for the use of transcranial magnetic stimulation in clinical practice and research. *Clinical Neurophysiology* **120**: 2008–2039.

CHAPTER 4

Biomagnetic Measurements

Sunao Iwaki

CONTENTS

4.1 INTRODUCTION

4.1.1 History of Magnetoencephalography

By the end of the twentieth century, there were several methods developed to measure human brain activities noninvasively. Magnetoencephalography (MEG) is one of the completely noninvasive techniques that records distribution of magnetic fields generated by the neuroelectrical activities of the human brain by using an array of ultrasensitive magnetic sensors located around the scalp (Hämäläinen et al., 1993). Since the first recording in 1929 (Berger, 1929), electroencephalography or EEG, which measures distributions and temporal changes in the electric potential generated by cerebral neural activities, was widely used both in clinical applications and in research on the working human brain. MEG is the magnetic counterpart of EEG that carries information complementary to that of EEG. More specifically, MEG mainly records neuroelectrical activity whose dipolar generator orients tangentially to the scalp, whereas EEG is sensitive to both tangentially and radially oriented sources. Also, MEG is thought to be useful to localize generators of the neural activities that lie in the cortical area. In EEG, accurate estimation of source location is often difficult because the distribution of electrical potentials on the scalp is severely affected by complex inhomogeneity of conductivity in the head and the poor electrical conductivity of the skull. On the other hand, spatial distribution of the magnetic field mainly reflects neural currents that flow in the macroscopically relatively homogeneous intracranial space (Hämäläinen et al., 1993). Under such favorable conditions where brain activity is modeled as a single current dipole, the location of the neural source in the brain can be determined within a few millimeters spatial accuracy.

Although the first MEG recordings were carried out by using an induction-coil magnetometer (Cohen, 1968), measurements of MEG signals typically require more sensitive magnetic sensors such as the superconducting quantum interference device (SQUID) (Jaklevic et al., 1964) because the magnetic fields generated by the human brain activity are very weak, on

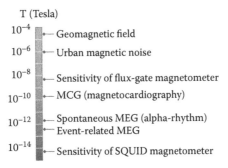

T (Tesla)

10^{-4} ← Geomagnetic field

10^{-6} ← Urban magnetic noise

10^{-8} ← Sensitivity of flux-gate magnetometer

10^{-10} ← MCG (magnetocardiography)

10^{-12} ← Spontaneous MEG (alpha-rhythm)
← Event-related MEG

10^{-14} ← Sensitivity of SQUID magnetometer

FIGURE 4.1   Amplitudes of magnetic fields generated from typical biomagnetic and environmental sources.

the order of 10–100 fT (femto-Tesla) (Figure 4.1). After the invention of the SQUID magnetometer (Zimmermann et al., 1970), which is based on the principle of superconductive tunneling (Josephson, 1972), and the first measurements of MEG signals corresponding to human alpha-band spontaneous brain activities (Cohen, 1972), SQUID systems have been widely used for practical MEG signal recordings. First, recordings of the event-related brain responses, which have smaller field strength compared to the spontaneous activities, were conducted a few years later by using SQUID magnetometers with signal averaging techniques (Brenner et al., 1975; Teyler et al., 1975).

The history of the development of SQUID systems is described in detail in Section 4.2.2.

4.1.2 Clinical Applications

Identifying the epileptic foci is the most prevalent use of MEG measurements in the clinical applications that had been studied since the early 1980s (Barth et al., 1982; Hari et al., 1993; Nakasato et al., 1994). Patients with recurrent and pharmacotherapy-resistant epilepsy could benefit from epilepsy surgery in which resection of epileptic foci is involved (Choi et al., 2008). In epileptic surgery, it is crucial to remove only the abnormal tissue and preserve normal functional tissue especially in the cortical regions. On the other hand, it is difficult to locate the epileptic foci even when the abnormal structure such as vascular malformation is detected by using other imaging modalities. Also, the abnormal tissues including epileptic foci are often surrounded by eloquent brain areas responsible for essential functions such as language. For these reasons, it is important to precisely locate the epileptic foci and dissociate the abnormal area from the surrounding intact tissues before surgery. MEG is one of the techniques used for the pre-surgical evaluation. Especially after the helmet-shaped SQUID magnetometer with whole-scalp coverage became commercially available, MEG has been routinely used as a tool to localize the epileptic foci (Stefan et al., 2003) and to optimize placement of intracranial electrodes to map brain functions around the foci for surgical planning (Knowlton et al., 2009).

Although pre-surgical evaluation for epileptic patients is the most common clinical application of MEG, there are several other clinical fields where MEG shows promise (Hari and Salmelin, 2012). There are accumulating evidences suggesting that MEG could be used for monitoring stroke recovery (Rossini et al., 2004; Forss et al., 2011) and objective evaluation of chronic pain (Ploner et al., 2002; Kakigi et al., 2003; Raij et al., 2004). Research on MEG application to human language functions are also being conducted. Some examples show possible use of MEG in determining language lateralization (Papanicolaou et al., 2004; Pirmoradi et al., 2010), which is originally performed by using an invasive method (Wada test) (Wada, 1949) before brain surgery, or in diagnosing language disabilities such as dyslexia (Salmelin et al., 1996; Helenius et al., 1999).

4.2 HARDWARE

4.2.1 SQUID

Although the first MEG recordings were carried out by using an induction-coil magnetometer (Cohen, 1968), measurements of MEG signals typically require more sensitive magnetic sensors such as a superconducting quantum interference device (SQUID) (Jaklevic et al., 1964). The SQUID is a superconducting ring with one (dc SQUID) or two (RF SQUID) insulating layers to form Josephson junction (Josephson, 1972). In the current instrumentation of MEG system, dc SQUID, which has more sensitivity to tiny changes in the magnetic field compared to RF SQUID (Clarke et al., 1976), is usually used. The dc SQUID has a couple of Josephson junctions in parallel in the superconducting ring (Figure 4.2). Without external magnetic flux passing through the SQUID ring, the input bias current I_b is split equally between both junctions. When a small external magnetic field is applied to the SQUID ring, Josephson junctions, which are characterized by the maximum critical current, limit the flow of induced currents without losing the superconductivity and magnetic flux quantization that occurs in the superconducting ring where magnetic flux through the ring

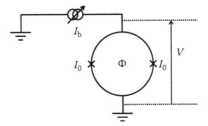

FIGURE 4.2 Schematic diagram of dc SQUID where I_b is the bias current, I_0 is the critical SQUID current, Φ is magnetic flux passing through the SQUID superconducting ring, and V is the output voltage to the flux Φ.

must be an integral multiple of the so-called flux quantum ($\Phi_0 = h/(2e)$ = 2.07×10^{-15} Wb). Under a constant biasing current I_b, measured voltage oscillates with the changes in phase at the two junctions, depending on the change in the magnetic flux, where one period of voltage variation corresponds to an increase of one flux quantum. Thus, the change in the magnetic flux is evaluated by counting the oscillations (Clarke, 1994). By attaching a flat coil of superconducting wire called flux transformer, the current induced in the SQUID ring can be increased, which results in the increased sensitivity to detect the magnetic fields smaller than 10^{-15} T, thus enabling measurements of brain magnetic fields.

4.2.2 Multichannel MEG System

Until the early 1980s, most of the MEG systems had single-channel SQUID coupled with gradiometric pickup coils to compensate for large changes in ambient magnetic fields or external magnetic noise (Figure 4.1). After the mid-1980s, there had been an extensive effort to build multichannel MEG systems. First, multichannel SQUID magnetometers, which have four to seven sensor channels covering an area of a few centimeters in diameter, were built between the mid- and late 1980s (Ilmoniemi et al., 1984; Romani et al., 1985; Williamson et al., 1985; Knuutila et al., 1987; Ohta et al., 1989).

From the late 1980s to the early 1990s, several private companies as well as academic institutes developed multichannel SQUID systems having larger spatial coverage of 10 cm in diameter or more, which usually covers most of the human hemisphere, with more than 20 SQUID magnetometers (Kajola et al., 1989; Hoenig et al., 1991; Pantev et al., 1991). In these systems, either axial or planar gradiometers were used, where pairs of oppositely wound coil is connected in series so that the sensor becomes

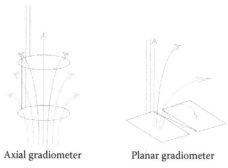

Axial gradiometer   Planar gradiometer

FIGURE 4.3   Gradiometric pickup coils to reduce disturbance from the external magnetic noise.

insensitive to uniform magnetic field generated far away from the sensor (Vrba et al., 1982; Duret and Karp, 1984) (Figure 4.3).

The first helmet-shaped MEG system with whole-scalp coverage using 122-channel planar-gradiometer SQUID sensor array was built in 1992 (Ahonen et al., 1993); it enabled users to monitor the entire cortical neural activities simultaneously. Since the early 1990s, several manufacturers made the helmet-shaped MEG systems commercially available (Vrba et al., 1993; Kado et al., 1999). Currently, systems with more than 300 SQUID sensors are available and are being used in many academic and clinical institutes.

4.3 PHYSIOLOGICAL BACKGROUND OF BIOMAGNETIC SIGNALS

4.3.1 Cortical Neural Activity

The human brain is composed of neurons and glial cells. The glial cells work primarily for the structural support for neurons and for regulating the internal environment by maintaining the concentration of cerebrospinal fluid and transporting substances between blood vessels and neurons to keep them working. Neurons are the electrically excitable cells that process and transmit information in the brain. The neuron is composed of a cell body (soma), dendrites, and an axon. The soma and dendrites are mostly located in the gray matter, which is distributed on the surface of the cerebral cortex. Inside the gray matter is a tissue called white matter that mostly consists of the glial cells and the myelinated axons, which connects between different brain structures. The distal terminal of the axon is connected to the dendrites of another neuron through synapse and transmits impulses by sending the neurotransmitters to postsynaptic receptors of the next neuron. When the neurotransmitters reach the post-synaptic neuron, electrochemical reaction changes the permeability of the postsynaptic membrane to specific ion, which causes a current along the interior of the postsynaptic neuron and results in the depolarization of the postsynaptic neuron and an excitatory postsynaptic potential (EPSP).

4.3.2 Current Dipole Modeling of Synaptic Activities

Because the cortical pyramidal neurons, which are major generators of EPSP, are spatially aligned parallel to each other and oriented perpendicu-lar to the cortical surface, the resultant direction of the electrical current flow in the dendrite becomes also perpendicular to the cortical surface. Observed from a distance, the local postsynaptic activity is well mod-eled by a current dipole oriented perpendicular to the cortical surface. Hämäläinen et al. discussed how the equivalent current dipole amplitude for a single PSP is comparable to approximately 20 fA m, and they con-clude that the estimated dimension of the "active" area required to gener-ate typical magnetic field strength outside of the head is tens of square millimeters or on the order of 10 nA m in terms of a local equivalent cur-rent dipole strength (Hämäläinen et al., 1993; Chapman et al., 1984) (see Figure 4.4).

Another major electrical activity of the neuron is the action poten-tial, which travels along the axon and is characterized by a wave of depo-larization followed by a repolarization and typically modeled by a pair of oppositely oriented dipoles in proximity whose magnitude is about 100 fA m. Because the two dipoles have opposite direction, they form a current quadrupole, whose magnetic field strength rapidly diminishes with distance as $1/r^3$ compared to the $1/r^2$-dependent dipolar field of

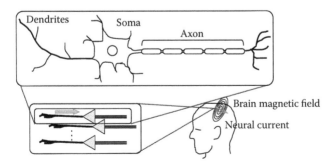

FIGURE 4.4   Equivalent current dipole model of the cortical neural activity.

synaptic activity. Further, temporal summation for action potentials that lasts only about 1 msec is less effective than the synaptic potentials that have a duration of tens of milliseconds. Thus, MEG signals are mostly generated by the synaptic activities (Hämäläinen et al., 1993).

4.4 FORMULATION OF MEG SIGNAL GENERATION

4.4.1 Maxwell's Equation and Primary Current

Section 4.3.2 described that the local cortical activities are modeled as an equivalent current dipole. Suppose that the macroscopic conductivity distribution $\sigma(\mathbf{r})$ and the spatial configurations of the source dipoles are known. Then MEG field distributions as they are measured outside of the head can be calculated from Maxwell's equation.

In the macroscopic point of view, the current density distribution $\mathbf{J}(\mathbf{r})$ generated by the electrical activities of the cortical neurons can be divided into two components, namely, the primary current $\mathbf{J}^{p}(\mathbf{r})$ and the volume current $\mathbf{J}^{v}(\mathbf{r})$. The volume current is a component that is produced by the electric field, which is formed by the primary electrical activities of the neurons, in the volume conductor.

$$\mathbf{J}(\mathbf{r}) = \mathbf{J}^{p}(\mathbf{r}) + \mathbf{J}^{v}(\mathbf{r}) = \mathbf{J}^{p}(\mathbf{r}) + \sigma(\mathbf{r})\mathbf{E}(\mathbf{r}) = \mathbf{J}^{p}(\mathbf{r}) - \sigma(\mathbf{r})\nabla V(\mathbf{r}) \qquad (4.1)$$

where $\mathbf{E}(\mathbf{r})$ is the distribution of the electric field, which is to be replaced by the gradient of the electric scalar potential $V(\mathbf{r})$. It is the distribution of primary current $\mathbf{J}^{p}(\mathbf{r})$ that should be estimated from the MEG data to locate the brain activity.

We can assume that the magnetic permeability is homogeneous throughout the biological tissues under consideration including scalp, skull, cerebrospinal fluid, and brain, and is the same as that of free space $\mu(\mathbf{r}) = \mu_0$. We can further assume that the quasi-static approximation on the Maxwell's equation in our frequencies of interest, which is typically less than 1 kHz, i.e., temporal derivatives of $\mathbf{E}$ and $\mathbf{B}$, can be ignored ($\partial E/\partial t = 0$, $\partial B/\partial t = 0$).

As we discussed in Section 4.3.2, the local synaptic activity is modeled as an equivalent current dipole. Suppose there is local activity at $\mathbf{r}_Q$; then, the primary current distribution can be described as

$$\mathbf{J}^{p}(\mathbf{r}) = \mathbf{Q} \cdot \delta(\mathbf{r} - \mathbf{r}_Q) \qquad (4.2)$$

where $\delta(\mathbf{r})$ is the Dirac delta function and $\mathbf{Q}$ is the amplitude of the equivalent current dipole.

4.4.2 Dipolar Current in Volume Conductor

From Maxwell's equation and Biot-Savart's law, the magnetic field outside of the head $\mathbf{B}(\mathbf{r})$ for given primary current distribution $\mathbf{J}^{P}(\mathbf{r}')$ is calculated. This is called the biomagnetic forward problem. Under quasi-static approximation, the magnetic field distribution is written as

$$\mathbf{B}(\mathbf{r}) = \frac{\mu_0}{4\pi} \int \frac{\mathbf{J}(\mathbf{r}') \times \mathbf{R}}{R^3} dv' \tag{4.3}$$

where $\mathbf{r}$ is the point where the magnetic field is calculated (outside of the head), $\mathbf{r}'$ is the primary current source location, and $\mathbf{R} = \mathbf{r} - \mathbf{r}'$ (see Figure 4.5).

Suppose the volume conductor, in which the primary and the volume currents are distributed, consists of several compartments G_i, $i = 1, \ldots, m$, each of them has homogeneous conductivity σ_i. The boundary between compartments G_i and G_j is defined by a surface S_{ij}. Then, the biomagnetic forward problem can be formulated as (Hämäläinen et al., 1993)

$$(\sigma_i + \sigma_j)V(\mathbf{r}) = 2\sigma_0 V_0(\mathbf{r}) + \frac{1}{2\pi} \sum_{ij} (\sigma_i - \sigma_j) \int_{S_{ij}} V(\mathbf{r}') d\Omega_r(\mathbf{r}') \tag{4.4}$$

$$\mathbf{B}(\mathbf{r}) = \mathbf{B}_0(\mathbf{r}) + \frac{\mu_0}{4\pi} \sum_{ij} (\sigma_i - \sigma_j) \int_{S_{ij}} V(\mathbf{r}') \frac{\mathbf{R}}{R^3} \times d\mathbf{S}'_{ij} \tag{4.5}$$

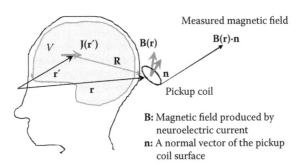

Measured magnetic field

$\mathbf{B}(\mathbf{r})\cdot\mathbf{n}$

Pickup coil

B: Magnetic field produced by neuroelectric current
n: A normal vector of the pickup coil surface

FIGURE 4.5 Configuration of the neural source and the magnetic field measurement.

where V_0 and $\mathbf{B}_0$ are the distributions of the electric potential and the magnetic field produced by $\mathrm{J^P}$ and are calculated as (Geselowitz, 1967)

$$V_0(\mathbf{r}) = -\frac{1}{4\pi\sigma_0}\int_G \mathbf{J^P}(\mathbf{r}')\cdot\nabla'\frac{1}{R}dv' = \frac{1}{4\pi\sigma_0}\int_G \frac{\nabla'\cdot\mathbf{J^P}}{R}dv' \tag{4.6}$$

$$\mathbf{B}_0(\mathbf{r}) = \frac{\mu_0}{4\pi}\int_G \mathbf{J^P}(\mathbf{r}')\times\frac{\mathbf{R}}{R^3}dv' \tag{4.7}$$

If the volume conductor is assumed to be spherically symmetric, where the distribution of the conductivity can be described as $\sigma(\mathbf{r}) = \sigma(r)$, the distribution of the magnetic field $\mathbf{B}$ outside of the conductor can be calculated analytically without solving partial differential equations (Ilmoniemi et al., 1985; Sarvas, 1987).

$$\mathbf{B}(\mathbf{r}) = \frac{\mu_0}{4\pi}\frac{F\mathbf{Q}\times\mathbf{r}_Q - (\mathbf{Q}\times\mathbf{r}_Q\cdot\mathbf{r})\nabla F(\mathbf{r},\mathbf{r}_Q)}{F(\mathbf{r},\mathbf{r}_Q)^2} \tag{4.8}$$

where

$$F(\mathbf{r},\mathbf{r}_Q) = a(ra + r^2 - \mathbf{r}_Q\cdot\mathbf{r}) \tag{4.9}$$

$$\nabla F(\mathbf{r},\mathbf{r}_Q) = (r^{-1}a^2 + a^{-1}\mathbf{a}\cdot\mathbf{r} + 2a + 2r)\mathbf{r} - (a + 2r + a^{-1}\mathbf{a}\cdot\mathbf{r})\mathbf{r}_Q \tag{4.10}$$

$$\mathbf{a} = (\mathbf{r} - \mathbf{r}_Q), a = |\mathbf{a}|, r = |\mathbf{r}| \tag{4.11}$$

The notable characteristics of the spherical volume conductor are that radially directed primary current does not generate any magnetic field and that a primary current located at the center of the sphere also does not produce any magnetic field outside of the conductor.

4.4.3 Boundary Element Method

Chapter 3 describes the calculation of the MEG distribution under the assumption that the human brain is modeled as a spherical volume conductor. When computing the MEG distribution for a volume conductor with arbitrary shape, we have to employ numerical techniques to solve Equation 4.4. One of the techniques to numerically solve the partial differential equation is to use the boundary element methods (BEM). In using BEM, we first need to create triangle mesh by segmenting the boundary surface S_i surrounding the compartments with different conductivities G_i (triangulation). Suppose the surface S_i is divided into n_i triangles $\Delta_k^i (k=1,\ldots,n_i)$, the continuous equation (Equation 4.4) can be discretized as follows (Barnard et al., 1967; Oostendorp and van Oosterom, 1989).

$$\mathbf{V}^i = \sum_{j=1}^{n_j} \mathbf{H}^{ij} \mathbf{V}^j + \mathbf{g}^i \tag{4.12}$$

where vector $\mathbf{g}^i$ and matrix $\mathbf{H}^{ij}$ are described as

$$g_k^i = \frac{1}{\mu_k^i} \frac{2}{\sigma_i^- + \sigma_i^+} \int_{\Delta_k^i} V_0(\mathbf{r}') \mathrm{d}S_i' \tag{4.13}$$

$$H_{kl}^{ij} = \frac{1}{2\pi} \frac{\sigma_j^- - \sigma_j^+}{\sigma_i^- + \sigma_i^+} \frac{1}{\mu_k^i} \int_{\Delta_k^i} \Omega_l^j(\mathbf{r}') \mathrm{d}S_l' \tag{4.14}$$

σ_i^- and σ_i^+ are conductivities of inside and outside of the surface S_i, μ_k^i is the area of the kth triangle on the surface S_i, and $\Omega_l^j(\mathbf{r})$ is a solid angle of the triangle Δ_k^i from the point $\mathbf{r}$ (van Oosterom and Strackee, 1983).

In solving the linear system Equation 4.12, the inverting matrix becomes singular reflecting the arbitrariness in determining the reference point to define the potential distribution. The mathematical technique called deflation can be used to avoid this singularity issue (Lynn and Timlake, 1968).

After obtaining the distribution of the potential distribution $\mathbf{V}$ on each surface, the magnetic field distribution can be calculated by directly integrating Equation 4.5 on the discretized surfaces.

4.4.4 Lead Field

Section 4.4.3 describes the numerical methods to calculate spatial distribution of the electrical potential on the scalp (i.e., EEG) and the magnetic field outside of the head (i.e., MEG) for given primary neural current distribution J^p. As shown in the equations, there is a linear relationship between MEG distribution $\mathbf{B}$ and amplitudes of the primary current distribution J^p. Suppose b_i is the amplitude of the magnetic field measured by the *i*th sensor. Then, b_i is written as a weighted integration of the sensitivity of the sensor in the volume conductor where the primary current is distributed.

$$b_i = \int \Lambda_i(\mathbf{r}) \cdot \mathbf{J}^P(\mathbf{r}) dv \qquad (4.15)$$

where vector field Λ_i is usually called the "lead field." The lead field is uniquely defined by the conductivity distribution $\sigma(\mathbf{r})$ and the configuration of the magnetic field sensor (i.e., location of the sensor and the direction of the pickup coil).

When employing a set of equivalent current dipoles shown in Equation 4.2 as a model of primary current distribution, the MEG signal generated by the *n* current dipoles with dipole moments $\mathbf{Q}_j$ located at $\mathbf{r}_{Q_j}$ ($j = 1, ..., n$) and measured by *i*th sensor is described as

$$b_i = \sum_{j=1}^{n} \Lambda_i\left(\mathbf{r}_{Q_j}\right) \cdot \mathbf{Q}_j \qquad (4.16)$$

4.5 NEURAL SOURCE RECONSTRUCTION FROM MEG SIGNALS

4.5.1 MEG Inverse Problem

The problem to estimate the neural current distribution from the spatio-temporal characteristics of measured MEG data is called the biomagnetic inverse problem (Figure 4.6).

In general, the problem to reconstruct the internal current distribution from the measurements of the electric potential or the magnetic field conducted outside of the conduction medium does not have a unique solution (an ill-posed problem). This is obvious from the fact that there are certain configurations of the primary current distribution that do not generate any magnetic fields outside the conducting medium (a magnetically "silent" source). One example of those magnetically silent sources is the radial dipolar current source in the spherically symmetric conductor. In solving the ill-posed inverse problem, we must impose additional mathematical and/or physiological constraints to obtain a unique solution.

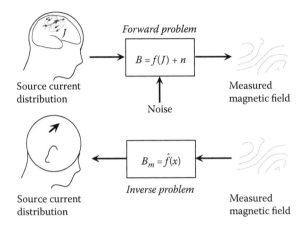

FIGURE 4.6 Forward and inverse problems in biomagnetism.

4.5.2 Equivalent Current Dipole

Modeling the primary current distribution by using a small number of equivalent current dipoles is a common technique to impose additional constraints to the ill-posed biomagnetic inverse problem. The dipole model was first introduced to biomagnetism to localize the primary current source of somatosensory evoked magnetic field (Brenner et al., 1978) and has been used until today to locate neural sources with relatively simple spatial configuration.

4.5.2.1 Single ECD Model

Single ECD is the simplest form of an equivalent dipole model to describe an internal neural generator. In the early days, the dipole parameters, i.e., location, amplitude, and orientation, had been estimated from the geometric characteristics of the field pattern, where the dipole lies directly below the midpoint between the two field extrema with its direction normal to the plane passing through the extrema. The depth of the dipole can be estimated from the distance between the extrema d as $d/\sqrt{2}$. The standard procedure to estimate the dipole parameters is to use the iterative nonlinear parameter optimization algorithm such as the Levenberg-Marquardt method (Levenberg, 1944; Marquardt, 1963). Details of the parameter estimation are described in a more general form in the next section.

4.5.2.2 Multidipole Model

In the multiple dipole model of the primary current distribution, we assume that the distribution of the measured magnetic field is generated by a set of n dipoles with their moments $\mathbf{Q}_j$ ($j = 1, ..., n$) and located at $\mathbf{r}_j$. Suppose the predicted magnetic field distribution generated by n ECDs as $\mathbf{b}_{model}$ and the measured field as $\mathbf{b}_{meas}$. Then, $\mathbf{b}_{model}$ is calculated by the forward equation (Equation 4.16). The nonlinear least squares parameter fitting technique (Marquardt, 1963) is applied to obtain a set of dipole parameters, which minimizes the difference between the predicted and the measured field distributions (Figure 4.7),

$$E = \left\| \mathbf{b}_{meas} - \mathbf{b}_{model} \right\|^2 \tag{4.17}$$

One way to evaluate the validity of the obtained model is to calculate the goodness-of-fit index,

$$g = 1 - \frac{\left\| \mathbf{b}_{meas} - \mathbf{b}_{model} \right\|^2}{\left\| \mathbf{b}_{meas} \right\|^2} \tag{4.18}$$

which indicates how well the field pattern predicted by the equivalent dipoles agrees with that measured in the experiment.

Another important point in using the multidipole model is how to determine the number of dipoles to explain the current field distribution. Typically, multiple sources with their distance larger than a few centimeters show field patterns that have only minor overlap if their orientations are favorable. In these cases, location, amplitude, and orientation of

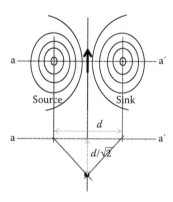

FIGURE 4.7 Simple geometric method to estimate location of the single ECD from magnetic field distribution.

each primary source may be estimated successfully by applying the single dipole model separately. On the other hand, if the field pattern does not show obvious separation between sources, spatiotemporal characteristics of the measured MEG data may be used to estimate the number of active source dipoles (Scherg, 1990 for review) to be assumed in the multidipole model. The standard method to estimate the number of active sources in the measured spatiotemporal MEG data is to use singular value decomposition (SVD) (de Munck, 1990).

$$\mathbf{B}_{meas} = \mathbf{U}\mathbf{S}\mathbf{V} \qquad (4.19)$$

where $\mathbf{B}_{meas}$ is a $m \times L$ spatiotemporal MEG data measured by m sensors at L time points $\mathbf{B}_{meas} = [\mathbf{b}_1, \mathbf{b}_2, ..., \mathbf{b}_L]$, $\mathbf{S}$ is a diagonal matrix of the singular values, and $\mathbf{U}$ and $\mathbf{V}$ are the unitary matrices containing left and right singular vectors, respectively. Suppose s_i $(i = 1, ..., m)$ are the singular values set in descending order. Then the square error between measured and predicted fields as expressed in Equation 4.17, where we assume r equivalent current dipoles to explain the spatiotemporal characteristics of the measured magnetic fields, can be calculated from the sum of squares of the least $m - r$ significant singular values,

$$E_{min} \geq \sum_{i=r+1}^{m} s_i^2 \qquad (4.20)$$

and the resulting goodness-of-fit index is estimated by

$$g \leq \frac{\displaystyle\sum_{i=1}^{r} s_i^2}{\displaystyle\sum_{i=r+1}^{m} s_i^2} \qquad (4.21)$$

The adequate number of equivalent current dipole r may be determined from the behavior of square error as shown in Equation 4.20 when r is changed. Some criteria in choosing a reasonable number of singular values r are reported in reference to the noise level of the measurements (Wax and Kailath, 1985) (see Figure 4.8).

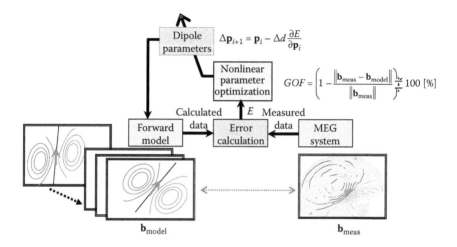

FIGURE 4.8 Schematic diagram of the standard technique to estimate dipole parameters by using the iterative nonlinear parameter optimization algorithm.

4.5.2.3 Multiple Signal Classification

In optimizing ECD parameters in the multidipole model, the global minimization of E_{min} sometimes cannot be achieved because the optimization step may get caught in a local minimum especially when the measured field patterns are relatively complex. To avoid the convergence to the local minimum, several optimization techniques such as the simulated annealing (Kirkpatrick et al., 1983) and the genetic algorithm (McNay et al., 1996) had been employed. Also, in these cases, it is usually difficult to determine the initial dipole locations where the optimization algorithm starts and is frequently dependent on heuristic or empirical methods. One approach to search optimal locations of multiple source dipoles is to use an algorithm called the multiple signal classification (MUSIC) (Mosher et al., 1992), which is originally developed to process signals from antenna array (Schmidt, 1986). Application of the MUSIC algorithm to the biomagnetic inverse problem is based on a separation of spatiotemporal measured data into signal and noise subspaces and on the three-dimensional search of the suboptimal estimates for multiple dipole locations.

First, eigendecomposition is applied to the covariance matrix of the spatiotemporal measured data, which is obtained by $\mathbf{R}_{meas} = \mathbf{B}_{meas}\mathbf{B}_{meas}^T$ to determine the separation between signal and noise subspaces.

$$\mathbf{R}_{meas} = \mathbf{VDV}^T = \begin{bmatrix} \mathbf{V}_s & \mathbf{V}_n \end{bmatrix} \begin{bmatrix} \mathbf{D}_s & \\ & \mathbf{D}_n \end{bmatrix} \begin{bmatrix} \mathbf{V}_s \\ \mathbf{V}_n \end{bmatrix} \qquad (4.22)$$

where $\mathbf{D}_s$ is a diagonal matrix containing r most significant eigenvalues of the signal subspace, $\mathbf{V}_s$ is corresponding eigenvectors that span a signal subspace, $\mathbf{D}_n$ is a diagonal matrix of eigenvalues of the noise subspace, and $\mathbf{V}_n$ is corresponding eigenvectors for noise subspace.

Next, MUSIC cost function, which has multiple maxima at the locations of the active sources, are calculated on the grid points covering the entire source space. Because the noise subspace is orthogonal to the lead field at the correct source location, the possible distribution of the active sources can be obtained by scanning the orthogonality S_i between lead field Λ_i and the noise subspace projector $\mathbf{P}_\perp = \mathbf{V}_n \mathbf{V}_n^T$ at each grid point i.

$$S_i = \left\| \mathbf{P}_\perp \cdot \hat{\Lambda}_i \right\|_F^2 \qquad (4.23)$$

where $\hat{\Lambda}_i$ is the normalized lead field at grid point i and is written as $\hat{\Lambda}_i = \Lambda_i / \|\Lambda_i\|$. Under ideal conditions where we can ignore noise, S_i approaches 0 at the exact dipolar source location. Contour plot of $1/S_i$, which is also called the MUSIC localizer (Sekihara et al., 1998), illustrates the distribution of possible source dipoles in the three-dimensional space. Because the scanning of the MUSIC localizer does not require an iterative optimization procedure, the search for the location of active source dipoles should not be time-consuming.

The basic assumption of the MUSIC algorithm is that the correlations between the active sources located in different positions are negligible. Under such conditions where this assumption does not hold, it is impossible to appropriately separate the signal and noise subspaces from the spatiotemporal measured data that may severely distort spatial distribution of the MUSIC localizer (see Figure 4.9).

4.5.3 Distributed Source Imaging

In the previous section, ECD models to solve biomagnetic inverse problems were discussed. The explicit assumption in using the ECD model is that the relevant neural activities are approximated by the set of small number of focal neural sources. There is an alternative approach to solve the biomagnetic inverse problem by using a more general form of the neural source configuration in which a large number of local ECDs are distributed in the entire brain volume or on the cortical surface (the distributed source model).

Suppose $\mathbf{Q} = [q_1, q_2, ..., q_N]$ as N local current dipoles to describe primary current distribution and $\mathbf{b}_{meas} = [b_1, b_2, ..., b_M]$ as the MEG field distribution

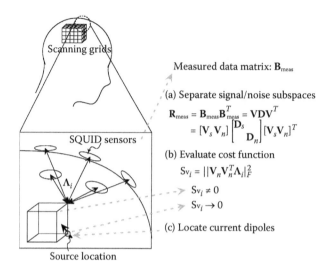

FIGURE 4.9 Schematic diagram of the multiple signal classification (MUSIC) algorithm applied to the biomagnetic inverse problem.

measured at M sensors, and $\Lambda = [\Lambda_1, \Lambda_2, ..., \Lambda_M]$ as the lead field matrix, the MEG forward model for distributed source model is written as

$$\mathbf{b}_{means} = \Lambda \cdot \mathbf{Q} + \mathbf{n} \tag{4.24}$$

where $\mathbf{n}$ is the noise in the measured data.

4.5.3.1 Minimum-Norm Estimates

The linear equation (Equation 4.24) does not generally have a unique solution because of the underdetermined nature of the system. The number of parameters to be estimated (N) is on the order of thousands, whereas the typical number of equations, which corresponds to the number of sensors (M), is on the order of hundreds at most. In the traditional inverse formulation, a "minimum-norm constraint" is commonly employed in which the least norm solution is selected from among the possible set of solutions (Hämäläinen and Ilmoniemi, 1984). The solution under minimum-norm constraint is given by

$$\hat{\mathbf{Q}} = \Lambda^- \mathbf{b}_{meas} \tag{4.25}$$

where Λ^- is the generalized inverse (Menke, 1984) of the lead field matrix and is written as

$$\Lambda^- = \left(\mathbf{W}_c \mathbf{W}_C^T \right)^{-1} \Lambda^T \left(\Lambda \left(\mathbf{W}_c \mathbf{W}_C^T \right)^{-1} \Lambda^T + \gamma^2 \mathbf{I} \right)^{-1} \tag{4.26}$$

The diagonal matrix $\mathbf{W}_c$ contains the weighting factors for "depth normalization" to avoid minimum-norm solution bias toward superficial locations (Jeffs et al., 1987), and γ is a regularization parameter that controls the degree of regularization and is determined in accordance with the signal-to-noise ratio of measured signal. Using this method, the solution $\hat{\mathbf{Q}}$ that minimizes the objective function,

$$E_c = \left\| \mathbf{b}_{\text{meas}} - \mathbf{b}_{\text{model}} \right\|^2 + \gamma^2 \left\| \mathbf{W}_c \hat{\mathbf{Q}} \right\|^2 \qquad (4.27)$$

is selected for the reconstructed current distribution (Menke, 1984). Here, $\mathbf{b}_{\text{model}}$ is the field distribution predicted by the reconstructed neural current distribution $\hat{\mathbf{Q}}$. Selecting the appropriate weighting factors $\mathbf{W}_c$ in Equations 4.26 and 4.27 (i.e., small weighting values for deep current locations and large values for shallow locations), the bias in the solution toward superficial sources can be equalized. However, in general, the optimal set of regularization parameters and weighting factors is difficult to derive theoretically and is highly dependent on detailed measurement geometry (Jeffs et al., 1987). Some methods that handle this difficulty with iterative algorithms have been proposed (Gorodnitsky et al., 1995; Srebro, 1996). The simplest form of obtaining the depth weighting factor $\mathbf{W}_c$ is derived directly from the lead field matrix for each possible source location i as

$$\mathbf{W}_c = \text{diag}\{\|\mathbf{\Lambda}_{i2}\|\} \qquad (4.28)$$

where $\|\mathbf{a}_2\|$ indicates the L_2 norm of vector $\mathbf{a}$.

The diagonal weighting matrix $\mathbf{W}_c$ can be also used to introduce prior information obtained by other neuroimaging modalities such as structural magnetic resonance imaging (MRI) and functional MRI (fMRI) into the source space when solving the biomagnetic inverse problem (Dale and Sereno, 1993; Dale et al., 2000; Iwaki et al., 2013).

4.5.3.2 Beamformer

Beamformer is an adaptive array signal processing technique originally developed for radar and seismic research (van Veen and Buckley, 1988). Beamforming techniques can be successfully applied to form an adaptive spatial filter to extract time series that arise from the neural activity at a specific brain location and block the signal from other parts of the brain with linear weighted sum of the MEG signals measured by the sensor

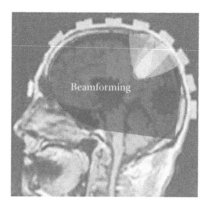

FIGURE 4.10 Beamforming technique to extract neural activity at a specific location by adaptive spatial filtering.

array. In this section, we discuss the technique to design linearly constrained minimum variance (LCMV) beamformer (van Veen et al., 1997) to be used to map the neural activities from MEG data (see Figure 4.10).

For the LCMV beamformer, we design a spatial filter for each point covering the whole cortical region by selecting the filter weights that minimize total signal power but at the same time do not suppress the MEG signal arising from the neural activity at the specific point. This is accomplished by solving the minimization problem below (van Veen et al., 1997).

$$\min_{F(r_0)}\left\{\text{trace}\left[\mathbf{C}\big(\mathbf{y}(\mathbf{r}_0)\big)\right]\right\}=\min_{F(r_0)}\left\{\left[\mathbf{F}^T(\mathbf{r}_0)\mathbf{C}(\mathbf{B})\mathbf{F}(\mathbf{r}_0)\right]\right\}$$

$$\text{subject to } \mathbf{F}^T(\mathbf{r}_0)\cdot\Lambda(\mathbf{r}_0)=\mathbf{I} \qquad (4.29)$$

where $\mathbf{F}(\mathbf{r}_0)$ is the filter weights for the cortical point of interest $\mathbf{r}_0$, $\mathbf{C}(\mathbf{y}(\mathbf{r}_0))$ is a covariance matrix of the neural activity at $\mathbf{r}_0$, which is not observable, $\mathbf{C}(\mathbf{B})$ is a covariance matrix of the spatiotemporal measure data $\mathbf{B}$, and $\Lambda(\mathbf{r}_0)$ is a lead field vector at $\mathbf{r}_0$. By using the methods of Lagrange multipliers, the solution for the minimization problem is obtained as

$$\mathbf{F}(\mathbf{r}_0) = [\Lambda^T(\mathbf{r}_0)\mathbf{C}^{-1}(\mathbf{B})\Lambda(\mathbf{r}_0)]^{-1}\Lambda^T(\mathbf{r}_0)\mathbf{C}^{-1}(\mathbf{B}) \qquad (4.30)$$

Applying linear filter $\mathbf{F}(\mathbf{r}_0)$ to the measured spatiotemporal MEG data $\mathbf{B}$, the time series of the neural activity at $\mathbf{r}_0$ are estimated as

$$\hat{\mathbf{y}}(\mathbf{r}_0)=\mathbf{F}(\mathbf{r}_0)\mathbf{B} \qquad (4.31)$$

4.6 EXAMPLES OF MEG MEASUREMENTS

This section describes some examples of MEG application on the cognitive neuroscience research.

4.6.1 Visual Evoked Responses

Here we present an example of visualizing human brain activity while performing a three-dimensional (3D) mental image manipulation task in which subjects were required to mentally rotate 3D objects (mental rotation) (Shepard and Metzler, 1971; Cooper, 1973). The mental rotation processing includes rotating and matching of a pair of 3D mental objects, which is known to activate both extrastriate visual areas and the parietal area as demonstrated by previous studies employing either positron emission tomography (PET) (Parsons et al., 1985) or functional MRI (Tagaris et al., 1996). Some studies have used EEG or MEG measures to investigate temporal characteristics of brain activity during mental rotation processing; however, owing to the difficulty in estimating the distribution of neural sources associated with higher-order visual functions, most of these studies could shed light only on the waveforms at the recording sites or on the single ECD modeling of the possible generators (Peronnet and Farah, 1989; Kaufman et al., 1990). To overcome this difficulty, the weighted minimum-norm approach as described in Equations 4.25 and 4.26 was used to map brain activities in the visual area and parietal areas with high temporal resolution from MEG data acquired during the mental rotation processing (see Figure 4.11).

Figure 4.12 shows the results of neural source imaging obtained by the weighted minimum-norm method (Iwaki et al., 2002). The neural activity in the primary visual area was followed by the activation in the lateral posterior temporal regions (Brodmann's area [BA] 19, 37) in the latency range between 190 and 220 msec after the onset of mental rotation stimuli. Activity in the intraparietal and inferior parietal areas (BA 7, 40) around

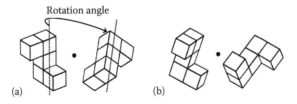

FIGURE 4.11   Examples of the visual stimuli used in the mental rotation task. (a) Rotationally-symmetric pair. (b) Minor-reversed pair.

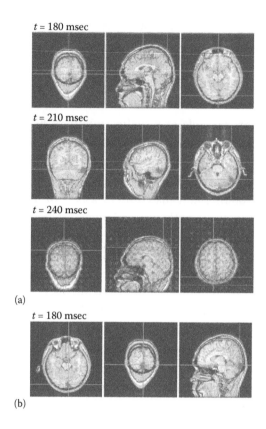

$t = 180$ msec

$t = 210$ msec

$t = 240$ msec

(a)

$t = 180$ msec

(b)

FIGURE 4.12 Results of the neural source imaging during the performance of the mental rotation task obtained by the weighted minimum-norm method. (a) Mental rotation. (b) Control.

240 msec was also observed. On the other hand, activity in the primary visual area was detected in the control condition in which subjects were not required to rotate the 3D objects (Iwaki et al., 1999).

This example suggests that the distributed source imaging techniques such as weighted minimum-norm estimates are capable of visualizing brain activities without *a priori* knowledge of the number of active sources when they are located a few centimeters apart from each other.

4.6.2 Auditory Evoked Responses

The most prominent evoked response to the auditory stimuli is called the N100 or N1 component that has a peak at about 100 msec after the onset of the stimuli. Its magnetic counterpart N100m or N1m is also recorded quite robustly by the MEG sensor covering the temporal cortex. Within each hemisphere, the MEG field pattern shows that the source of N1m can be well explained by a single ECD. The results of the single ECD estimation in each hemisphere show that the neural sources of N1m located in the primary auditory are in the Heschl's gyrus on the superior temporal gyrus (Figure 4.13).

The N1m component is evoked in response not only to the onset of the auditory stimuli but also to the offset of the stimuli. This offset response $N1m_{off}$ is also measured for the auditory language stream while the subjects listened to (1) a series of spoken sentences and (2) white noise whose amplitude was modulated by the envelope of the spoken sentences as control stimuli, and both included sudden and brief interruptions (Figure 4.14a). Auditory evoked magnetic fields corresponding to the interruption and resumption of the speech were measured. Statistically significant enhancement was observed only in the N1m component of the auditory evoked offset response ($N1m_{off}$) to the interruptions of the verbal stream compared to the onset response ($N1m_{on}$) to the resumptions of the stream (Figure 4.14b) (Iwaki et al., 2012). These results suggest that the human auditory cortex plays an important role in perceiving interruptions of verbal streams.

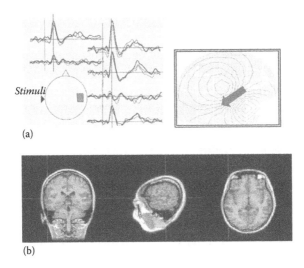

(a)

(b)

FIGURE 4.13 (a) Waveforms and the spatial distribution of the N1m auditory evoked magnetic fields. (b) Results of the neural source estimation from N1m MEG distribution. The dipoles are located in the bilateral primary auditory cortex.

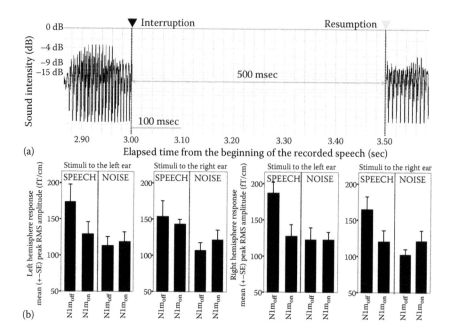

(a)

(b)

FIGURE 4.14　(a) Typical acoustic waveforms of the sudden interruptions with 500 msec duration embedded in the natural spoken language stream. (b) Mean and standard error of the $N1m_{off}$ and $N1m_{on}$ peak amplitudes elicited by the interruption of speech stream (SPEECH) and corresponding white noise (NOISE).

4.6.3 Multimodal Neuroimaging

Although MEG can record temporal changes of post-synaptic neural activity with good temporal resolution (on the order of milliseconds), the principal difficulty in interpreting MEG data is in reconstructing the spatial distribution of the neural activity in the brain from MEG spatiotemporal distributions measured at sensors located at a distance from the brain area from which activity is being assessed. On the other hand, fMRI makes it possible to precisely visualize the spatial distribution of human brain activity with a resolution that is typically approximately a few millimeters; however, fMRI measures hemodynamic changes (blood oxygenation level dependent [BOLD] signals) (Ogawa et al., 1990; Belliveau et al., 1991), which are an indirect measure of neural activity. These changes peak a few seconds after the onset of neural firing in an area; therefore, this technique has limited temporal resolution. On the basis of these inherent limitations in each of the measurement modalities, it is natural to combine data from both techniques to enhance both the spatial and temporal resolutions. As stated in Section 4.5.3.1, the weighted minimum-norm technique can be used to impose physiologically plausible constraints on the biomagnetic inverse problem by introducing structural and functional MRI priors into the weighting matrix when solving the underdetermined linear system (Equations 4.25 and 4.26).

The example shown here uses a combination of MEG and fMRI data as well as a brain structure obtained from high-resolution structural MRI to visualize human brain activity while perceiving 3D objects in 2D random-dot motion images, which is usually called 3D structure perception from motion or 3D-SFM (Wallach and O'Connell, 1953). Perception of a three-dimensional object shape from visually presented moving dots requires integration of the embedded movement and spatial information that acti-vates both dorsal (i.e., the parieto-occipital [PO] and the intraparietal [IP] regions) and ventral (i.e., the posterior inferior temporal [IT] region) visual systems (Orban et al., 1999; Paradis et al., 2000; Beers et al., 2009); however, the neural dynamics underlying the reconstruction of a 3D object from a 2D optic flow is not known. The results of the fMRI statistical parametric map-ping (SPM) analysis (Friston, 1998) were used to introduce plausible weights in the weighted minimum-norm estimates (Liu et al., 1998) to obtain time-resolved images of the brain activity during 3D-SFM (Figure 4.15).

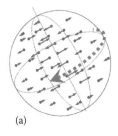

(a)

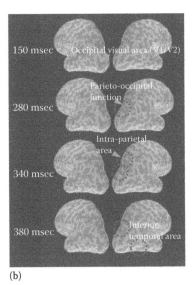

150 msec — Occipital visual area (V1/V2)

Parieto-occipital junction

280 msec

Intra-parietal area

340 msec

380 msec — Inferior temporal area

(b)

FIGURE 4.15   (a) Example of the visual stimuli used in the 3D-SFM experiment. (b) Results of the fMRI-weighted MEG inverse analysis in the typical subject performing the 3D-SFM task. The blue circles denote the sites where we observed the major neural activity during 3D-SFM processing.

The results show that the neural activity in the parieto-occipital, the intraparietal, and the inferior-temporal regions as well as primary visual area were increased during the 3D SFM at the latencies between 200 and 450 msec after the onset of visual motion (Iwaki et al., 2013). These results demonstrate that the noninvasive multimodal neuroimaging, in which structural and functional MRI data are incorporated into the MEG inverse problem, is capable of visualizing neural dynamics of the human brain with spatial accuracy comparable to fMRI without compromising the excellent temporal accuracy of MEG (Iwaki, 2012).

REFERENCES

Ahonen, A. I., Hämäläinen, M. S., Kajola, M. J., Knuutila, J. E. T., Laine, P. L., Lounasmaa, O. V., Parkkonen, L. T., Simola, J. T., and Tesche, C. D. 1993. 122-channel SQUID instrument for investigating the magnetic signals from the human brain. *Physica Scripta* **T49A**: 198–205.

Barnard, A. C. L., Duck, I. M., Lynn, M. S., and Timlake, W. P. 1967. The application of electromagnetic theory to electrocardiography. II. Numerical solution of the integral equations. *Biophysics Journal* **7**: 463–491.

Barth, D., Sutherling, W., Engel, J. J., and Beatty, J. 1982. Neuromagnetic localization of epileptiform spike activity in the human brain. *Science* **218**: 891–894.

Beers, A. L., Watanabe, T., Ni, R., Sasaki, Y., and Anderson, G. J. 2009. 3D surface perception from motion involves a temporal–parietal network. *European Journal of Neuroscience* **30**: 703–713.

Belliveau, J. W., Kennedy, D. N., McKinstry, R. C., Buchbinder, B. R., Weisskoff, R. M., Cohen, M. S., Vevea, J. M., Brady, T. J., and Rosen, B. R. 1991. Functional mapping of the human visual cortex by magnetic resonance imaging. *Science* **254**: 716–719.

Berger, H. 1929. Über das Elektroenkephalogramm des Menschen. *Archiv für Psychiatrie und Nervenkrankheiten* **87**: 527–570.

Brenner, D., Williamson, S. J., and Kaufman, L. 1975. Visually evoked magnetic fields of the human brain. *Science* **190**: 480–481.

Brenner, D., Lipton, J., Kaufman, L., and Williamson, S. J. 1978. Somatically evoked magnetic fields of the human brain. *Science* **199**: 81–83.

Chapman, R. M., Ilmoniemi, R. J., Barbanera, S., and Romani, G. L. 1984. Selective localization of alpha brain activity with neuromagnetic measurements. *Electroencephalography and Clinical Neurophysiology* **58**: 569–572.

Choi, H., Sell, R. L., Lenert, L., Muennig, P., Goodman, R. R., Gilliam, F. G., and Wong, J. B. 2008. Epilepsy surgery for pharmacoresistant temporal lobe epilepsy: A decision analysis. *JAMA* **300**: 2497–2505.

Clarke, J. 1994. SQUIDs. *Scientific American* **271**: 46.

Clarke, J., Goubau, W. M., and Ketchen, M. B. 1976. Tunnel junction dc SQUID fabrication, operation, and performance. *Journal of Low Temperature Physics* **25**: 99–144.

Cohen, D. 1968. Magnetoencephalography: Evidence of magnetic fields produced by alpha-rhythm currents. *Science* 161: 784–786.

Cohen, D. 1972. Magnetoencephalography: Detection of the brain's electrical activity with a superconducting magnetometer. *Science* 175: 664–666.

Cooper, L. A. 1973. Chronometric studies of the rotation of mental images. In *Visual Information Processing*. Academic Press, New York.

Dale, A. M., and Sereno, M. I. 1993. Improved localization of cortical activity by combining EEG and MEG with MRI cortical surface reconstruction: A linear approach. *Journal of Cognitive Neuroscience* 5: 162–176.

Dale, A. M., Liu, A. K., Fischl, B., Buckner, R., Belliveau, J. W., Lewine, J. D., and Halgren, E. 2000. Dynamic statistical parametric mapping: Combining fMRI and MEG for high-resolution imaging of cortical activity. *Neuron* 26: 55–67.

de Munck, J. C. 1990. The estimation of time-varying dipoles on the basis of evoked potentials. *Electroencephalography and Clinical Neurophysiology* 77: 156–160.

Duret, D., and Karp, P. 1984. Figure of merit and spatial resolution of superconducting flux transformers. *Journal of Applied Physics* 56: 1762–1768.

Forss, N., Mustanoja, S., Roiha, K., Kirveskari, E., Makela, J. P., Salonen, O., Tatlisumak, T., and Kaste, M. 2011. Activation in parietal operculum parallels motor recovery in stroke. *Human Brain Mapping* 33: 534–541.

Friston, K. J. 1998. Imaging neuroscience: Principles or maps? *Proceedings of the National Academy of Sciences of the United States of America* 95: 796–802.

Geselowitz, D. B. 1967. On bioelectric potentials in an inhomogeneous volume conductor. *Biophysical Journal* 7: 1–11.

Gorodnitsky, I. F., George, J. S., and Rao, B. D. 1995. Neuromagnetic source imaging with FOCUSS: A recursive weighted minimum norm algorithm. *Electroencephagraphy and Clinical Neurophysiology* 95: 231–251.

Hämäläinen, M., and Ilmoniemi, R. J. 1984. Interpreting magnetic fields of the brain: Minimum norm estimates. *Medical & Biological Engineering & Computing* 32: 35–42.

Hämäläinen, M., Hari, R., Ilmoniemi, R. J., Knuutila, J., and Lounasmaa, O. V. 1993. Magnetoencephalography—Theory, instrumentation, and applications to noninvasive studies of the working human brain. *Reviews of Modern Physics* 65: 413–497.

Hari, R., and Salmelin, R. 2012. Magnetoencephalography: From SQUIDs to neuroscience, Neuroimage 20th anniversary special edition. *Neuroimage* 61: 386–396.

Hari, R., Ahonen, A., Forss, N., Granström, M. L., Hämäläinen, M., Kajola, M., Knuutila, J., Lounasmaa, O. V., Mäkelä, J. P., Paetau, R., Salmelin, R., and Simola, J. 1993. Parietal epileptic mirror focus detected with a whole-head neuromagnetometer. *Neuroreport* 5: 45–48.

Helenius, P., Salmelin, R., Service, E., and Connolly, J. F. 1999. Semantic cortical activation in dyslexic readers. *Journal of Cognitive Neuroscience* 11: 535–550.

Hoenig, H. E., Daalmans, G. M., Bär, L., Bömmel, F., Paulus, S., Uhl, D., Weisse, H. J., Schneider, S., Seifert, H., Reichenberger, H., and Abraham-Fuchs, K. 1991. Multi channel dc SQUID sensor array for biomagnetic applications. *IEEE Transactions on Magnetics* 27: 2777–2785.

Ilmoniemi, R., Hari, R., and Reinikainen, K. 1984. A four-channel SQUID magnetometer for brain research. *Electroencephalography and Clinical Neurophysiology* **58**: 467–473.

Ilmoniemi, R. J., Hämäläinen, M. S., and Knuutila, J. 1985. The forward and inverse problems in the spherical model. In *Biomagnetism: Application & Theory*, Weinberg, H., Stroink, G., and Katila, T., eds. Pergamon, New York: 278–282.

Iwaki, S. 2012. Multimodal neuroimaging to visualize human visual processing. In *Biomedical Engineering and Cognitive Neuroscience for Healthcare*. IGI Press, Hershey, PA: 273–281.

Iwaki, S., Ueno, S., Imada, T., and Tonoike, M. 1999. Dynamic cortical activation in mental image processing revealed by biomagnetic measurement. *Neuroreport* **10**: 1793–1797.

Iwaki, S., Tonoike, M., and Ueno, S. 2002. Visualization of the brain activity during mental rotation processing using MUSIC-weighted lead-field synthetic filtering. *IEICE Transactions on Information and Systems* **85-D**: 175–183.

Iwaki, S., Hamada, T., and Kawano, T. 2012. Differential offset and onset responses to interruptions and resumptions of verbal streams in the human auditory cortex. *International Journal of Bioelectromagnetism* **14**: 74–79.

Iwaki, S., Bonmassar, G., and Belliveau, J. W. 2013. Dynamic cortical activity during the perception of three-dimensional object shape from two-dimensional random-dot motion. *Journal of Integrative Neuroscience* **12**: 355–367.

Jaklevic, R. C., Lambe, J., Silver, A. H., and Mercereau, J. E. 1964. Quantum interference effects in Josephson tunneling. *Physical Review Letters* **12**(7): 159–160.

Jeffs, B., Leahy, R. M., and Singh, M. 1987. An evaluation of methods for neuromagnetic image reconstruction. *IEEE Transactions on Biomedical Engineering* **34**: 713–723.

Josephson, B. D. 1972. Possible new effects in superconductive tunneling. *Physics Letters* **1**: 251–253.

Kado, H., Higuchi, M., Shimogawara, M., Haruta, Y., Adachi, Y., Kawai, J., Ogata, H., and Uehara, G. 1999. Magnetoencephalogram systems developed at KIT. *IEEE Transactions of Applied Superconductivity* **9**: 4057–4062.

Kajola, M., Ahlfors, S., Ehnholm, G. J., Hällström, J., Hämäläinen, M. S., Ilmoniemi, R. J., Kirviranta, M., Knuutila, J., Lounasmaa, O. V., Tesche, C. D., and Vilkman, V. 1989. A 24-channel magnetometer for brain research. In *Advances in Biomagnetism*, Williamson, S. J., Hoke, M., Stroink, G., and Kotani, M., eds. Plenum, New York: 673–676.

Kakigi, R., Tran, T. D., Qiu, Y., Wang, X., Nguyen, T. B., Inui, K., Watanabe, S., and Hoshiyama, M. 2003. Cerebral responses following stimulation of unmyelinated C-fibers in humans: Electro- and magneto-encephalographic study. *Neuroscience Research* **45**: 255–275.

Kaufman, L., Schwartz, B., Salustri, C., and Williamson, S. J. 1990. Modulation of spontaneous brain activity during mental imagery. *Journal of Cognitive Neuroscience* **3**: 124–132.

Kirkpatrick, S., Gelatt Jr., C. D., and Vecchi, M. P. 1983. Optimization by simulated annealing. *Science* **220**: 671–680.

Knowlton, R. C., Razdan, S. N., Limdi, N., Elgavish, R. A., Killen, J., Blount, J., Burneo, J. G., Ver Hoef, L., Paige, L., Faught, E., Kankirawatana, P., Bartolucci, A., Riley, K., and Kuzniecky, R. 2009. Effect of epilepsy magnetic source imaging on intracranial electrode placement. *Annals of Neurology* 65: 716–723.

Knuutila, J., Ahlfors, S., Ahonen, A., Hällström, J., Kajola, M., Lounasmaa, O. V., Vilkman, V., and Tesche, C. 1987. Large-area low-noise seven-channel dc SQUID magnetometer for brain research. *Review of Scientific Instruments* 58: 2145–2156.

Levenberg, K. 1944. A method for the solution of certain problems in least squares. *Quarterly of Applied Mathematics* 2: 164–168.

Liu, A. K., Belliveau, J. W., and Dale, A. M. 1998. Spatiotemporal imaging of human brain activity using functional MRI constrained magnetoencephalography data: Monte Carlo simulations. *Proceedings of the National Academy of Sciences of the United States of America* 95: 8945–8950.

Lynn, M. S., and Timlake, W. P. 1968. The use of multiple deflations in the numerical solution of singular systems of equations to potential theory. *SIAM Journal on Numerical Analysis* 5: 303–322.

Marquardt, D. 1963. An algorithm for least-squares estimation of nonlinear parameters. *SIAM Journal of Applied Mathematics* 11: 431–441.

McNay, D., Michielssen, E., Rogers, R. L., Taylor, S. A., Akhtari, M., and Sutherling, W. W. 1996. Multiple source localization using genetic algorithms. *Journal of Neuroscience Methods* 64: 163–182.

Menke, W. 1984. *Geophysical Data Analysis: Discrete Inverse Theory*. Academic, New York.

Mosher, J. C., Lewis, P. S., and Leahy R. 1992. Multiple dipole modeling and localization from spatio-temporal MEG data. *IEEE Transactions on Biomedical Engineering* 39: 541–557.

Nakasato, N., Levesque, M. F., Barth, D. S., Baumgartner, C., Rogers, R. L., and Sutherling W. W. 1994. Comparisons of MEG, EEG, and ECoG source localization in neocortical partial epilepsy in humans. *Electroencephalography and Clinical Neurophysiology* 91: 171–178.

Ogawa, S., Lee, T. M., Nayak, A. S., and Glynn, P. 1990. Oxygenation-sensitive contrast in magnetic resonance image or rodent brain at high magnetic field. *Magnetic Resonance in Medicine* 14: 68–78.

Ohta, H., Takahata, M., Takahashi, Y., Shinada, K., Yamada, Y., Hanasaka, T., Uchikawa, Y., Kotani, M., Matsui, T., and Komiyama, B. 1989. Seven-channel rf SQUID with 1/f noises only at very low frequencies. *IEEE Transactions on Magnetics* 25: 1018–1021.

Oostendorp, T. F., and van Oosterom, A. 1989. Source parameter estimation in inhomogeneous volume conductors of arbitrary shape. *IEEE Transactions on Biomedical Engineering* 36: 382–391.

Orban, G. A., Sunaert, S., Todd, J. T., Hecke, P. V., and Marchal, G. 1999. Human cortical regions involved in extracting depth from motion. *Neuron* 24: 929–940.

Pantev, C., Makaig, S., Hoke, M., Galambos, R., Hampson, S., and Gallan, C. 1991. Human auditory evoked gamma-band magnetic fields. *Proceedings of the National Academy of Sciences of the United States of America* 88: 8996–9000.

Papanicolaou, A. C., Simos, P. G., Castillo, E. M., Breier, J. I., Sarkari, S., Pataraia, E., Billingsley, R. L., Buchanan, S., Wheless, J., Maggio, V., and Maggio, W. W. 2004. Magnetocephalography: A noninvasive alternative to the Wada procedure. *Journal of Neurosurgery* **100**: 867–876.

Paradis, A. L., Cornilleau-Peres, V., Droulez, J., Van de Moortele, P. F., Lobel, E., Berthoz, A., Le Bihan, D., and Poline, J. B. 2000. Visual perception of motion and 3-D structure from motion: an fMRI study. *Cerebral Cortex* **10**: 772–783.

Parsons, L. M., Fox, P. T., Downs, J. H., Glass, T., Hirsch, T. B., Martin, C. C., Jerabek, P. A., and Lancaster, J. 1985. Use of implicit motor imagery for visual shape discrimination as revealed by PET. *Nature* **375**: 54–58.

Peronnet, F., and Farah, M. J. 1989. Mental rotation: An event-related potential study with a validated mental rotation task. *Brain and Cognition* **9**: 279–288.

Pirmoradi, M., Beland, R., Nguyen, D. K., Bacon, B. A., and Lassonde, M. 2010. Language tasks used for the presurgical assessment of epileptic patients with MEG. *Epileptic Disorders* **12**: 97–108.

Ploner, M., Gross, J., Timmermann, L., and Schnitzler, A. 2002. Cortical representation of first and second pain sensation in humans. *Proceedings of the National Academy of Sciences of the United States of America* **99**: 12444–12448.

Raij, T. T., Forss, N., Stancak, A., and Hari, R. 2004. Modulation of motor-cortex oscillatory activity by painful Adelta- and C-fiber stimuli. *NeuroImage* **23**: 569–573.

Romani, G. L., Leoni, R., and Salustri C. 1985. Multichannel instrumentation for biomagnetism. In *SQUID '85: Superconducting Quantum Interference Devices and their Applications*, Hahlbohm, H. D. and Lübbig, H., eds. Walter de Gruyter, Berlin: 919–932.

Rossini, P. M., Altamura, C., Ferretti, A., Vernieri, F., Zappasodi, F., Caulo, M., Pizzella, V., Del Gratta, C., Romani, G. L., and Tecchio, F. 2004. Does cerebrovascular disease affect the coupling between neuronal activity and local haemodynamics? *Brain* **127**: 99–110.

Salmelin, R., Service, E., Kiesilä, P., Uutela, K., and Salonen, O. 1996. Impaired visual word processing in dyslexia revealed with magnetoencephalography. *Annals of Neurology* **40**: 157–162.

Sarvas, J. 1987. Basic mathematical and electromagnetic concepts of the biomagnetic inverse problem. *Physics in Medicine and Biology* **32**: 11–22.

Scherg, M. 1990. Fundamentals of dipole source potential analysis. In *Auditory Evoked Magnetic Fields and Electric Potentials, Advances in Audiology*, vol. 6. Grandori, F., Hoke, M., and Romani, G. L., eds. Karger, Basel: 40–69.

Schmidt, R. O. 1986. Multiple emitter location and signal parameter estimation. *IEEE Transactions on Antennas and Propagation* AP-**34**: 276–280.

Sekihara, K., Poeppel, D., Marantz, A., Phillips, C., Koizumi, H., and Miyashita, Y. 1998. MEG covariance difference analysis: A method to extract target source activities by using task and control measurements. *IEEE Transactions on Biomedical Engineering* **45**: 87–97.

Shepard, R. N., and Metzler, J. 1971. Mental rotation of three dimensional objects. *Science* **171**: 701–703.

Srebro, R. 1996. An iterative approach to the solution of the inverse problem. *Electroencephagraphy and Clinical Neurophysiology* 98: 349–362.

Stefan, H., Hummel, C., Scheler, G., Genow, A., Druschky, K., Tilz, C., Kaltenhäuser, M., Hopfengärtner, R., Buchfelder, M., and Romstöck, J. 2003. Magnetic brain source imaging of focal epileptic activity: A synopsis of 455 cases. *Brain* 126: 2396–2405.

Tagaris, G. A., Kim, S. G., Strupp, J. P., Andersen, P., Ugurbil, K., and Georgopoulos, A. P. 1996. Quantitative relation between parietal activation and performance in mental rotation. *Neuroreport* 7: 773–776.

Teyler, T. J., Cuffin, B. N., and Cohen, D. 1975. The visual evoked magnetoencephalogram. *Life Sciences* 17: 683–691.

van Oosterom, A., and Strackee, J. 1983. The solid angle of a plane triangle. *IEEE Transactions on Biomedical Engineering* 30: 125–126.

van Veen, B. D., and Buckley, K. M. 1988. Beamforming: a versatile approach to spatial filtering. *IEEE ASSP Magazine* 5: 4–24.

van Veen, B. D., van Drongelen, W., Yuchtman, M., and Suzuki, A. 1997. Localization of brain electrical activity via linearly constrained minimum variance spatial filter. *IEEE Transactions on Biomedical Engineering* 44: 867–880.

Vrba, J., Fife, A. A., Burbank, M. B., Weinberg, H., and Brickett, P. A. 1982. Spatial discrimination in SQUID gradiometers and 3rd order gradiometer performance. *Canadian Journal of Physics* 60: 1060–1073.

Vrba, J., Betts, K., Burbank, M., Cheung, T., Fife, A. A., Haid, G., Kubik, P. R., Lee, S., McCubbin, J., McKay, J., McKenzie, D., Spear, P., Taylor, B., Tillotson, M., Cheyne, D., and Weinberg, H. 1993. Whole cortex, 64 channel SQUID biomagnetometer system. *IEEE Transactions on Applied Superconductivity* 3: 1878–1882.

Wada, J. 1949. A new method for the determination of the side of cerebral speech dominance. A preliminary report of the intra-carotid injection of sodium amytal in man. *Igaku to Seibutsugaku* 14: 221–222.

Wallach, H., and O'Connell, D. N. 1953. The kinetic depth effect. *Journal of Experimental Psychology* 45: 205–217.

Wax, M., and Kailath, T. 1985. Detection of signals by information theoretic criteria. *IEEE Transactions on Acoustics, Speech, and Signal Processing* ASSP-33: 387–392.

Williamson, S. J., Pelizzone, M., Okada, Y., Kaufman, L., Crum, D. B., and Marsden, J. R. 1985. Five channel SQUID installation for unshielded neuromagnetic measurements. In *Biomagnetism: Applications and Theory*, Weinberg, H., Stroink, G., and Katila, T., eds. Pergamon, New York: 46–51.

Zimmermann, J. E., Thiene, P., and Harding, J. T. 1970. Design and operation of stable rf-biased superconducting point-contact quantum devices and a note on the properties of perfectly clean metal contacts. *Journal of Applied Physics* 41: 1572–1580.

CHAPTER 5

Principles of Magnetic Resonance Imaging

Masaki Sekino, Norio Iriguchi, and Shoogo Ueno

CONTENTS

5.1 INTRODUCTION

Magnetic resonance imaging (MRI) is a way of producing cross-sectional images of the human body noninvasively. A proton (or hydrogen nucleus) is spinning with a positive charge, and it is generating a magnetic flux in the direction of the Ampere's law. Therefore, protons can be regarded as small, spinning magnets with north poles and south poles in Newtonian physics. When a part of the body is exposed to a static magnetic field, each spinning magnet makes a precession around the axis of the static magnetic field. If a radiofrequency (rf) wave is then transmitted, some of the magnets begin to make a precession with the same phase of the transmitted rf wave. This is the magnetic resonance (MR). In the resonating state, a large number of magnets establish a large magnetic flux rotating around the axis of the static magnetic field at the frequency of the transmitted rf wave. The rf wave is then turned off, and subsequently an electrical coil tuned at the proper rf frequency picks up the electromotive force (EMF) before the rotating magnetic flux decays in amplitude. Resonance frequency is proportional to the external magnetic field. Hence, modifying the static magnetic field by using a gradient magnetic field, the position of each proton can be distinguished, and a computer can reconstruct an image from the EMF signal.

The first successful nuclear magnetic resonance (NMR) experiments were reported by two independent groups in the United States (Bloch et al., 1946; Purcell et al., 1946). In 1952, Purcell and Bloch received the Nobel Prize for their observations. Since chemical shift phenomena were discovered in the 1950s, NMR techniques were developed as important tools for chemical analyses. The early experiments were limited in scope by the relatively poor instrumentation available at that time. In the late 1960s, superconducting magnets were introduced to NMR experiments and revolutionized the scope of NMR together with the emergence of Fourier transform NMR (Ernst and Anderson, 1966). Lauterbur (1973) published the first NMR image of two small tubes of water. By 1980, the clinical evaluation of MRI had begun, and since that time, there have been continuing developments in instrumentation and applications that have led to where we are today. MRI is now established as an important modality in medical practice. The application techniques of MRI have diversified in recent years. Those techniques are, for example, magnetic resonance angiography (MRA), perfusion and diffusion imaging, and functional neuroimaging (fMRI). Scanning time has been dramatically shortened. Because there are clear attractions in using noninvasive methods for the study of living systems, those MRI techniques that investigate the brain and tissue characteristics appear to have a very bright future. Therefore, in this chapter, we give an overview of the principles of MRI techniques.

5.2 MAGNETIC RESONANCE SIGNAL AND RELAXATION TIMES

5.2.1 Electromotive Force

When protons are placed in a static magnetic field B_0, the spinning nuclei begin to make a precession. The phase of the precession is primarily governed by uncertainty, and the phases of each precession are at random. Therefore, the direction of the net macroscopic vector of magnetization M_0 is expected to be that of the z axis, the axis of the external static magnetic field B_0. When the radiofrequency (rf) field of the magnetic flux density B_1 and the Larmor angular frequency ω_0, which is equal to γB_0, is applied to the nuclei for a duration Δt and in the direction of the x axis perpendicular to the z axis, the spinning nuclei make a precession around the x axis by the angle $\theta = \gamma B_1 \Delta t$, where γ is the gyromagnetic ratio of the nuclei. This occurs because the B_1 field is the only apparent field experienced by the nuclei in the rotation frame (Figure 5.1). Thus, the vector of magnetization M_0 makes precession around the x axis, and the flip angle θ is the phase of the macroscopic precession of the magnetization M_0.

The sample is generally conductive. When a magnetic flux tries to pass through the sample, eddy currents are generated in the sample to produce

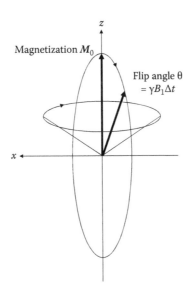

FIGURE 5.1 Precession of the magnetization M_0 around the x axis.

a counter magnetic flux, and this is experienced by the rf coil as the resistance of the coil itself. When the electrical rf coil and the sample are supplied with the rf power P, the current i^* of the coil is $i^* = (P/R)^{1/2}$, where R is the equivalent, gross resistance of the rf coil and the sample. If B_1 is defined as the magnetic flux density produced by a coil carrying unit current, the magnetic flux density $B_1^* = B_1 (P/R)^{1/2}$. The B_1 value of the rf coil can be determined also by the geometrical configuration of the rf coil.

Figure 5.2 shows the magnetization M_0^* of a volume of interest (VOI) of a sample at the time when the lip angle θ has been made 90 degree by the rf field ($B_1 \cos \omega_0 t$) generated by the coil, carrying unit current $i = \cos \omega_0 t$. Magnetic resonance signals can be acquired with the same coil, and the electromotive force (EMF) ξ_s induced by the magnetization M_0^* is given by

$$\xi_s = -d\left(B_1^* M_0^*\right)/dt$$

Because the rf field is $B_1 \cos \omega_0 t$, the EMF is exhibited by

$$\xi_s = \omega_0 B_1 M_0^* \sin \omega_0 t$$

The number N^* of nuclei in the VOI is $N^* = 10^3 \, cvA$, where A is Avogadro's number (6.02×10^{23}), v is the volume of the VOI in m^3, and c is the concentration of nuclei in M. Then, $M_0^* = 10^3 \, cvA\gamma^2 \, (h/2\pi)^2 \, B_0 \, I \, (I + 1)/3kT$, where

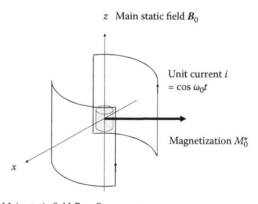

FIGURE 5.2 Magnetization M_0^* just after the application of a 90° pulse.

h is Planck's constant, I is the spin quantum number, k is Boltzmann's constant, and T is the temperature in Kelvin. If the concentration c, the rf field B_1 and the volume v of VOI are determined, then the EMF ξ_s can be numerically calculated, because all other constants are known.

Proton density is an important source of the NMR signal that conveys much information about the subject tissue. But the total NMR signal has more information than the number of nuclei reflected in the amplitude of the signal. Relaxation is the term given to a process of transition from an unstable, high-energy state to a stable, low-energy state.

When proton nuclei are exposed to a static magnetic field B_0, the individual magnetic moments align parallel or antiparallel with the static magnetic field B_0. Spinning nuclei whose expected axes of precessions are parallel with the static magnetic field B_0 are called α nuclei and are in a low-energy, stable state. Spinning nuclei whose expected axes of precessions are antiparallel with the static magnetic field B_0 are called β nuclei and are in a high-energy, unstable state. In the state of thermal equilibrium, nuclei are in only one of those two states, and the number of α nuclei is slightly greater than the number of β nuclei. As small as this slight excess is, it accounts for the small macroscopic net magnetization M_0 directing parallel with the static magnetic field B_0, and it is this differential that accounts for the on which MRI is based.

When a 90 degree rf pulse is applied, α nuclei are excited and elevated to the high-energy state, and the number of α nuclei becomes the same as that of β nuclei. When a 180 degree rf pulse is applied, the number of β nuclei becomes greater than that of α nuclei. If excited nuclei are left in the thermal equilibrium, they will be allowed to lose energy to become low-energy, stable α state. This is the process of T_1 or longitudinal relaxation. T_1 relaxation is an exponential process, and T_1 is the time constant of the exponential process. T_1 relaxation can be expressed as a macroscopic phenomenon, $dM_z/dt = (M_0 - M_z)/T_1$, where M_z is the z axis component of the transitional magnetization. T_1 relaxation is also the macroscopic phenomenon where the magnetization is allowed to return to the initial state of the thermal equilibrium. For example, the proton T_1 value of the white matter of the human brain is typically 690 msec at 20 MHz. When molecules are restricted in movement as in solids, T_1 relaxation hardly occurs, and T_1 values are even as long as several hours. The nuclei are often the elements of large molecules, and T_1 values reflect the molecular structures.

Spinning nuclei, small magnets with north poles and south poles, affect each other. Here, the magnetic flux density of the static magnetic field B_0

actually experienced by a spinning nucleus is altered by other spinning nuclei. For this reason, spinning nuclei placed in the same static magnetic field B_0 make precessions with slightly different frequencies, and phases of precessions become incoherent. Thus, T_2 relaxation takes place. T_2 relaxation is also an exponential process, and T_2 is the time constant of the exponential process. T_2 relaxation can be expressed as a macroscopic phenomenon, $dM_{xy}/dt = -M_{xy}/T_2$, where M_{xy} is the component of transitional magnetization perpendicular to the direction of the static magnetic field B_0. T_2 relaxation is also the macroscopic phenomenon where the magnetization decays in amplitude in the xy plane. By T_2 relaxation, the amplitude of EMF ξ_s becomes smaller and exhibits free induction decay (FID). The FID signal ξ_s can be expressed as

$$\xi_s = \xi_{s0} \exp(i\omega_0 t) \exp(-t/T_2)$$

and the initial value of the amplitude $\xi_{s0} = \omega_0 B_1 M_0^*$. In contrast to T_1 relaxation, the exchange of energy in T_2 relaxation does not take place between spinning nuclei and lattices but only among nuclei. For example, the proton T_2 value of the white matter of the human brain is typically 110 msec.

T_2 relaxation is accelerated in large molecules, and T_2 values of the nuclei of large molecules are generally short. T_2 values of nuclei of small molecules are also short when small molecules are combined to large molecules. For example, an anticancer drug fluorourasil solution exhibits a sharp ^{19}F resonance peak and has a longer T_2 value than 200 msec, but once it is incorporated into the liver cells, the T_2 value becomes shorter than 10 msec.

When the static magnetic field B_0 is inhomogeneous, phases of nuclei precessions become incoherent. This process appears like a relaxation and is called T_2 star $\left(T_2^*\right)$ relaxation.

By shimming, or improving the homogeneity of the static magnetic ratio B_0, the T_2^* value is made long. The FID signal ξ_s from nuclei in an inhomogeneous static magnetic field can be expressed as

$$\xi_s = \xi_{s0} \exp(i\omega_0 t)\exp\left(-t/T_2^*\right)$$

After nuclei are excited by a 90 degree pulse, the vector of magnetization M_0 on the xy plane begins to decay exponentially in amplitude by T_2^* relaxation. However, if a 180 degree pulse follows immediately, the phases

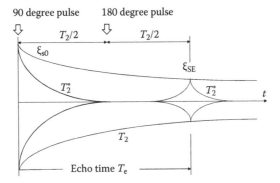

FIGURE 5.3   Spin echo (SE) signal.

are refocused to be coherent and the nuclei generate an echo (Hahn, 1950). This is the spin echo (SE) signal, and the 180 degree pulse is often called a time-reversing pulse, because it refocuses the phases of nuclear precessions as if time were reversed (Figure 5.3). The spin echo signal amplitude ξ_{SE} is affected by T_1 relaxation as a matter of course during the repetition time T_r and the T_2 relaxation during the echo time T_e. ξ_{SE} is expressed as $\xi_{SE} = \xi_{s0}\{1 - \exp(-T_r/T_1)\}\exp(-T_e/T_2)$. The SE signal ξ_s can be hence expressed as

$$\xi_s = \xi_{s0}\{1-\exp(-T_r/T_1)\}\exp(-T_e/T_2)\exp(i\omega_0 t)\exp\left(-\left|t-T_e\right|/T_2^*\right)$$

The SE signal ξ_s has a feature where the amplitude is symmetrical in regard to the echo time T_e. Therefore, if the excitation pulse and refocusing pulse are θ_1 and $2\theta_2$ instead of 90 and 180 degree, respectively, then the signal ξ_s can be expressed as

$$\xi_s=\xi_{s0}\{1-\exp(-T_r/T_1)\}\exp(-T_e/T_2)\exp(i\omega_0 t)\exp\left(-\left|t-T_e\right|/T_2^*\right)\sin\theta_1\sin^2\theta_2$$

5.3 OVERVIEW OF THE SPIN-WARP IMAGING TECHNIQUE

In MRI scanners, the scanning space inside the magnet has three intersecting planes corresponding to three anatomical planes of the body (axial, coronal, and sagittal). The three planes are indicated by three intersecting axes, with the z axis of the static magnetic field B_0.

A gradient magnetic field can be used to assign nuclei to a position in space within the body. The gradient field is generated by a gradient coil (Turner, 1993). The gradient coils create opposing magnetic fields by carrying currents in opposite directions. The field from one coil opposes the field from the other, creating a net field that serves as a gradient field. Gradient field can be applied in each of the three orthogonal directions (x, y, and z). A gradient coil of any direction changes the magnetic flux density of the z direction. When a gradient coil of z direction is applied, the magnetic flux density in the z direction is the static magnetic field B_0 itself at the center of the magnet, but it is higher at a position apart from the center in the positive z direction. When a gradient coil of x direction is applied, the magnetic flux density in the z direction is higher than B_0 at a position apart from the center in the positive x direction.

Slice selection is performed by applying a gradient magnetic field and an excitation rf pulse simultaneously. Also, the recognition of nuclear distribution in the second and third directions is performed by employing gradient coils. The spin-warp imaging technique is the traditional term for such popular methods of MRI as field echo (FE) imaging, spin echo (SE) imaging, and echo-planar imaging (EPI). The main feature of spin-warp imaging technique is the process of detecting the location of a spinning nucleus in the phase of precession by varying a linear, gradient magnetic field (Edelstein et al., 1980).

5.3.1 Recognition of Nuclei Distributed in the First Direction

When a sample is placed in a static magnetic field B_0 and a linear gradient magnetic field G_z is superimposed to the static magnetic field B_0, nuclei in the sample are excited by an rf field only in the region where the magnetic flux density is specific to the nuclear magnetic resonance frequency ω_0. This process is the slice-selective excitation of nuclei, and the selection is in the z axis direction of the applied gradient field G_z. By modulating the excitation power, the profile of the slice selection can be determined. An excitation power with a sinc, or (sin t)/t, envelope function gives an excitation of a square profile. Because a Fourier-transformed sinc function in time is a square function in frequency, the slice profile resulted from a sinc excitation is square. An excitation power with a Gaussian envelope function gives a Gaussian excitation. The effective frequency bandwidth of an excitation is the reciprocal of the effective duration of the applied rf pulse. When an excitation pulse has a sinc envelope function with a main lobe of 1 msec in duration at half maximum, the resulting slice has a square profile of about 1 kHz in bandwidth. The longer the excitation duration is, the thinner the resulting slice thickness is.

As shown in Figure 5.4, when the gradient field is 1 mT/m, the proton resonance frequency is made distributed in 42.6 Hz/mm, because the

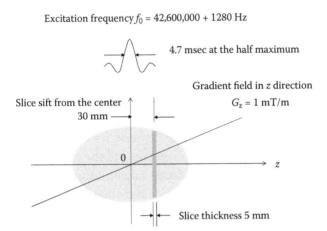

Excitation frequency f_0 = 42,600,000 + 1280 Hz

4.7 msec at the half maximum

Gradient field in z direction

Slice sift from the center

30 mm

G_z = 1 mT/m

0

z

Slice thickness 5 mm

FIGURE 5.4   Slice selection with a gradient field.

resonance frequency of protons is about 42,600,000 Hz/T. When a pulse is applied with a sinc envelope of the main lobe of 4.7 msec in duration at half maximum, the Fourier-transformed bandwidth is 213 Hz. Hence, a slice is excited with a thickness of 5 mm. If the frequency f_0 of the of the rf pulse is f_0 = 42,600,000 + 1280 Hz, then the position of the slice selection will be shifted by 30 mm in the +z axis direction.

5.3.2 Recognition of Nuclei Distributed in the Second Direction

The recognition of nuclei distributed in the second, or x axis direction, can be performed by acquiring signals simultaneously applying a linear, gradient magnetic field G_x in the x axis direction. With the gradient field G_x, the resonance frequency of nuclei distributed in the x axis is linearly distributed in the direction of x axis from lower frequencies to higher frequencies than that of nuclei without the gradient field. For example, if the gradient field G_x is 1 mT/m, then in the region of −0.1 m to +0.1 m inclusively in the x axis direction, the frequency deviation of resonance will be linearly distributed −4260 Hz to +4260 Hz inclusively. This frequency bandwidth is small enough to be processed by an ordinary computer.

The signal acquired by an rf coil has various frequencies specific to the locations of nuclei distributed in the x axis direction, and the special distribution of nuclei is restored by Fourier transformation. This is the process of encoding the locations of nuclei in the x axis direction and simply called reading-out of the signal. By the Fourier transformation, a projection curve of the sample in the x axis direction can be obtained.

Projection data are often exhibited in an absolute value $\left(R_e^2 + I_m^2\right)^{1/2}$ with the real amplitude R_e and imaginary I_m both detected by quadrature demodulation. For example, if the number of acquired data points is 512 and half of those points are real and the rest are imaginary, then the projection can be exhibited in a 256-point line graph.

With the application of a reading-out gradient field G_x, the signal intensity rapidly diminishes because frequencies of nuclear precessions are made distributed in the x axis direction and the phases of precessions rapidly become incoherent. This decay can be recovered by applying a negative, dephasing gradient field $-G_{xd}$ and then reversing the gradient field to the positive gradient field $+G_x$. By this, a field echo (FE) signal is generated. As shown in Figure 5.5, the FE is generated at the time when the gray areas are the same. When the gradient field G_x is 1 mT/m, proton frequency deviation occurring because of the gradient field G_x is 42,600 Hz/m in the x axis direction. If the field of view (FOV) of the MR image is, for example, 30 cm in the x axis direction, then the frequency bandwidth covering the FOV range of 30 cm will be 12,800 Hz, and according to the Sampling theorem, the sampling rate of the AD converter must be greater than 25,600 Hz.

In spin echo (SE) imaging, most of acquired signals are the combined echoes of the SE signal generated by a 180 degree pulse and the FE signal generated by reversing the reading-out gradient field G_x. Usually, the moment of SE refocusing and the moment of FE refocusing are made coincident. The acquired data contain various frequencies of precessions of nuclei, and the frequencies are restored to the original state of locations of nuclei by the Fourier transformation.

Figure 5.6 is an example of an echo signal acquired during the time t_s. The axis is time in second. Figures 5.7 and 5.8 show the real part R_e and

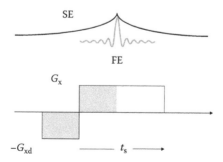

FIGURE 5.5 Reversing the gradient magnetic field and a field echo (FE) signal.

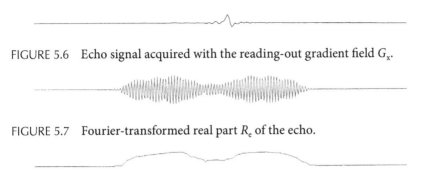

FIGURE 5.6 Echo signal acquired with the reading-out gradient field G_x.

FIGURE 5.7 Fourier-transformed real part R_e of the echo.

FIGURE 5.8 Fourier-transformed absolute value $A_v = \left(R_e^2 + I_m^2 \right)^{1/2}$ of the echo.

absolute value $A_v = \left(R_e^2 + I_m^2 \right)^{1/2}$ of the signal data, respectively. The axes are frequency in hertz. The curve of the absolute value A_v is similar to the envelope shape of the real part R_e curve.

5.3.3 Recognition of Nuclei Distributed in the Third Direction

Recognition of nuclei distributed in the third direction is performed by a technique called phase encoding. It begins after the slice-selective excitation. First, a small gradient field G_y in the third, y axis direction is applied for a certain period Δt. By the application of the gradient G_y, a nucleus located far in the $+y$ direction from the original, central point is exposed to a higher static magnetic field than that of the nucleus located at the original, central point. In the period Δt, the phase of precession of the nucleus far from the center becomes larger than that of the nucleus at the center. On the other hand, a nucleus located far in the $-y$ direction from the original, central point is exposed to a lower static magnetic field than that of the nucleus located at the original, central point. In the period Δt, the phase of precession of the nucleus far from the center becomes smaller than that of the nucleus at the center. Second, a gradient field G_y which is 2 times larger in amplitude is applied in the same y axis direction for the same period Δt and also after the second slice-selective excitation. Thus, the gradient field is successively changed stepwise in amplitude and applied, for example, in 256 steps. By this, the farther nuclei are located from the original point, the greater change of the gradient field G_y in amplitude is experienced by the nuclei, and hence the greater change of the precession phase is experienced by the nuclei during the period Δt. By Fourier transformation of the acquired 256 data set, it is recognized how far the nuclei are located from the original point.

When the gradient field G_y is varied in amplitude 256 times, for example, from −2.54 mT/m to +2.56 mT/m by step of 0.02 mT/m, a nucleus at the position 10 cm apart from the original point in the y axis direction, experiences a deviation of the static magnetic field in amplitude by 0.002 mT each time, and the proton resonance frequency changes 256 times by 85.2 Hz each time. If the duration Δt is 3 msec, for example, at the position 10 cm apart from the original point, the phase of the precession changes 256 times by $2\pi \times 85.2 \times 0.003 = 1.6$ rad every time. By Fourier transformation, the nuclear distribution in the y axis direction is restored. A typical FE imaging sequence is, as shown in Figure 5.9, a combination of a slice-selective excitation in the z axis direction, a frequency encoding in the x axis direction, and a phase encoding in the y axis direction. By two-dimensional Fourier transformation of a set of acquired signal data, the two-dimensional distribution of nuclei is restored. Hence, the resulting image is on the plane of one axis in frequency in Hz and another axis in θ in rad per encoding step, as shown in Figure 5.10.

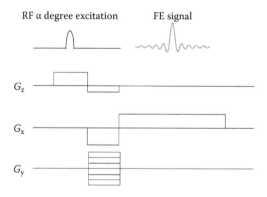

FIGURE 5.9   FE imaging sequence.

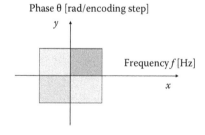

FIGURE 5.10   Frequency-phase plane.

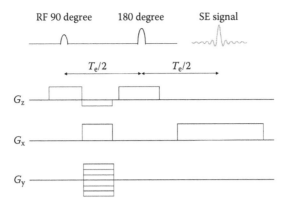

FIGURE 5.11   SE imaging sequence.

A SE imaging sequence employs a 180 degree pulse, as shown in Figure 5.11. The 180 degree pulse is applied in the period $T_e/2$ after the 90 degree selective excitation. By this, SE signal is detected in the echo time T_e after the slice-selective excitation.

5.3.4 k Space

By varying the amplitude of the gradient field in the phase-encoding direction, nuclei are labeled according to their phases of precessions. The acquired data are then put in successive order on a plane traditionally called k space (Ljunggren, 1983). The echo signals put on the k space are those demodulated and subtracted by the primary frequency $\omega_0 = 2\pi\nu_0$. Obviously, each signal data appear to have waves containing different frequencies. Those waves are specific to the horizontal x axis distribution of nuclei and generated by the reading-out gradient field G_x.

The waves in the vertical direction of the k space are also specific to the vertical y axis distribution of nuclei and generated by varying the amplitude of the phase-encoding gradient field G_y. When a quadrature demodulation is employed, each data has a real part and imaginary part. For example, if the number of data points of a signal is 512, then 256 are real data points and the remaining 256 are imaginary data points. Hence, when the number of phase-encoding steps is 256, the vertical phase-encoded data have 256 real data points and imaginary data points. Therefore, on the k space, there are a real data set of 256×256 points and an imaginary data set of 256×256 points. Figure 5.12 shows an example of a real data set of 256×256.

FIGURE 5.12   Acquired real part data on the k space.

Figure 5.13 shows 256 one-dimensional Fourier-transformed real data on the k space. Each horizontal line is a line graph consisting of 256 points. The vertical array is not Fourier transformed yet. Figure 5.14 shows one-dimensional Fourier-transformed absolute value $A_v = \left(R_e^2 + I_m^2 \right)^{1/2}$ data on the k space. Each horizontal line is a line graph consisting of 256 points, while the vertical array is not Fourier transformed yet.

A two-dimensional nuclear distribution of an orange is shown in Figure 5.15. This was restored by a two-dimensional Fourier transformation of the horizontal and vertical wave lines on the k space. The nuclear distribution is displayed in absolute values. The matrix size is 256 × 256. The unit of the horizontal axis is frequency f in hertz and corresponds linearly to the distance l in meters from the central point. The unit of the vertical axis is phase θ in rad/step and also corresponds linearly to the

FIGURE 5.13   One-dimensional Fourier-transformed real part data on the k space.

FIGURE 5.14   One-dimensional Fourier-transformed absolute value data on the k space.

FIGURE 5.15   Two-dimensional Fourier-transformed absolute value data on the k space.

distance l in meters from the point. If the intensity diagram is gray scaled and displayed on a CRT, then we obtain an ordinary MR image.

5.3.5 Image Contrast

By applying an excitation rf pulse, the magnetization vector is given a component perpendicular to the z axis and the component decays in amplitude by T_2^* relaxation. When a FE image sequence is used, and if both the repetition time T_r and echo time T_e are long, then the resulting FE image is a T_2^*-weighted image affected by T_2^* relaxation. The T_2^* relaxation of the brain tissue often reflects the brain activity, and T_2^*-weighted images are important in the field of fMRI.

With a SE imaging sequence, the magnetization vector is given a component perpendicular to the z axis and the component decays in amplitude by T_2 relaxation. If the repetition time T_r is longer than T_1 and the echo time T_e is shorter than T_2, then the resulting SE image reflects the nuclear density distribution.

If the echo time T_e remains short and the repletion time T_r is shortened, the sample of long T_1 undergoes successive excitations before the magnetization recovery by T_1 relaxation and generates small signal. The resulting image is a T_1-weighted image and has a constant reflecting the T_1 distribution. If both the repetition time T_r and echo time T_e are long, then the resulting SE image is a T_2-weighted image affected by T_2 relaxation. Compared with FE images, the influence of inhomogeneity of the static magnetic field is reduced in SE imaging.

T_1-weighted images give higher-intensity signals from the white matter than the gray matter. T_2-weighted images give higher-intensity signals from the gray matter than the white matter. The contrast arises because the water protons of white matter have shorter T_1 and T_2 values than those of gray matter. The shorter T_1 value causes the white matter protons to undergo less saturation, and hence, to generate higher signal intensity in the T_1-weighted image. On the other hand, because of their shorter T_2 value, the white matter protons undergo a greater degree of signal loss during the echo time T_e of the T_2-weighted image.

Figure 5.16 shows a coronal T_1-weighted SE image (Figure 5.16a) and T_2-weighted SE image (Figure 5.16b) of a 46-year-old male patient

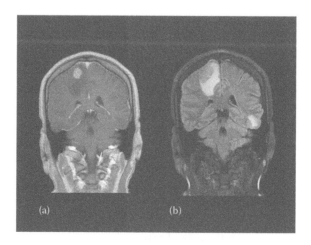

FIGURE 5.16 (a) T_1-weighted and (b) T_2-weighted images.

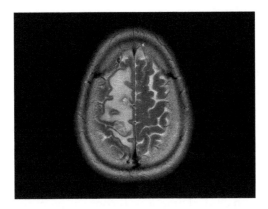

FIGURE 5.17 T_2-weighted image of the same patient of Figure 5.16.

with multiple tumor lesions. Those images were obtained at IKR-TRIC University Hospital, Muenster in Germany and offered by Siemens Japan K.K. The MRI scanner was a Siemens Prisma 3Tesla machine. The T_1-weighted image (Figure 5.16a) is a TSE image with a slice thickness of 4 mm and matrix size of 320. TSE is a technique to obtain an SE image very fast. The total acquisition time was 84 sec. The T_2-weighted image (Figure 5.16b) is a TIRM image with a slice thickness of 4 mm and matrix size of 640. TIRM is a technique to obtain an inversion-recovery (IR) image very fast. The total acquisition time was 180 sec. Both images were obtained by a Siemens-specific implementation called iPAT[2] (standing for integrated parallel acquisition techniques square), which allows iPAT in two directions simultaneously (phase-encoding direction and three-dimensional direction for three-dimensional sequences).

Figure 5.17 shows an axial T_2-weighted SE image of the same patient of Figure 5.16. The image was obtained also at IKR-TRIC University Hospital, Muenster in Germany and offered by Siemens Japan K.K. The MRI scanner was a Siemens Prisma 3Tesla machine. The T_2-weighted image (left) is a TSE image with a slice thickness of 4 mm and matrix size of 512. The total acquisition time was 132 sec. The image was obtained also by the iPAT[2].

5.4 DIVERSIFICATION OF MRI APPLICATION TECHNIQUES

5.4.1 Magnetic Resonance Angiography

Magnetic resonance angiography (MRA) is developing a powerful technique to obtain flow images. It does not necessarily require the injection of a contrast agent and permits the measurement of flow velocities. It is becoming increasingly used for the assessment of abnormalities in the cerebrovascular system and peripheral vasculature.

There are two important methods for MRA. Those are the time-of-flight (TOF) method and the phased-contrast (PC) method. In a TOF method, the water protons in a slice of a certain thickness are first labeled. The labeling is performed by applying an excitation rf pulse. Because saturation makes the numbers of α and β nuclei equal, few signals are generated from the saturated region. In a time t after the saturation, fresh and unsaturated nuclei flow into the saturated region. As a result, contrast is generated between stationary and flowing water protons. Three-dimensional, TOF MRA is used in clinical practices. If consecutive excitation pulses are applied at intervals that are much shorter than T_1, there is an increase in signal intensity, because the inflowing spins generate more signals than the stationary nuclei, which undergo the saturation. FE, FLASH (fast low-angle shot) sequences are commonly used for the three-dimensional, TOF MRA. The three-dimensional data sets contain information about both stationary and flowing nuclei. Vascular structures can be extracted out of the full data set by virtue of their increased intensity via a post-processing routine called maximum intensity projection (MIP).

Flow can induce phase change in MR signals. As water protons move along a gradient magnetic field, they accumulate a phase shift that is proportional to their velocity. In the PC method, a gradient field is applied in a certain direction. The faster the nuclei flow in the gradient direction, the more the transverse magnetization changes phase. By subtraction of data sets with and without the gradient magnetic field, an angiogram is obtained. Three-dimensional, PC MRA is also used in clinical practices. Compared to the TOF method, the PC method takes a longer time for signal acquisition, but the background tissue signals are small.

5.4.2 Perfusion and Diffusion Imaging

Once nuclei in the neck region are excited, arterial flow will carry the labeled nuclei into the brain tissue and influence the observed signal from brain water. Subtractions of images with and without the labeling permit measurements of regional perfusion (Edelstein et al., 1980; Ljunggren, 1983). A technique of signal targeting with alternating radiofrequency (EPISTER) is also used (Williams et al., 1992; Zhang et al., 1993). This is an echo-planar imaging (EPI) technique combined with TOF technique, and generates a perfusion image by subtraction of images with and without saturation pulses (Edelman et al., 1994). The EPI is a rapid imaging technique to obtain a two-dimensional data set by filling the k space with signals generated by one excitation pulse.

Diffusion measurements by NMR were introduced in the 1960s (Stejskal, 1965; Stejskal and Tanner, 1965). In water, all molecules move randomly. It is often referred to as Brownian motion and involves displacement distances that are small and comparable to the cell sizes. Hence, the water diffusion can be affected by the cell structure and diffusion-weighted images can reflect the changes of the cell structure. In the tissue of cerebral infarction, diffusion of water is restricted and the apparent diffusion coefficient (ADC) of water becomes small. This is allegedly caused by the water shift from the extracellular spaces into the intracellular spaces (Kohno et al., 1995; Mosley et al., 1990). Nuclei moving randomly under a gradient magnetic field lose signals, and the diffusion can be detected by MRI. Neither a T_1-weighted nor T_2-weighted image reveals the lesions of cerebral infarction obtained in a few hours after the onset, but a diffusion-weighted image often reveals the lesions in the early stage of infarction (Mintrovich et al., 1991; Pennock et al., 1994; Warach et al., 1992). Diffusion-weighted images are often linked to the anisotropy of the tissue structure. The anisotropy can be detected by the relations between the direction of the gradient field and the extent of signal loss by the water movement (LeBihan and Turner, 1992).

Figure 5.18 shows axial diffusion-weighted images (left and center) and an ADC image (right) of the same patient of Figures 5.16 and 5.17. The attenuation b factors were $b = 0$ (left) and 1000 (center). The images were obtained also at IKR-TRIC University Hospital, Muenster in Germany

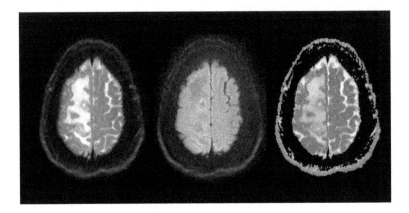

FIGURE 5.18 Diffusion-weighted and apparent diffusion coefficient images of the same patient in Figure 5.16.

and offered by Siemens Japan K.K. The MRI scanner was a Siemens Prisma 3Tesla machine. The images are with a slice thickness of 4 mm. The total acquisition time was 63 sec. The images were obtained by GRAPPA, standing for generalized autocalibrating partially parallel acquisition, a Siemens-specific technique.

5.4.3 Functional Neuroimaging

Functional neuroimaging (fMRI) is one of the most important MRI techniques developed recent years and visualizes the active regions of the brain noninvasively. Functional imaging uses FE imaging sequences of long echo time T_e as 50 msec and reflects T_2^* characteristics of the tissues. Electrons of Fe^{2+} of deoxy-hemoglobin (Deoxy-Hb) have much greater magnetic momentum than those of proton nuclei, and proton resonance spectra of deoxy-hemoglobin (Deoxy-Hb) solution exhibit broad peaks caused by the effects of paramagnetic features of Deoxy-Hb. Also, the water resonance frequency is broadened by the effects of the gradient magnetic field due to the Fe^{2+} electrons. Therefore, T_2^*-weighted images of water around Deoxy-Hb generate few signals. In the living brain, the gradient magnetic fields caused by the Fe^{2+} electrons strongly influence not only the water in the blood vessels but also, to a much greater extent, the outer neural tissues of the water volume of the blood vessels (Ogawa and Lee, 1990; Ogawa et al., 1990; Turner et al., 1991).

When a region in the brain is active, the blood flow increases in the region, oxy-hemoglobin (Oxy-Hb) increases, and Deoxy-Hb decreases in relative quantity. By the blood oxygenation level dependent effects, the signal intensities from the active region in the obtained T_2^*-weighted images increases (Bandettini et al., 1992; Frahm et al., 1992). Also, fMRI reflects internal verbal generations and recalling of sensory or motor stimulation. (LeBihan et al., 1993; McCauthy et al., 1993; Rueckert et al., 1994).

Figure 5.19 shows a functional neuroimaging routine. A simultaneous time-course analysis for multiple regions of interest (ROIs) is carried out in routine. The image was also offered by Siemens Japan K.K.

5.4.4 Magnetic Resonance Spectroscopy

The ability to obtain localized spectra offers great potential for detecting human disease at the early stage of biochemical alteration. The proton signal from water is much greater than signals from metabolites. However, the suppression of water is expanding the use of MRI in metabolite evaluation. In the field of magnetic resonance spectroscopy, metabolite maps of specific chemicals or chemical shift images are also obtained.

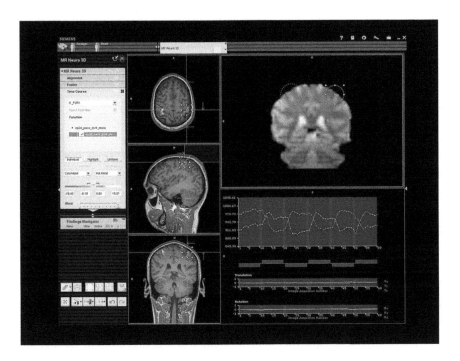

FIGURE 5.19   Functional neuroimaging routine.

5.4.5 MRI of Other Nuclei than Protons

There are important nuclei other than protons for MRI. Figure 5.20 shows a natural abundance carbon-13 MRI of a human arm reflects the distribution of $-CH_2-$ chain of the subcutaneous fatty tissues and bone marrow (Iriguchi and Hasegawa, 1993). Phosphorus-31 chemical shift images,

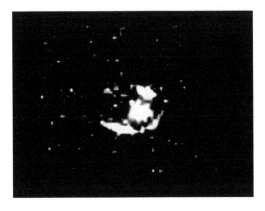

FIGURE 5.20   Natural abundance carbon-13 MRI of a human upper arm.

which are images of adenosine triphosphate (ATP), phosphocreatine (PCr), and inorganic phosphate (Pi) of a human brain, are obtained. Those images reflect the energy metabolism of the brain. Deuterium MRI and fluorine MRI are carried out in experiments with those nuclei as contrast media.

5.5 CONCLUDING REMARKS

In the last century, MRI scanners were classified simply by their field strength, namely, low-field (0.5 T and below), mid-field (1.0 T), and high-field (1.5 T and higher) scanners. Continued progress in high-performance gradient coil technologies have made high-quality and fast imaging a reality. Techniques have been diversified dramatically. MRI scanners are classified in a more sophisticated way, i.e., productivity in scanning reality and real-time post-processing procedures and friendliness in design concepts for clinical use. In the field of basic technology, it is apparent that MRI is not simply a method of obtaining radiological images, but it can also provide a great deal of physiological information. Some MRI techniques will be developed for noninvasive mapping of physical, chemical, and physiological properties such as temperature, mechanical elasticity, pressure, electrical conductivity, permittivity, and metabolite concentration. Development will be continued in this century.

REFERENCES

Bandettini, P. A., Wong, E. C., Hinks, R. S., Tikofsky, R. S., and Hyde, J. S. 1992. Time course EPI of human brain function during task activation. *Magnetic Resonance Medicine* **25**: 390.

Bloch, F., Hansen, W. W., and Packard, M. 1946. The nuclear induction experiment. *Physical Review* **70**: 474.

Edelman, R. R., Siewert, B., Darby, D. G., Thangaraj, V., Nobre, A. C., Mesulam, M. M., and Warach, S. 1994. Qualitative mapping of cerebral blood flow and functional localization with echo-planar MR imaging and signal targeting with alternating radiofrequency. *Radiology* **192**: 513.

Edelstein, W. A., Hutchison, J. M. S., Johnson, G., and Redpath, T. 1980. Spin warp NMR imaging and applications to whole body imaging. *Physics in Medicine and Biology* **25**: 751.

Ernst, R. R., and Anderson, W. A. 1966. Application of Fourier transform spectroscopy to magnetic resonance. *Review of Scientific Instruments* **37**: 93.

Frahm, J., Bruhn, H., Merboldt, K. D., and Haenicke, W. 1992. Dynamic MR imaging of human brain oxygenation during rest and photic stimulation. *Journal of Magnetic Resonance Imaging* **2**: 501.

Hahn, E. L. 1950. Spin echoes. *Physical Review* **80**: 580.

Iriguchi, N., and Hasegawa, J. 1993. Carbon-13 magnetic resonance imaging of a human arm. *Magnetic Resonance Imaging* **11**: 269.

Kohno, K., Hoehn-Berlage, M., Mies, G., Back, T., and Hossmann, K. A. 1995. Relationship between diffusion-weighted MR images, cerebral blood flow, and energy state in experimental brain infarction. *Magnetic Resonance Imaging* **13**: 73.

Lauterbur, P. C. 1973. Image formation by induced local interactions: Examples employing nuclear magnetic resonance. *Nature* **242**: 190.

LeBihan D., and Turner, R. 1992. Diffusion and perfusion. In *Magnetic Resonance Imaging*. Mosby Year Books, St. Louis.

LeBihan, D., Turner, R., Zeffiro, T. A., Cuenod, C. A., Jezzard, P., and Bonnerot, V. 1993. Activation of human primary visual cortex during visual recall: An MRI study. *Proceedings of the National Academy of Sciences of the United States of America* **90**: 1802.

Ljunggren, S. 1983. A simple graphical representation of Fourier-based imaging method. *Journal of Magnetic Resonance* **54**: 338.

McCauthy, G., Blamire, A. M., Rothman, D. L., Gruetter, R., and Schulman, R. G. 1993. Echo-planar MRI studies of frontal cortex activation during word generation in humans. *Proceedings of the National Academy of Sciences of the United States of America* **90**: 4952.

Mintrovich, J., Mosley, M. E., Chileuitt, L., Shimizu, H., Cohen, Y., and Weinstein, R. R. 1991. Comparison of diffusion- and T_2-weighted MRI for early detection of cerebral ischemia and reperfusion in rats. *Magnetic Resonance Medicine* **18**: 39.

Mosley, M. E., Cohen, Y., Mintorovich, J., Chileuitt, L., Shimizu, H., Kucharczyk, J., Wendland, F. M., and Weinstein, P. R. 1990. Early detection of regional cerebral ischemia in cats: Comparison of diffusion- and T_2-weighted MRI and spectroscopy. *Magnetic Resonance in Medicine* **14**: 330.

Ogawa, S., and Lee, T-M. 1990. Magnetic resonance imaging of blood vessels at high field: In vivo and in in vitro measurements and image simulation. *Magnetic Resonance Medicine* **16**: 9.

Ogawa, S., Lee, T-M., Nayak, A. S., and Glynn, P. 1990. Oxygenation-sensitive contrast in magnetic resonance image of rodent brain at high magnetic field. *Magnetic Resonance Medicine* **14**: 68.

Pennock, J. M., Cowan, F. W., Schweiso, J. E., Oatridge, A., Rutherfard, M. A., Pubowitz, L. M. S., and Bydder, G. M. 1994. Clinical role of diffusion-weighted imaging: Neonatal studies. *Magnetic Resonance Materials in Physics, Biology and Medicine* **2**: 273.

Purcell, E. M., Torroy, H. C., and Pound, R. V. 1946. Resonance absorption by nuclear magnetic moments in a solid. *Physical Review* **69**: 37.

Rueckert, L., Appollonio, I., Grafmaii, J., Jezzard, P., Johnson, Jr., R., LeBihan, D., and Turner, R. 1994. MRI functional activation of the left frontal cortex during covert word production. *Journal of Neuroimaging* **4**: 67.

Stejskal, E. O., and Tanner, J. E. 1965. Spin diffusion measurements: Spin echoes in the presence of time-dependent field gradient. *Journal of Chemical Phys.* **42**: 288.

Stejskal, E. O. 1965. Use of spin echo in a pulsed magnetic-field gradient to study anisotropic, restricted diffusion and flow. *Journal of Chemical Physics* **43**: 3597.

Turner, R. 1993. Gradient coil design: A review of method. *Magnetic Resonance Imaging* 11: 903.

Turner, R., LeBihan, D., Moonen, C. T. W., Despres, D., and Frank, J. 1991. Echo-planar time course MRI of cat brain by oxygenation changes. *Magnetic Resonance Medicine* **22**: 159.

Warach, S., Chien, D., Li, W., Ronthal, M., and Edelman, R. R. 1992. Fast magnetic resonance diffusion-weighted imaging of acute human stroke. *Neurology* **42**: 1717.

Williams, D. S., Detre, J.A., Leigh, J. S., and Koretsky, A. P. 1992. Magnetic resonance imaging of perfusion using spin inversion of arterial water. *Proceedings of the National Academy of Sciences of the United States of America* **89**: 212.

Zhang, W., Williams, D. S., and Koretsky, A. P. 1993. Measurement of rat brain perfusion by NMR using spin labeling of arterial water: In vivo determination of the degree of spin labeling. *Magnetic Resonance in Medicine* **29**: 416.

CHAPTER 6

Prospects for Magnetic Resonance Imaging of Impedance and Electric Currents

Masaki Sekino, Norio Iriguchi, and Shoogo Ueno

CONTENTS

6.1 INTRODUCTION

6.1.1 Advantages of Magnetic Resonance Imaging in Impedance Imaging

Analyses of electromagnetic fields play an important role in understanding electromagnetic phenomena in living bodies. Worldwide, several groups develop and distribute standard models of the human body,[1,2] and these models benefit research into electromagnetic fields. However, the present methodology for developing these standard models imposes certain limitations on the scope and accuracy of these analyses. The models consist of multiple tissues or organs segmented from cross-sectional images of the body. Each tissue has specific permittivity and conductivity values. Each organ, such as the brain, is modeled as a uniform medium. The electric properties of actual organs and tissues, however, are not necessarily uniform because of a variety of substructures. Moreover, these standard models do not replicate the anatomy of an individual subject and are not applicable to subject-specific analyses. One solution to the aforementioned issues will be the development of new methodology for impedance imaging of living bodies. Imaging of electric impedance will enable us to determine both the local distribution of electric properties within an organ and the detailed anatomy of an individual subject.

Electrical impedance tomography (EIT), in which multiple electrodes are attached to the subject's skin, has widely been used to obtain the three-dimensional impedance distribution in a body.[3] The basic principle of EIT is to apply electric currents between a pair of selected electrodes and to simultaneously measure the resulting potentials at other electrodes. Another technique for impedance imaging is to apply microwaves, instead of currents, from surrounding antennas to the body.[4] The spatial resolution of these techniques is limited, in principle, by the number of electrodes or antennas. In addition, it is difficult to obtain EIT of the brain, which is surrounded by a resistive skull.

To improve spatial resolution, studies have been conducted to investigate the use of magnetic resonance imaging (MRI) to obtain impedance distributions. MRI is inherently sensitive to the magnetic fields generated in the body during measurement. When an electric current is externally applied to the body, the magnetic field generated by the current affects the amplitude and phase angle of the MRI signal. The distribution of conductivity and permittivity inside the body can be estimated from this effect. Compared with EIT, the use of MRI in impedance imaging significantly improves spatial resolution. The use of MRI can also improve the accuracy of estimates for the electric properties of organs surrounded by bones or other resistive tissues. The first half of this chapter provides an overview of several major techniques for estimating the distribution of conductivity and permittivity using MRI.

6.1.2 Advantages of MRI in Electric Current Imaging

Electric currents flowing in a body can be detected by observing surface potentials or surrounding magnetic fields. This idea forms the foundation for principal measurement techniques in biomagnetics. Although the current distribution inside a body can be estimated by inversely analyzing the measured fields, such analysis contains a mathematical ill-posedness. Because MRI is sensitive to magnetic fields, MRI can be a potent tool for mapping electric currents flowing in the body. The second half of this chapter provides an overview of MRI-based imaging of electric currents and discusses techniques that are especially applicable to neuronal electric activities.

As explained in Chapter 5, the frequency of the MRI signal is proportional to the strength of the magnetic field. The magnetic field produced by current in the body is added to the main static magnetic field, which results in a slight local shift in the signal frequency and phase angle in and around the current pathway. When the current is externally controllable, a typical method for electric current imaging is to obtain a pair of phase angle maps with the current switched on and off and to then subtract the two phase angle maps to estimate the current distribution. Another effect of electric current is a local decrease in the MRI signal intensity. Because the magnetic fields resulting from an internal current are usually inhomogeneous, the magnetic fields cause dephasing of nuclear magnetization, which leads to the decreased signal intensity. One can select either the phase angle or signal intensity to obtain current distributions.

One of the ultimate goals of MRI-based electric current imaging is to detect extremely weak magnetic fields arising from neuronal electric activities. Visualization of the spatiotemporal dynamics of the brain's neuronal activities is essential for understanding functional networks of neurons. Detailed spatial distributions of neuronal activities have been investigated using functional MRI (fMRI), in which a local change in blood flow resulting from neuronal activities affects the MRI signal intensity. However, using fMRI, detailed measurements of the temporal evolution of neuronal activities are difficult to obtain because the change in blood flow lags a second behind the electrical activity. During the last 15 years, there have been attempts to detect both weak magnetic fields generated in phantoms and neuronal magnetic fields produced in cultured cells or animals. This methodology has a potential temporal resolution of tens of milliseconds and a spatial resolution below 1 mm. Neuronal magnetic fields are very weak, as low as 1 pT or smaller in the human brain. Evaluating the sensitivity to magnetic fields is crucial for discussing this methodology's feasibility.

6.2 INFLUENCE OF SAMPLE IMPEDANCE ON MRI

6.2.1 Electric Properties of Biological Tissues

6.2.1.1 Permittivity and Conductivity

An MRI signal is emitted from the body as a radiofrequency (RF) electromagnetic wave, and propagating through space, this wave is received by an RF coil. As the Larmor equation in Chapter 5 indicates, frequency

f of the RF wave is proportional to strength B of the applied magnetic field. The frequencies for 1.5- and 3-T field strengths are 64 and 128 MHz, respectively.

The propagation velocity of RF waves is 3.0×10^8 m/sec in a vacuum. The velocity in biological tissue depends on electric permittivity ε and magnetic permeability μ of the tissue, as expressed by

$$v = \frac{1}{\sqrt{\varepsilon\mu}} \tag{6.1}$$

The ratio of tissue permittivity ε to vacuum permittivity $\varepsilon_0 = 8.9 \times 10^{-12}$ F/m is denoted by the relative permittivity: $\varepsilon_r = \varepsilon/\varepsilon_0$. Because the relative permittivities of biological tissues are higher than 1, propagation velocities in tissues are smaller than velocities in vacuum.

A traveling wave generally satisfies the relation

$$\lambda = \frac{v}{f} \tag{6.2}$$

where λ is the wavelength, v is the velocity, and f is the frequency. Combining Equations 6.1 and 6.2 with the Larmor equation demonstrates the following important behavior: the wavelength of the RF wave decreases when the strength of the magnetic field increases or when the permittivity is high.

In many molecules contained in biological tissues, the centers of positive and negative electric charges are not in the same position. Such molecules are called polar molecules. One of the major components of biological tissues, a water molecule has two O-H bonds at an angle of 104°. The distribution of electrons within a water molecule shifts toward the oxygen atom, and consequently, water is a polar molecule. The relative permittivity of water is as high as 80. Permittivity becomes high when the material contains a considerable amount of polar molecules. On the basis of Equation 6.1, the velocity of an RF wave traveling in water is estimated to be as low as 1/9 of the velocity in vacuum.

Conductivity is another important electric property for understanding MRI physics. Conductivity is a measure of a material's ability to conduct electric currents when an electric field is applied. Conductivity σ is defined as a constant of proportionality between electric field E and current density j:

$$j = \sigma E \tag{6.3}$$

Conductivity has an extremely broad range of values. Metals are representative conductors with conductivities around 10^7 S/m. Conductivities of typical insulators are as low as 10^{-15} S/m. Although biological tissues are usually modeled as conductors, their conductivity is much lower than metals' conductivity.

To analyze an electromagnetic wave, it is useful to define the following complex permittivity, in which the permittivity and conductivity are included in the equation's real and imaginary parts, respectively:

$$\varepsilon^* = \varepsilon - \frac{i\sigma}{\omega} \tag{6.4}$$

where ω is the angular frequency. In some cases, the sign of the imaginary part may be positive instead of negative. When analyzing the distribution of an RF wave, the use of complex permittivity enables us to write the wave equation in a simple form.

6.2.1.2 Frequency Characteristics and Anisotropy

A material's permittivity and conductivity generally depend on the frequency of an applied field, and this phenomenon is called dispersion. Because biological tissues have multiple mechanisms for electric conduction, such as the drift of ions and the rotation of polar molecules, tissues exhibit relatively complicated dispersion. As follows, Debye's equation of complex permittivity is widely accepted as a model of dispersion:

$$\varepsilon^*(\omega) = \varepsilon_\infty + \frac{\varepsilon_0 - \varepsilon_\infty}{1 + i\omega\tau} \tag{6.5}$$

where τ is the relaxation time and where ε_0 and ε_∞ indicate the respective permittivities at sufficiently low and high frequencies (compared with $1/\tau$). With ω ranging from 0 to ∞, a plot of Equation 6.5 on a complex plane produces a semicircle, which is called a Cole-Cole plot.[5] There are multiple sources of dielectric polarization in a tissue, such as charges on cell membranes and polar molecules (including water), and these polarizations have different relaxation times. With four relaxation times, an extended version of Equation 6.5 is used for modeling the tissue's complex permittivity:

$$\varepsilon^*(\omega) = \varepsilon_\infty + \sum_{m=1}^{4} \frac{\Delta\varepsilon_m}{1+(i\omega\tau_m)^{1-\alpha_m}} + \frac{\sigma_i}{j\omega\varepsilon_0} \tag{6.6}$$

where $\Delta\varepsilon_m$ indicates the drop in permittivity for a frequency range across $\omega = 1/\tau_m$, which corresponds to each relaxation time τ_m; σ_i indicates the ionic conductivity. The above empirically defined equation is well fitted to the measured complex permittivities in a broad frequency range, from 10 Hz to 100 GHz.[6,7] Figure 6.1 shows the frequency characteristics of gray

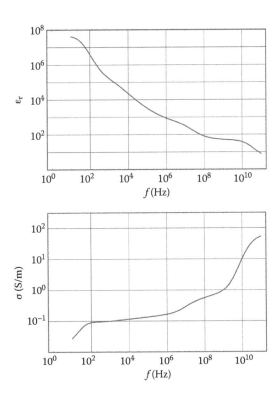

FIGURE 6.1 Frequency characteristics of electric permittivity ε_r and conductivity σ of gray matter. (From Safety Committee of the Japan Society of Magnetic Resonance in Medicine. 2010. *MR Safety—Principles, Standards and Clinical Concerns*. Gakken Medical Shujunsha, Tokyo.)

matter's permittivity and conductivity, obtained by fitting Equation 6.6 to the measured values.[8] An increase in frequency causes a decrease in permittivity and an increase in conductivity. A Cole-Cole plot of a tissue differs from a semicircle because of the tissue's relatively complicated frequency characteristics. Table 6.1 shows the permittivities and conductivities of several tissues for 63.9 and 128 MHz.

In addition to the frequency characteristics, another unique property of biological tissues' permittivity and conductivity is their anisotropy. The permittivity and conductivity depend on the direction of the applied electric field. In nerves and muscles, cells form fibrous microscopic structures, and the cells align in a specific direction. Because the cell membranes hinder ions' ability to pass through the membranes, ions easily move in the direction of fibers but are mostly unable to move orthogonally. Consequently, conductivity exhibits higher values in the direction of fibrous structures. Because of cell membranes' resistance, low-frequency currents flow mainly through extracellular spaces. Furthermore, because of surface charges on cell membranes, a tissue's effective permittivity becomes high when an electric field is applied in the direction perpendicular to the cell membrane. In a measurement of the conductivity of the cerebral cortex, the conductivity varied by a

TABLE 6.1 Electric Permittivity and Conductivity of Body Tissues

Tissue	Relative Permittivity		Conductivity (S/m)	
	63.9 MHz	128 MHz	63.9 MHz	128 MHz
Brain gray matter	97.5	73.5	0.511	0.587
Brain white matter	67.9	52.5	0.291	0.342
Cerebellum	116.4	79.7	0.719	0.829
Muscle	72.3	63.5	0.688	0.719
Heart	106.6	84.3	0.678	0.766
Fat	6.5	5.9	0.035	0.037
Bone cortical	16.7	14.7	0.060	0.067
Bone cancellous	30.9	26.3	0.161	0.180
Cerebrospinal fluid	97.3	84.0	2.066	2.143
Blood	86.5	73.2	1.207	1.249
Liver	80.6	64.3	0.448	0.511

Source: Gabriel, C. et al., *Physics in Medicine and Biology* **41**: 2231–2249, 1996; Gabriel, S. et al., *Physics in Medicine and Biology* **41**: 2251–2269, 1996.

factor of 1.7, depending on the measurement direction.[9] The anisotropies originating from the membrane structures become prominent when
considerable quantities of ions oscillating in an AC field collide with
membranes. These anisotropies become negligibly small at frequencies
higher than a megahertz.

6.2.1.3 Human Models for Numerical Simulations of Electromagnetic Fields

Numerical simulations of electromagnetic fields in the human body are
effective for safety assessments of electromagnetic fields. In addition to
computer performance and algorithms, modeling the distributions of permittivity and conductivity in the body is an important endeavor.

Several institutions develop and distribute numerical human models
without charge for noncommercial users.[1,2] A few software developers
provide numerical human models specifically for use with their software.
These models are built by segmenting an anatomical data set (obtained
from MRI, CT, or dissected slices of a cadaver) into multiple tissues. Each
voxel in the data set has a labeling number to indicate a particular tissue
type, such as bone or muscle. A model's typical spatial resolution is 1 mm.
Because there is no method to automatically segment the entire body,
building a model necessitates extensive manual procedures that require
significant effort and time. The aim of such a model is to provide a risk
assessment for exposing the human body to equipment's electromagnetic
fields. This model is also useful for medical applications, such as electric
and magnetic stimulations, microwave hyperthermia, and electromagnetic fields in MRI systems.

The National Institute of Information and Communications Technology
(NICT), Japan, develops and distributes numerical human models.[1] This
group provides standard male and female models with average body sizes.
These models were built by segmenting a 3D MRI data set into 51 different
tissues with a spatial resolution of 2 mm. The models can be deformed to
assume a variety of postures. A model of a pregnant woman is also available for evaluating a fetus's field exposure.[10]

The above models relate each voxel in the data set to a tissue type. Other
databases should be utilized to ascertain permittivity and conductivity
values of different tissues. The values are estimated for a target frequency
using Equation 6.6.[6,7] Because typical models of the human body consist of cubic voxels, the model has immediate applicability to simulation

methods that use cubic cells, such as the finite difference time domain method, the impedance method, and the scalar-potential finite difference method. The finite element method is also available.

6.2.2 RF Distribution in a Cylindrical Sample

Analyses of RF fields in a cylindrical sample provide a comprehensive view of the influence of electric permittivity and conductivity on RF field distributions. Here, we consider two major effects: dielectric resonance and the skin effect. Both effects are prominent in ultrahigh-field MRI systems.

A distinctive signal inhomogeneity arises in a sample whose dimensions are comparable to or smaller than the RF fields' wavelength. This phenomenon, dielectric resonance, complicates the quantification of MRI. A typical dielectric resonance effect appears as a brightening of an image at the center of a sample. MRI frequencies of presently available scanners range from 8.5 to 340 MHz, which corresponds to wavelengths in water ranging from 10 to 400 cm. The degree of signal inhomogeneity resulting from dielectric resonance highly depends on the sample's permittivity because the wavelength is directly related to the permittivity.

The skin effect is the attenuation of the electromagnetic field in the deep part of a conductive sample. The following equation gives the representative skin depth (the penetration depth in the sample):

$$l = \sqrt{\frac{2}{\mu \sigma \omega}} \tag{6.7}$$

At frequencies applicable to ultrahigh-field MRI systems, the skin depth becomes comparable to the size of the human head, which results in a clear occurrence of the skin effect.

Figure 6.2 shows the magnitudes and phase angles of magnetic resonance signals obtained from cylindrical phantoms with different radii and electric properties.[11] The measuring frequency of 200 MHz corresponds to a wavelength of 170 mm in water. Because of dielectric resonance, the phantom with $a = 50$ mm (compared to a phantom with $a = 30$ mm) exhibited more significant inhomogeneity in the RF magnitude. An increase in conductivity, from 0 to 1.9 S/m (comparable to saline solution), resulted in the suppression of inhomogeneity. As shown in Figure 6.2, the skin effect partly counteracted the dielectric resonance.

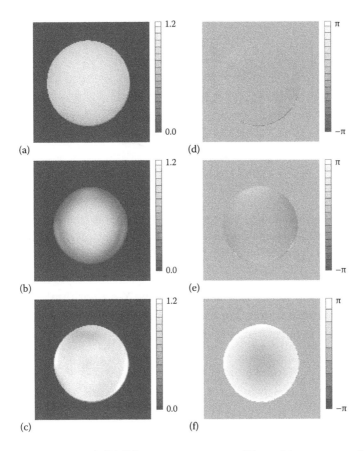

FIGURE 6.2 Magnitude $|\xi|$ ([a] $a = 30$ mm, water; [b] $a = 50$ mm, water; [c] $a = 50$ mm, saline) and phase angle $\arg(\xi)$ ([d] $a = 30$ mm, water; [e] $a = 50$ mm, water; [f] $a = 50$ mm, saline) of magnetic resonance signals obtained from cylindrical phantoms with different radii and electric properties. The signals are normalized by the values at the center of the phantom. (From Sekino, M. et al., *Journal of Applied Physics* **97**: 10R303, 2005.)

6.2.3 RF Distribution in the Human Head

6.2.3.1 Finite Difference Time Domain Method

The finite difference time domain (FDTD) method is a major computational technique for analyzing high-frequency electromagnetic fields. Maxwell's equations are directly discretized and sequentially solved to obtain the time evolution and spatial distribution of electromagnetic fields. Electromagnetic field components are assigned to each computational cell, as shown in Figure 6.3. The electromagnetic field distributions

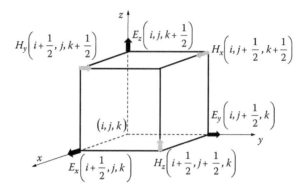

FIGURE 6.3 Computational cell for the FDTD method. The electromagnetic field components are assigned on the edges or faces of each cell. (From Sekino, M. et al., *Journal of Applied Physics* **103**: 07A318, 2008.)

for each time step can be calculated using the Yee algorithm[12,13] when an RF coil is the wave source. For example, the z component of the electromagnetic field is calculated as follows:

$$E_z^n\left(i,j,k+\frac{1}{2}\right) = \frac{1-\dfrac{\sigma(i,j,k+1/2)\ t}{2\varepsilon(i,j,k+1/2)}}{1+\dfrac{\sigma(i,j,k+1/2)\ t}{2\varepsilon(i,j,k+1/2)}} E_z^{n-1}\left(i,j,k+\frac{1}{2}\right)$$

$$+\frac{\dfrac{\Delta t}{\varepsilon(i,j,k+1/2)}}{1+\dfrac{\sigma(i,j,k+1/2)\Delta t}{2\varepsilon(i,j,k+1/2)}}\frac{1}{\Delta x}$$

$$\times\left\{H_y^{n-1/2}\left(i+\frac{1}{2},j,k+\frac{1}{2}\right)-H_y^{n-1/2}\left(i-\frac{1}{2},j,k+\frac{1}{2}\right)\right\}$$

$$-\frac{\dfrac{\Delta t}{\varepsilon(i,j,k+1/2)}}{1+\dfrac{\sigma(i,j,k+1/2)\Delta t}{2\varepsilon(i,j,k+1/2)}}\frac{1}{\Delta y}$$

$$\times\left\{H_x^{n-1/2}\left(i,j+\frac{1}{2},k+\frac{1}{2}\right)-H_x^{n-1/2}\left(i,j-\frac{1}{2},k+\frac{1}{2}\right)\right\} \quad (6.8)$$

$$H_z^{n+1/2}\left(i+\frac{1}{2},j+\frac{1}{2},k\right) = H_z^{n-1/2}\left(i+\frac{1}{2},j+\frac{1}{2},k\right)$$

$$+\frac{\Delta t}{\mu_0}\frac{E_y^n(i+1,j+1/2,k)-E_y^n(i,j+1/2,k)}{\Delta x} \quad (6.9)$$

$$+\frac{\Delta t}{\mu_0}\frac{E_x^n(i+1/2,j+1,k)-E_x^n(i+1/2,j,k)}{\Delta y}$$

where E_z^n and H_z^n are, respectively, the electric and magnetic field components at the nth time step, i, j, and k are the index numbers of the computational cell, and Δt, Δx, and Δy are the length of a time step and the lengths of the computational cell in the x and y directions, respectively. At the outer boundaries of a computational model, absorbing boundary conditions are applied to eliminate the influence of the boundaries. There is a variety of formulations for the boundary conditions.

6.2.3.2 RF Distribution in the Human Head at Varied Frequencies
Recent advances in superconducting magnets have led to the realization of ultrahigh-field MRI systems of around 10 T.[14] These ultrahigh-field systems have numerous advantages, such as the acquisition of precise anatomical and functional images in a short acquisition time, quantification of complicated biomolecules, and imaging of nonproton nuclei. As high as 500 MHz, the frequencies of such systems cause dielectric resonance and the skin effect in the human brain. These two effects result in a complicated RF electromagnetic field distribution. The inhomogeneity of the RF fields may lead to difficulty quantifying images and a regional increase in the specific absorption rate (SAR). A quantitative estimate of the signal distribution and SAR is essential for initial evaluation of the advantages and characteristics of ultrahigh-field MRI systems. The signal inhomogeneities and SAR distribution in ultrahigh-field MRI systems have been evaluated using numerical simulations.[13]

Figure 6.4 shows the distributions of the electric field and magnetic field in an axial section for frequencies of 64, 128, 200, and 500 MHz (corresponding to static magnetic field intensities of 1.5, 3.0, 4.7, and 11.7 T). A numerical model of the human head was constructed using the NICT database.[1] The analyses were conducted using the FDTD method. At 64 MHz (the frequency commonly used in clinical MRI

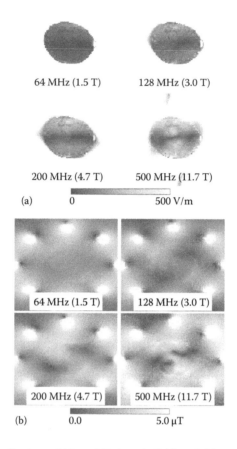

64 MHz (1.5 T) 128 MHz (3.0 T)

200 MHz (4.7 T) 500 MHz (11.7 T)

(a) 0 500 V/m

64 MHz (1.5 T) 128 MHz (3.0 T)

200 MHz (4.7 T) 500 MHz (11.7 T)

(b) 0.0 5.0 µT

FIGURE 6.4 Distribution of (a) an RF electric field and (b) an RF magnetic field in a horizontal section of the human head with different magnetic resonance frequencies. The increase in frequency causes the occurrence of hot spots that exhibit high RF absorption. (From Sekino, M. et al., *Journal of Applied Physics* **103**: 07A318, 2008.)

systems), the magnetic field exhibited homogeneous distributions, and the electric field increased its intensity as the distance from the center of the head increased. An increase in frequency resulted in considerable inhomogeneities in the RF electromagnetic fields. The magnetic field at 200 MHz had a high intensity at the center of the head. The electromagnetic fields at 500 MHz exhibited a complicated amplitude distribution. Although the calculations were conducted under the same coil current intensity, the intensity of the electric field increased with an increase in frequency.

6.3 MAGNETIC RESONANCE IMAGING OF IMPEDANCE

6.3.1 Large Flip Angle Method

When conductive tissues are subjected to an excitation RF field in MRI, eddy currents are induced in the tissues. This phenomenon can be used to obtain the impedance distribution of the tissue.[15] The basic principle is to apply the shielding effects of the eddy currents induced in the body to spin precession. The occurrence of eddy currents results in a reduction in the net RF fields penetrating the tissues. Because of shielding effects, the flip angles are reduced by various degrees, depending on the electrical properties of the tissues.

When a precise 180°, 360°, or 540° excitation pulse is applied to conductive tissues, the tissues do not yield a signal because magnetization's transversal components are absent. Conversely, resistive tissues yield signals because they are less electrically shielded than conducting tissues and because they simultaneously undergo different flip angles. In addition, the resistive tissues produce transversal components with magnitudes determined by the sine wave functions of the flip angles. The difference in signal, therefore, reflects the tissues' conductivity. By applying very large flip angles, conductivity-enhanced MRI can be obtained. The RF fields are applied at the given Larmor frequency and in the direction perpendicular to the main static field.

Figure 6.5 shows images of a mouse head with excitation flip angles of 160°, 180°, and 200°. The 180° image shows a slight signal from the brain and muscle tissues because there were almost no transversal components of magnetization. Conversely, in the same 180° image, the resistive fatty tissues, which were transparent to the RF field, yielded a specific signal. By applying 180° pulses to the conducting cerebrospinal fluid and muscle tissues, resistive fatty tissues simultaneously received excitation for flip angles larger than 180° and produced an image signal.

6.3.2 Use of an External AC Field

To obtain conductivity-enhanced images at an arbitrary frequency, an additional time-varying field parallel to the main static field, B_0, is introduced.[15,16] The perturbing field is produced by the third coil, which is hereby denoted the B_c coil. The sample is located inside this coil. The perturbing field (the B_c field) affects the image's slice positioning, and the slice selection fluctuates. Because of shielding effects, conducting tissues are less affected by the B_c field. Because the frequency of the AC field is independent of the

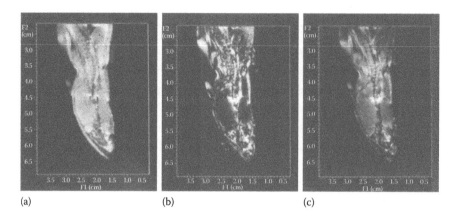

(a) (b) (c)

FIGURE 6.5 Impedance imaging of the mouse head using the large flip angle method. The images were obtained with different excitation flip angles and were evaluated in the cerebrospinal fluid. The flips angles were (a) 160°, (b) 180°, and (c) 200°. (From Ueno, S., and Iriguchi, N., *Journal of Applied Physics* **83**: 6450–6452, 1998.)

given Larmor frequency, conductivity-enhanced images can be obtained at any frequency (but only in the direction perpendicular to the AC field). The AC field is added to the main static field only during the period of slice selection, and the artifacts in the reading-out and phase-encoding directions can be eliminated. A series of pre-pulses are applied before slice selection to dephase the sample's nuclear spins in the neighboring regions. When the AC field shifts the sliced position into the neighboring regions, fewer nuclei are excited and the resulting image intensity weakens.

When two parallel columns with different conductivities are subjected to imaging, as shown in Figure 6.6, two different regions of each column are excited by the application of two pre-pulses. The subsequent application of slice-selective 90° pulse superimposed on the AC field excites a thicker slice involving the pre-excited regions in the resistive material. This excitation is inefficient because spins in the re-excited regions have lost coherency during the precession phase. Conversely, a thinner slice that excludes the pre-excited regions excites the conductive material. This excitation is efficient because the excitation is exclusive to the selected slice. One approach to evaluate an impedance-enhanced image is to compare images with and without the AC field. Impedance imaging of a phantom with a sinusoidal 100-Hz AC field has been reported.[16]

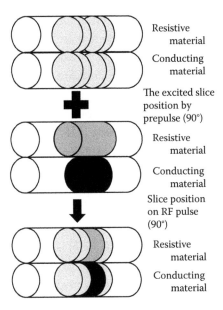

Resistive material

Conducting material

The excited slice position by prepulse (90°)

Resistive material

Conducting material

Slice position on RF pulse (90°)

Resistive material

Conducting material

FIGURE 6.6 Principles of impedance imaging using external AC fields and prepulses. Because the slice position fluctuates in response to AC fields, a slice-selective 90° pulse excites a thicker slice infiltrating into the preexcited regions in the resistive material. The signal intensity of the reexcited regions is very small. The signal intensity of the region excited by the 90° pulse in the conductive material is less affected by the pre-pulses. (From Yukawa, Y. et al., *IEEE Transactions on Magnetics* **35**: 4121–4123, 1999.)

6.3.3 Conductivity Imaging Using Diffusion MRI

6.3.3.1 Principles

Low-frequency electric currents in living tissues are mainly conducted by the migration of ions. Because the migrating ions encounter high resistance from cell membranes, most currents flow through extracellular fluid. Conductivity depends on the viscosity of extracellular fluid because the balance between the electrostatic force and viscous drag governs an ion's drift velocity. The diffusion coefficient of extracellular fluid is also related to its viscosity, which is described by the Stokes-Einstein equation. On the basis of these relationships, tissue conductivity can be obtained from extracellular fluid's diffusion coefficient. Several research groups have reported the use of diffusion MRI for mapping tissue conductivity.[17–20]

Biological tissues have multiple diffusion components with different diffusion coefficients. In many studies, such diffusion is simply divided into the fast component and the slow component. As an approximation, the fast component and the slow component are attributed to extracellular fluid and intracellular fluid, respectively. The diffusion coefficient and the fractional volume of the extracellular fluid are approximated by the fast diffusion coefficient and the fraction of the fast component, respectively. For each pixel in the image, the fast and slow diffusion components for the direction of the motion-probing gradient (MPG) were obtained by fitting the following function to the measured signals:

$$\frac{S_i(b)}{S_i(0)} = f_{\text{fast},i} \exp(-bD_{\text{fast},i}) + f_{\text{slow},i} \exp(-bD_{\text{slow},i}) \tag{6.10}$$

where i is the index number of the MPG, $S_i(b)$ is the signal intensity of the diffusion-weighted images, b is the b factor of the MPG, $D_{\text{fast},i}$ and $D_{\text{slow},i}$ are the fast and slow diffusion components, respectively, and $f_{\text{fast},i}$ and $f_{\text{slow},i}$ are the respective fractions of the fast and slow components.

The conductivity of the extracellular fluid is obtained from the fast diffusion coefficient based on the proportionality between conductivity and the diffusion coefficient. Finally, the effective conductivity of the tissue is estimated from the extracellular fluid's conductivity and volume fraction. The effective conductivity σ_i for direction i is given by[17]

$$\sigma_i = \frac{2f_{\text{fast},i} \times (8.1 \times 10^8)}{3 - f_{\text{fast},i}} D_{\text{fast},i} \tag{6.11}$$

6.3.3.2 Experimental Results

Figure 6.7 shows the relationships between the b factor and the signal intensities of diffusion-weighted images of the human brain.[18] The regions of interest were located on the putamen (gray matter), the posterior limb of the internal capsule (white matter), and the genu of the corpus callosum (white matter). The plots indicate the signal intensities measured with the following MPG directions: anterior-posterior, right-left, and superior-inferior. The signals gradually decreased with an increase in the b factor. The putamen did not demonstrate clear anisotropy in signal attenuation. In the internal capsule, the application of the MPG in the superior-inferior direction caused the most rapid signal attenuation. In the corpus callosum, the application of the MPG in the right-left direction caused the most rapid signal attenuation.

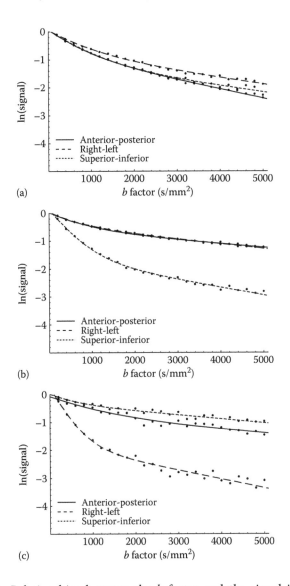

(a)

(b)

(c)

FIGURE 6.7   Relationships between the b factor and the signal intensities of diffusion-weighted images in the putamen. The three plots were obtained with different MPG directions. (a) The putamen did not exhibit clear anisotropy in the signal attenuations. (b) Signal attenuations in the internal capsule. The application of the MPG in the superior-inferior direction caused the most rapid signal attenuation. (c) Signal attenuations in the corpus callosum. The application of the MPG in the right-left direction caused the most rapid signal attenuation. (From Sekino, M. et al., *IEEE Transactions on Magnetics* **41**: 4203–4205, 2005.)

Figure 6.8a through c show images of conductivities estimated using Equation 6.11. The gray matter's conductivity does not have a clear dependence on the MPG direction. The white matter contains several regions where conductivities are highly dependent on the MPG direction. Figure 6.8d and e show images of the mean conductivity (MC) and the anisotropy index (AI). Regions with high AI values are found in the white matter.

The signals of diffusion-weighted images are derived from both the intracellular and extracellular fluids. This method provides the tissue conductivity based on only the fast diffusion component, which corresponds to diffusion in the extracellular fluid. Marked anisotropy in brain regions is also demonstrated for tissue conductivity. The anatomical structures of

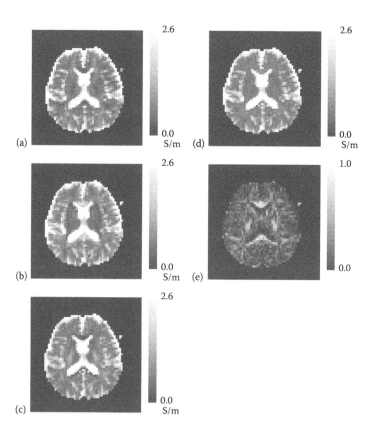

FIGURE 6.8 Conductivity imaging of the human brain using diffusion MRI. (a through c) Images of the estimated conductivities in the anterior-posterior, right-left, and superior-inferior directions, respectively. (d, e) Images of the MC and AI. (From Sekino, M. et al., *IEEE Transactions on Magnetics* **41**: 4203–4205, 2005.)

the brain's principal neuronal fiber tracts are well established from previous anatomical and histological studies. The corpus callosum connects the right hemisphere and the left hemisphere with several neuronal fibers that run in the right-left direction. The posterior limb of the internal capsule lies on the pyramidal tract, which mainly consists of neuronal fibers running in the superior-inferior direction. The diffusion of water molecules in the extracellular fluid is disturbed by the cell membranes, and diffusibility is higher in the direction of neuronal fibers than in the other directions. The high AI values in the internal capsule and corpus callosum are attributable to the anatomical structures of these regions.

The inhomogeneity and anisotropy of the conductivity in each tissue is not normally considered in conductivity models for current source estimations of electroencephalography and magnetoencephalography. However, the results of current source estimations depend on the spatial distribution of conductivity in the models. The results of conductivity imaging show significant inhomogeneity and anisotropy in the white matter. Therefore, the use of an inhomogeneous and anisotropic conductivity model is desirable, particularly for estimating current sources in the deep regions of the brain.

6.3.4 Impedance Imaging Using RF Current

6.3.4.1 Principles

When electric current is applied to a sample during MRI acquisition, the current causes a change in the MRI signal, which depends on the sample's permittivity and conductivity. This phenomenon enables us to estimate the distribution of tissue impedance. Because of increased attention to the RF absorption of bodies in ultrahigh-field MRI, methods for mapping tissue impedance and the resulting RF absorption have been developed.[21-24] Figure 6.9a shows a schematic of MRI hardware for an impedance imaging method using RF electromagnetic fields.[21] In addition to the conventional components, a pair of electrodes and an RF transmitter are introduced to apply electric currents to the sample. Figure 6.9b shows the operational diagram of the RF transmitters and receiver. The two transmitters operate at the same frequency and maintain a difference in the phase angle at 0 or $\pi/2$. Two consecutive RF pulses are applied from transmitter A with the same pulse duration $(\tau/2)$ and opposite polarity. An RF pulse is simultaneously applied from transmitter B with a duration of τ. The magnetic field generated from the RF coil rotates around the z axis with the Larmor frequency. In addition, the application of an RF electric current to the sample

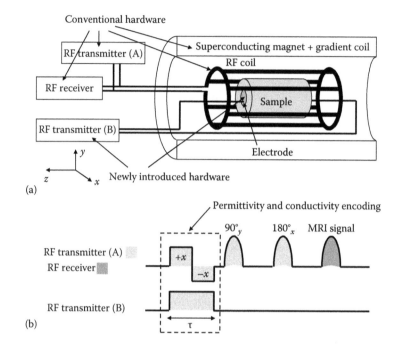

FIGURE 6.9 Principle of impedance imaging using an RF current. (a) A weak electric current at the Larmor frequency was applied to the sample through a pair of surface electrodes. The current was supplied from an external RF transmitter. (b) The operational diagram of the RF transmitters and receiver. Transmitter A produces RF pulses with inverted polarities for a duration of $\tau/2$. Transmitter B simultaneously produces an RF pulse for a duration of τ. (From Sekino, M. et al., *IEEE Transactions on Magnetics* 44: 4460–4463, 2008.)

gives rise to an oscillating magnetic field. An x'-y'-z' rotating frame with the x' and y' axes rotating at the magnetic resonance frequency is introduced for the following discussion. The phase angle of transmitter A is adjusted such that the magnetic field generated from the coil (B_1) turns toward the x' direction. Magnetization M rotates around the externally applied magnetic field. Assuming that B_1 is much stronger than B'_x and B'_y, the direction of the total RF magnetic field becomes approximately x'. During application of B_1 in a positive polarity, the magnetization vector rotates with an angle of $(B_1 + B'_x)\gamma\tau/2$. The B_1 magnetic field is subsequently applied in a negative polarity, and the magnetization vector rotates with an angle of $(-B_1 + B'_x)\gamma\tau/2$. Consequently, the angle between the magnetization and the z axis becomes $\theta_0 = B'_x\gamma\tau$. The remaining RF pulses in Figure 6.9b enable us to measure θ_0 based on the phase angle of the magnetic resonance signals. In the next step, B_1 turns toward the y' direction, and $\theta_{\pi/2} = B'_y \gamma \tau$ is measured. On the basis of the wave equation of the electromagnetic fields in the sample, the following equation is derived to estimate permittivity ε_r and conductivity σ:

$$\omega_0^2 \varepsilon_0 \varepsilon_r \mu_0 - i\omega_0 \mu_0 \sigma = -\frac{\nabla^2(\theta_0 - i\theta_{\pi/2})}{\theta_0 - i\theta_{\pi/2}} \tag{6.12}$$

6.3.4.2 Experimental Results

Preliminary results of phantom experiments have been reported in a previous paper.[21] An acrylic tube was filled with a solution comprised of saline (30%) and ethanol (70%). The permittivity and conductivity of the solution were approximately 40 and 0.14 S/m, respectively. A pair of platinum electrodes was attached to both ends of the sample. Using a coaxial cable, the electrodes were connected to an RF transmitter. Signals were detected using a birdcage-type RF coil.

As shown in Figure 6.10, images of magnetization angles were obtained from magnetic resonance signals, and the sample's permittivity and conductivity were estimated. The lower left part of the sample exhibited a relatively large error in the estimated permittivity and conductivity. The lead wire caused inhomogeneity in the RF electromagnetic field, which resulted in an error around the wire. At the center of the sample, we obtained an estimated relative permittivity of 40 and an estimated conductivity of 2.0 S/m, which are reasonable results. In a part of the image, however, the results showed errors that are mainly attributable to inhomogeneity in the field generated by the RF coil.

This methodology has potential advantages over conventional impedance imaging techniques because of its high spatial resolution and its capacity to measure permittivity. In addition, this method is easily applicable to both biological tissues and human subjects.

6.3.5 Magnetic Resonance Electrical Impedance Tomography

As discussed in the next section, MRI quantitatively visualizes the spatial distribution of electric current.[25] The distribution of conductivity can then be estimated from the measured current.[26-28] Compared with conventional EIT, the advantages of this approach are its high spatial distributions and high robustness with respect to the contact resistance between the electrodes and the body. Equation 6.13 is an algorithm proposed

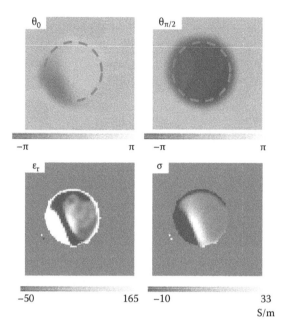

FIGURE 6.10   Experimentally obtained images of the phase angle of magnetic resonance signals θ_0 (upper left) and $\theta_{\pi/2}$ (upper right), estimated electric permittivity ε_r (lower left), and estimated conductivity σ (lower right). (From Sekino, M. et al., *IEEE Transactions on Magnetics* 44: 4460–4463, 2008.)

in the primary stage of MRI-based EIT to estimate the conductivity distribution.[26]

The current distribution in the sample measured using MRI is denoted by J^*, and the respective current and voltage applied to the electrodes attached to the surface are denoted by I and V^*. The aim of the algorithm is to estimate unknown resistivity distribution ρ^* using the given values of J^*, I, and V^*. The area inside the sample is divided by a computational grid, and initial resistivity ρ^0 is given to each computational grid. Given initial resistivity ρ^0 and measured electrode current I, current density J^0 at each grid and electrode voltage V^0 is calculated using the finite-element method. Revised resistivity ρ^1 for each node is then obtained by substituting $k = 0$ into

$$\rho^{k+1} = \rho^k \frac{J^k}{J^*} \frac{V^*}{V^k} \qquad (6.13)$$

Iterative calculations for $k = 1, 2, 3, \ldots$, make resistivity ρ^k of each node converge to the true resistivity ρ^*.

Magnetic resonance electrical impedance tomography has been demonstrated in both computer simulations and experiments for phantoms, animals, and humans.

6.4 MAGNETIC RESONANCE IMAGING OF ELECTRIC CURRENTS

6.4.1 Changes in Magnetic Resonance Signals Resulting from the Magnetic Field

As the Larmor equation indicates, the precession velocity of magnetization is proportional to the strength of the externally applied magnetic field. Magnetic fields generated by electric currents flowing in a body are generally much weaker than the main static magnetic field applied in the z direction. Consequently, the strength of the total magnetic field is the sum of the main static field and the z component of the additional magnetic field.[29] When an electric current pulse is applied to the body with a pulse width of τ, the magnetic field generated by the current causes the following phase shift in the magnetic resonance signal:

$$\phi = \gamma B_z \tau \tag{6.14}$$

To map the generated magnetic field, two phase angle images are acquired with and without the application of an electric current, and the two images are subtracted.[25] Spontaneously generated magnetic fields, such as neuronal fields, generally have time-varying currents. In this case, Equation 6.14 is modified as

$$\phi = \gamma \int_0^\tau B_z(t)\,dt \tag{6.15}$$

The above discussion indicates that only the z component of the magnetic field can be measured using MRI. The process of rotating the sample inside the MRI scanner such that the sample is oriented along the other two orthogonal directions and then measuring the corresponding magnetic field maps provides the three components of the magnetic field vector. Taking the rotation of the magnetic field gives the electric current distribution:

$$j = \frac{1}{\mu_0} \nabla \times B \tag{6.16}$$

However, mechanical rotation is possible only in the case of small samples. Using mathematical approaches to incorporate certain physical constraints on the field distribution, the current density distribution can be estimated from the measured magnetic field distribution.[30,31] For example, the Fourier transform of the z component of Equation 6.16 is given by

$$j_z' = -\frac{i}{\mu_0}\frac{k_x^2 + k_y^2}{k_y}b_x'$$

(6.17)

where j_z' and b_x' are the Fourier transforms of the z component of the current density and the x component of the magnetic field, respectively. According to Equation 6.17, estimating one component of the current density requires one orthogonal component of the magnetic field to be mapped.

When the generated magnetic field is strongly inhomogeneous, the field causes the loss of signal coherence in a voxel, which results in a decrease in signal magnitude. This effect may occur in close proximity to the current source. While the measurement based on the phase angle has relatively high sensitivity to the magnetic field, the image analyses include a somewhat complicated process for phase unwrapping. The measurement using the signal magnitude allows for simple post processing.

When the intensity of the magnetic field is relatively high, the influence of the magnetic field can also be observed as a shift in the Larmor frequency. Numerical simulations based on the Bloch equation provide a quantitative relationship between the generated field and the signal intensity.[31]

6.4.2 Sensitivity to a Weak Electric Current

6.4.2.1 Neuronal Current Dipole

Neuronal activities give rise to ionic currents inside and outside the neuron. The realistic current distribution and the resulting magnetic field distribution are expected to be complicated. However, the following formula for magnetic fields produced by a current dipole gives a first-order approximation for neuronal magnetic fields:[32]

$$B(r) = \frac{\mu_0}{4\pi}Q(r') \times \frac{r - r'}{|r - r'|^3}$$

(6.18)

where $B(r)$ is the neuronal magnetic field at location r, $Q(r')$ is the current dipole at location r', and μ_0 is the permittivity of free space. Here, we assume that the main static magnetic field, B_0, is applied in $+z$ direction and that the current dipole is located at $r' = (0, 0, 0)$, pointing in the $+x$ direction. When $B(r)$ is much smaller than B_0, the magnitude of the total magnetic field can be approximated as[33]

$$\left|\mathbf{B_0} + \mathbf{B}(r)\right| \approx \left|\mathbf{B_0}\right| + B_z(r) = \left|\mathbf{B_0}\right| + \frac{\mu_0}{4\pi} \frac{Qy}{(x^2 + y^2 + z^2)^{3/2}} \qquad (6.19)$$

where $B_z(r)$ is the z component of $B(r)$ and Q is the strength of the current dipole. The frequency shift caused by the neuronal magnetic field is

$$\Delta\omega = \gamma\left(\left|\mathbf{B_0} - \mathbf{B}(r)\right| - \left|\mathbf{B_0}\right|\right) \approx \frac{\mu_0\gamma}{4\pi} \frac{Qy}{(x^2 + y^2 + z^2)^{3/2}} \qquad (6.20)$$

For an image voxel centered at (x_0, y_0, z_0), the following equation gives the signal intensity normalized by the intensity without a neuronal magnetic field:

$$S = \frac{1}{h_x h_y h_z} \left| \int_{x_0-h_x/2}^{x_0+h_x/2} dx \int_{y_0-h_y/2}^{y_0+h_y/2} dy \int_{z_0-h_z/2}^{z_0+h_z/2} dz \exp(iT_E\Delta\omega) \right| \qquad (6.21)$$

where h_x, h_y, and h_z are the dimensions of the voxel in the x, y, and z directions, respectively, and T_E is the echo time for MRI acquisition. Examples of calculating the above signal intensity are reported elsewhere.[33]

6.4.2.2 Theoretical Evaluation of Sensitivity

Magnetic fields arising from neuronal electrical activity are extremely weak. A magnetic field of 10 pT generated during an echo time of 20 msec, for example, leads to a phase angle shift of only 5.3×10^{-5} rad. To discuss the feasibility of detecting neuronal magnetic fields, theoretical evaluation of sensitivity is essential. Several studies have reported the sensitivity of MRI to magnetic fields.[34-36]

Here, we assume that the magnetic fields are detected as a particular phase angle shift, as explained above. Resulting from noise in the MRI signal, uncertainty σ_B in the estimated value of the magnetic field is given by

$$\sigma_B = \frac{N}{S\gamma T_E} \qquad (6.22)$$

where N is the intensity of noise in the MRI signal, S is the signal intensity, and T_E is the echo time.[34] The magnetic field can be detected when intensity β is higher than uncertainty σ_B. Therefore, σ_B gives the theoretical sensitivity for magnetic fields.

Magnetization M_0, which is induced in a sample by the main static field, B_0, is given by

$$M_0 = N_s\gamma^2\hbar^2 I(I+1)B_0/3k_B T_s \qquad (6.23)$$

where N_s is the ^{1}H density of the sample, $\hbar$ is Planck's constant, I is the spin quantum number, k_B is the Boltzmann constant, and T_s is the sample temperature. When an MRI is obtained with a field of view of $L \times L$ and a slice thickness of h, signal intensity S per voxel is

$$S = \gamma B_0 B_1 M_0 L^2 h \frac{[1-\exp(-T_R/T_1)]\exp(-T_E/T_2^*)}{1-\cos\theta\exp(-T_R/T_1)}\sin\theta \qquad (6.24)$$

where T_R is the repetition time, T_1 and T_2 are the sample's relaxation times, and θ is the flip angle. B_1 is the RF field intensity that the unit current flowing through the RF receiver coil produces at the voxel.

In an MRI consisting of $n \times n$ voxels, noise N per voxel is

$$\Delta N = n\sqrt{4k_B T_s \Delta f R} \qquad (6.25)$$

where Δf is the spectral width of the receiver circuit and R is the effective resistance in the receiver circuit. The resistance is partly caused by the conductors in the receiver coil and partly caused by the sample. Under typical MRI conditions in the human head, the resistance caused by the sample is dominant. Produced by the sample, equivalent resistance R in the receiver circuit is related to the absorption, which the unit current flowing in the receiver coil at the Larmor frequency causes in the sample; namely,

$$R = 2 \int \frac{|\mathbf{j}|^2}{\sigma} dV_s \qquad (6.26)$$

where $\mathbf{j}$ is the induced current distribution in the sample, which is caused by the unit current in the receiver coil, and σ is the conductivity.

A previous study reported the theoretical sensitivity for detecting neuronal magnetic fields in the human brain.[35] The effective resistance in the receiver circuit caused by the human head was evaluated using numerical simulations on an anatomically realistic model, as shown in Figure 6.11. The electromagnetic field distributions were calculated using the finite element method. The results show that the theoretical sensitivity for magnetic fields in the brain is approximately 10^{-8} T.

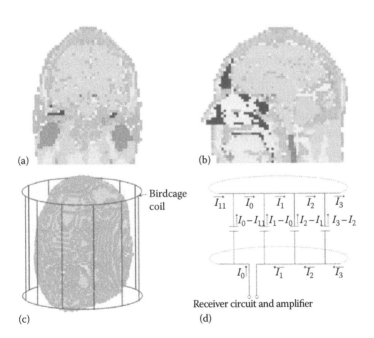

FIGURE 6.11 Evaluation of the sensitivity for detecting weak magnetic fields generated in the human brain. Numerical analyses were conducted on a realistic human head model and a birdcage-type receiver coil. (a) Coronal and (b) sagittal slices of the model. (c) Three-dimensional view of the model and the receiver coil. (d) A circuit schematic of the receiver coil. (From Hatada, T. et al., *Journal of Applied Physics* **97**: 10E109, 2005.)

6.4.2.3 Experimental Evaluation of Sensitivity

Sensitivity to weak magnetic fields has been experimentally evaluated for phantoms and animals.[36–38] One of our studies used columnar phantoms with electrodes attached to the ends of the column.[36] MRI was obtained from externally applied currents with intensities ranging from 0.75 to 150 A/m². The experimentally determined sensitivity was approximately 10^{-8} T, which agreed well with the theoretical prediction.

Measurements of neuronal magnetic fields have been conducted using sensors located outside the brain. However, MRI detects magnetic fields generated in close proximity to the current source. Local mappings of neuronal magnetic fields in the rat brain have been reported when discussing the possibility of detecting neuronal fields.[38] A 16-channel electrode with a needle length of 3 mm was prepared to measure three-dimensional potential distributions. The electrode was inserted into the brain through a cranial window with varied depths ranging from 0.5 to 1.4 mm. Electric stimulations were delivered to the left hind paw. Somatosensory-evoked potentials were measured through the electrode, and with a 300-μm spatial resolution, three-dimensional distributions of the current density and magnetic field were calculated from the potentials.

The peak magnitude of the evoked potential was 807 μV. The maximum calculated electric current density was 168 μA/cm², which occurred 21 msec after stimulation. The time and location of the peak current density were similar to the time and location of electric potentials. The maximum magnetic flux density within the calculated region was 50 pT. The neuronal magnetic field caused a phase angle shift in the MRI signal. The estimated phase angle shift in the rat brain was 5×10^{-5} rad. Without signal averaging, the experimentally evaluated signal-to-noise ratio for the rat brain was 200 in a 7-T MRI. The resulting sensitivity for magnetic field detection was 940 pT. Sensitivity is inversely proportional to the square root of the number of averaged signal averaging. Data averaging should be performed with more than 10,000 iterations to detect neuronal magnetic fields.

6.4.3 Animal Studies

6.4.3.1 Brain Slices

In addition to the magnetic field arising from neuronal electric activities, associated changes in blood oxygenation cause considerable fluctuation in the magnetic field and result in baseline signal fluctuations. This situation makes it difficult to detect neuronal magnetic fields. In order to eliminate baseline signal fluctuations for hemodynamics, experimental setups under hemoglobin-free conditions have been reported using cell cultures, turtle brain slices, and rat brain slices.[39–41] Among these methods, an experimental setup was developed to maintain the rat brain slice in a hemoglobin-free medium and to measure the neuronal electric activity in an MRI system.[39] The developed nonmagnetic sample holder consisted of a multielectrode array (MEA) and a perfusing system, as shown in Figure 6.12a. The eight-by-eight-channel MEA was fabricated on a

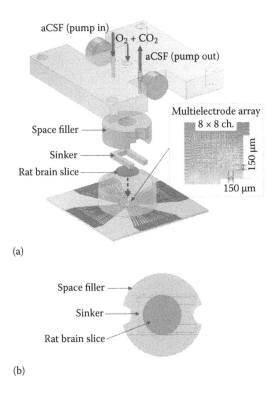

(a)

(b)

FIGURE 6.12 Experimental setup for detecting magnetic fields arising from an extracted brain slice. A multielectrode array measures and stimulates the living brain slice. The perfusion of artificial cerebrospinal fluid washes the blood from the brain slice. (a) Construction of perfusion chamber and perfusing system. (b) Position relation in perfusion chamber. (From Kim, D. et al., *Proceedings of the Annual International Conference of the IEEE Engineering in Medicine and Biology Society* 1370–1373, 2013.)

glass substrate and recorded the evoked extracellular electrical potential during electrical stimulation of the brain slice. The perfusion system circulated the artificial cerebrospinal fluid (aCSF) and the mixture gas [O_2 (95%) + CO_2 (5%)] to maintain cell activities in the slice. The MEA was equipped with a space filler controlling the aCSF flow and a nonmagnetic material sinker to achieve the desired contact condition for the MEA. The relative positions of the space filler, the sinker, and the rat brain slice in the perfusion chamber are shown in Figure 6.12b. The field excitatory postsynaptic potentials (fEPSP) were measured with the MEA by electrically stimulating the brain slice. Assuming that the brain slice had uniform conductivity, the distribution of current density could be calculated from the measured distribution of the evoked electric potential. Using the distribution of current density, the magnetic field intensity arising from neuronal electrical activity in the brain slice was analyzed. The possibility of directly detecting the weak neuronal magnetic field could be experimentally evaluated according to the sensitivity of the MRI system. Moreover, an MRI acquisition condition for direct detection of neuronal activity was investigated. However, validation is required to determine whether the sensitivity of MRI is sufficiently high to detect these neuronal activities.

6.4.3.2 In Vivo Experiments

In vivo experiments have also been performed to measure transient changes in the signal intensity resulting from neuronal activities in the rat brain.[33] An implantable platinum electrode was attached to the sciatic nerve for electric stimulation. Pulsed electric stimulations were applied to the electrodes with a repetition rate of 3 pulses/sec. Measurements were performed using a 4.7-T MRI system. Functional images were obtained using a gradient echo sequence. The excitation pulses were applied every 30 msec at ten time points (0, 30, 60, …, and 270 msec) after electric stimulation. The signal intensities of functional images obtained with and without stimulation were statistically compared using a Student's t-test. In addition, the signal intensities of the images acquired at adjacent time points were compared.

Figure 6.13 shows temporal changes in the signal intensity of functional images. The signals were averaged in regions of interest (ROIs) located on the left and right somatosensory cortices. In the left ROI, no clear difference was found between the signals obtained with and without

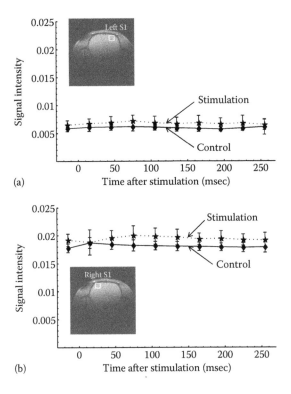

(a)

(b)

FIGURE 6.13 Time courses of the signal intensity in the (a) left and (b) right somatosensory areas in the rat brain after electric stimulation of the left sciatic nerve. (From Sekino, M. et al., *IEEE Transactions on Magnetics* 45: 4841–4844, 2009.)

stimulation. In the right ROI, the signal intensity with stimulation was higher than the intensity without stimulation at most time points. This increase in the signal intensity's baseline was caused by the blood oxygenation level-dependent effect associated with increased blood flow in the right somatosensory cortex. A transient decrease in the signal intensity was found at 0 to 30 msec after stimulation. This result is consistent with a local change in the magnetic field resulting from neuronal activities. The images were compared at adjacent time points to investigate temporal changes, as shown in Figure 6.14. A significant difference was observed between 30 to 60 msec and 60 to 90 msec in the right somatosensory area, which suggests that these effects arose temporarily from neuronal magnetic fields. However, both the somatosensory cortex and other areas of

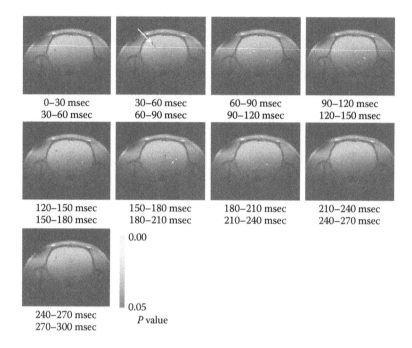

FIGURE 6.14   Detection of magnetic fields arising from neuronal activities. The signal intensities from the images obtained at adjacent time points were compared after electric stimulation. The time "0–30 msec, 30–60 msec" means a comparison between the image excited at 0 msec (and acquired at 30 msec) and the image excited at 30 msec (and acquired at 60 msec). Color indicates the level of statistical significance (P value) in a Student's t-test. (From Sekino, M. et al., *IEEE Transactions on Magnetics* 45: 4841–4844, 2009.)

the brain showed activations. This findings is perhaps caused by noise in images because, compared with the strength of neuronal magnetic fields, the sensitivity of present MRI is not sufficiently high. The signal-to-noise ratio should be improved in future works.

6.4.4 Human Studies

Several papers have reported human studies seeking to detect neuronal currents in the brain or the median nerve.[42–44] A pioneering work in this topic used gradient magnetic fields with different polarities to detect neuronal fields and to eliminate signal variations resulting from other effects.[42] Measurements were conducted using a 1.5-T MRI system. In addition to neuronal magnetic fields, the blood oxygenation level and blood volume affect images. As shown in Table 6.2, functional images were obtained from gradient fields with different polarities, and the images were edited to eliminate intensity changes resulting from causes other than neuronal currents normal to the static magnetic field. Identical effects on I_p and I_n are produced by changes in blood's magnetic susceptibility. Signal intensity changes resulting from causes other than neuronal currents are eliminated in subtracted images of I_p and I_n. Neuronal currents cause different effects on I_p and I_n, and signal intensity changes are observed in subtracted images ($I_p - I_n$ and $I_n - I_p$). The component of currents normal to the readout gradient causes symmetrical signal intensity changes about the component's axis. The parallel component produces asymmetrical intensity changes. Areas with symmetrical intensity changes were extracted to avoid misrepresentation by artifacts.

A double-subtracted image was produced for the activated area of the human brain during repeated tapping of the right hand's middle finger and thumb. Figure 6.15 shows a neuronal current map based on the double-subtracted image. The y component (normal to the static magnetic field and the readout gradient) of the currents in the activated sensory area was clearly observed. The currents in the motor area, however, were not detected because the y component of the currents in the motor area was too small.

TABLE 6.2   Procedures for Obtaining Neural Current Distribution Images

Image	State	Polarity of the Read-Out Gradient
		I
I_{pr}	Rest	p
I_{pa}	Activated	p
I_{nr}	Rest	n
I_{na}	Activated	n
		II
$I_{pa} - I_{pr} = I_p \rightarrow$ fMRI		
$I_{na} - I_{nr} = I_n \rightarrow$ fMRI		
		III
$I_n - I_p \rightarrow$ Current distribution image		
$I_p - I_n \rightarrow$ Current distribution image		

Source: Kamei, H. et al., *IEEE Transactions on Magnetics* **35**: 4109–4111, 1999.

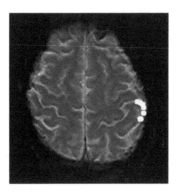

FIGURE 6.15 Neuronal current distribution imaging of the human brain using readout gradients with reversed polarities. The detected signal from the activated sensory area has been enhanced bright enough to be clearly visible on the image. (From Kamei, H. et al., *IEEE Transactions on Magnetics* 35: 4109–4111, 1999.)

6.5 SUMMARY AND FUTURE PROSPECTS

As discussed in this chapter, the use of MRI in impedance imaging realizes millimeter-scale resolutions and measurement of anisotropic conductivity. Currently, the results of MRI-based impedance imaging have been put into practical use in electromagnetic field analyses. Recent progress in ultrahigh-field MRI has raised a safety issue associated with high RF absorption in the patient's body. Because of differences in body anatomy among individuals and variable postures of the MRI scanner, there is a strong need to map the impedance of the subject's body. One of the technical challenges for future impedance imaging is a measurement with arbitrary frequency. When a quantitative measurement of impedance at an arbitrary frequency is realized, the results will be used in even broader technological fields.

Detection of a neuronal magnetic field using MRI is an attractive but challenging endeavor. As introduced in this chapter, some previous studies have provided positive evidence supporting the possibility of detection. However, the methods for detecting neuronal fields remain to be established for practical use. Conventional methods of functional imaging provide some brain activity information; for example, fMRI measures the spatial distribution of the activities, and MEG measures the temporal changes in the activities. The neuronal-field MRI potentially visualizes the spatiotemporal dynamics of brain activities and may become a new technique leading to a deeper understanding of brain function.

Finally, authors would appreciate if readers understand that the contents of this chapter belong to future prospects, and authors have tried to introduce whatever small or even incomplete data and results of experiments toward MR imaging of impedance or electric currents to be verified in the future.

REFERENCES

1. Nagaoka, T., Watanabe, S., Sakurai, K., Kunieda, E., Watanabe, S., Taki, M., and Yamanaka, Y. 2004. Development of realistic high-resolution whole-body voxel models of Japanese adult males and females of average height and weight, and application of models to radio-frequency electromagnetic-field dosimetry. *Physics in Medicine and Biology* **49**: 1–15.
2. Christ, A., Kainz, W., Hahn, E. G., Honegger, K., Zefferer, M., Neufeld, E., Rascher, W., Janka, R., Bautz, W., Chen, J., Kiefer, B., Schmitt, P., Hollenbach, H. P., Shen, J., Oberle, M., Szczerba, D., Kam, A., Guag, J. W., and Kuster, N. 2010. The virtual family—Development of surface-based anatomical models of two adults and two children for dosimetric simulations. *Physics in Medicine and Biology* **55**: 23–38.
3. Metherall, P., Barber, D. C., Smallwood, R. H., and Brown, B. H. 1996. Three-dimensional electrical impedance tomography. *Nature* **380**: 509–512.
4. Hashemzadeh, P., Fhager, A., and Persson, M., 2006. Experimental investigation of an optimization approach to microwave tomography. *Electromagnetic Biology and Medicine* **25**: 1–12.
5. Cole, K. S., and Cole, R. H. 1941. Dispersion and absorption in dielectrics: I. Alternating current characteristics. *Journal of Chemical Physics* **9**: 341–351.
6. Gabriel, C., Gabriel, S., and Corthout, E. 1996. The dielectric properties of biological tissues: I. Literature survey. *Physics in Medicine and Biology* **41**: 2231–2249.
7. Gabriel, S., Lau, R. W., and Gabriel, C. 1996. The dielectric properties of biological tissues: II. Measurements in the frequency range 10 Hz to 20 GHz. *Physics in Medicine and Biology* **41**: 2251–2269.
8. Safety Committee of the Japan Society of Magnetic Resonance in Medicine. 2010. *MR Safety—Principles, Standards and Clinical Concerns.* Gakken Medical Shujunsha, Tokyo.
9. Hoeltzell, P. B., and Dykes, R. W. 1979. Conductivity in the somatosensory cortex of the cat—Evidence for cortical anisotropy. *Brain Research* **177**: 61–82.
10. Nagaoka, T., Togashi, T., Saito, K., Takahashi, M., Ito, K., and Watanabe, S. 2007. An anatomically realistic whole-body pregnant-woman model and specific absorption rates for pregnant-woman exposure to electromagnetic plane waves from 10 MHz to 2 GHz. *Physics in Medicine and Biology* **52**: 6731–6745.

11. Sekino, M., Mihara, H., Iriguchi, N., and Ueno, S. 2005. Dielectric resonance in magnetic resonance imaging: Signal inhomogeneities in samples of high permittivity. *Journal of Applied Physics* **97**: 10R303.
12. Yee, K. 1966. Numerical solution of initial boundary value problems involving Maxwell's equations in isotropic media. *IEEE Transactions on Antennas and Propagation* **14**: 302–307.
13. Wada, H., Sekino, M., Ohsaki, H., Hisatsune, T., Ikehira, H., and Kiyoshi, T. 2010. Prospect of high-field MRI. *IEEE Transactions on Applied Superconductivity* **20**: 115–122.
14. Sekino, M., Kim, D., and Ohsaki, H. 2008. FDTD simulations of RF electromagnetic fields and signal inhomogeneities in ultrahigh-field MRI systems. *Journal of Applied Physics* **103**: 07A318.
15. Ueno, S., and Iriguchi, N. 1998. Impedance magnetic resonance imaging: A method for imaging of impedance distributions based on magnetic resonance imaging. *Journal of Applied Physics* **83**: 6450–6452.
16. Yukawa, Y., Iriguchi, N., and Ueno, S. 1999. Impedance magnetic resonance imaging with external AC field added to main static field. *IEEE Transactions on Magnetics* **35**: 4121–4123.
17. Sekino, M., Yamaguchi, K., Iriguchi, N., and Ueno, S. 2003. Conductivity tensor imaging of the brain using diffusion-weighted magnetic resonance imaging. *Journal of Applied Physics* **93**: 6730–6732.
18. Sekino, M., Inoue, Y., and Ueno, S. 2005. Magnetic resonance imaging of electrical conductivity in the human brain. *IEEE Transactions on Magnetics* **41**: 4203–4205.
19. Sekino, M., Ohsaki, H., Yamaguchi-Sekino, S., Iriguchi, N., and Ueno, S. 2009. Low-frequency conductivity tensor of rat brain tissues inferred from diffusion MRI. *Bioelectromagnetics* **30**: 489–499.
20. Tuch, D. S., Wedeen, V. J., Dale, A. M., George, J. S., and Belliveau, J. W. 2001. Conductivity tensor mapping of the human brain using diffusion tensor MRI. *Proceedings of the National Academy of Sciences of the United States of America* **98**: 11,697–11,701.
21. Sekino, M., Tatara, S., and Ohsaki, H. 2008. Imaging of electric permittivity and conductivity using MRI. *IEEE Transactions on Magnetics* **44**: 4460–4463.
22. Katscher, U., Voigt, T., Findeklee, C., Vernickel, P., Nehrke, K., and Dössel, O. 2009. Determination of electric conductivity and local SAR via B1 mapping. *IEEE Transactions on Medical Imaging* **28**: 1365–1374.
23. Voigt, T., Katscher, U., and Doessel, O. 2011. Quantitative conductivity and permittivity imaging of the human brain using electric properties tomography. *Magnetic Resonance in Medicine* **66**: 456–466.
24. van Lier, A. L. H. M. W., Brunner, D. O., Pruessmann, K. P., Klomp, D. W. J., Luijten, P. R,. Lagendijk, J. J. W., and van den Berg, C. A. T. 2012. B1+ phase mapping at 7T and its application for in vivo electrical conductivity mapping. *Magnetic Resonance in Medicine* **67**: 552–561.
25. Joy, M., Scott, G., and Henkelman, M. 1989. *In vivo* detection of applied electric currents by magnetic resonance imaging. *Magnetic Resonance Imaging* **7**: 89–94.

26. Khang, H. S., Lee, B. I., Oh, S. H., Woo, E. J., Lee, S. Y., Cho, M. H., Kwon, O., Yoon, J. R., and Seo, J. K. 2002. J-substitution algorithm in magnetic resonance electrical impedance tomography (MREIT): Phantom experiments for static resistivity images. *IEEE Transactions on Medical Imaging* **21**: 695–702.

27. Lee, C. O., Jeon, K., Ahn, S., Kim, H. J., and Woo, E. J. 2011. Ramp-preserving denoising for conductivity image reconstruction in magnetic resonance electrical impedance tomography. *IEEE Transactions on Biomedical Engineering* **58**: 2038–2050.

28. Oh, T. I., Jeong, W. C., McEwan, A., Park, H. M., Kim, H. J., Kwon, O. I., and Woo, E. J. 2013. Feasibility of magnetic resonance electrical impedance tomography (MREIT) conductivity imaging to evaluate brain abscess lesion: In vivo canine model. *Journal of Magnetic Resonance Imaging* **38**: 189–197.

29. Seo, J. K., Yoon, J. R., Woo, E. J., and Kwon, O. 2003. Reconstruction of conductivity and current density images using only one component of magnetic field measurements. *IEEE Transactions on Biomedical Engineering* **50**: 1121–1124.

30. Ider, Y. Z. Onart, S., and Lionheart, W. R. 2003. Uniqueness and reconstruction in magnetic resonance-electrical impedance tomography (MR-EIT). *Physiological Measurement* **24**: 591–604.

31. Sekino, M., Matsumoto, T., Yamaguchi, K., Iriguchi, N., and Ueno, S. 2004. A method for NMR imaging of a magnetic field generated by electric current. *IEEE Transactions on Magnetics* **40**: 2188–2190.

32. Demachi, K., Rybalko, S., and Fujita, M. 2008. Inverse analysis of the current dipoles distribution in a human brain applied with the shifting-aperture method. *IEEE Transactions on Magnetics* **44**: 1426–1429.

33. Sekino, M., Ohsaki, H., Yamaguchi-Sekino, S., and Ueno, S. 2009. Toward detection of transient changes in magnetic resonance signal intensity arising from neuronal electrical activities. *IEEE Transactions on Magnetics* **45**: 4841–4844.

34. Scott, G. C., Joy, M. L. G., Armstrong, R. L., and Henkelman, R. M. 1992. Sensitivity of magnetic-resonance current-density imaging. *Journal of Magnetic Resonance* **97**: 235–254.

35. Hatada, T., Sekino, M., and Ueno, S. 2005. FEM-based calculation of the theoretical limit of sensitivity for detecting weak magnetic fields in the human brain using magnetic resonance imaging. *Journal of Applied Physics* **97**: 10E109.

36. Hatada, T., Sekino, M., and Ueno, S. 2004. Detection of weak magnetic fields induced by electrical currents with MRI: Theoretical and practical limits of sensitivity. *Magnetic Resonance in Medical Sciences* **3**: 159–163.

37. Halpern-Manners, N. W., Bajaj, V. S., Teisseyre, T. Z., and Pines, A. 2010. Magnetic resonance imaging of oscillating electrical currents. *Proceedings of the National Academy of Sciences of the United States of America* **107**: 8519–8524.

38. Sekino, M., Chin, Y., Takewa, T., Kim, D., and Someya, T. 2014. Submillimeter-scale mapping of evoked magnetic fields in the rat brain for neuronal current MRI. *IEEE International Magnetics Conference*.

39. Petridou, N., Plenz, D., Silva, A. C., Loew, M., Bodurka, J., and Bandettini, P. A. 2006. Direct magnetic resonance detection of neuronal electrical activity. *Proceedings of the National Academy of Sciences of the United States of America* **103**: 16,015–16,020.

40. Luo, Q., Lu, H., Lu, H., Senseman, D., Worsley, K., Yang, Y., and Gao, J. H. 2009. Physiologically evoked neuronal current MRI in a bloodless turtle brain: Detectable or not? *NeuroImage* **47**: 1268–1276.

41. Kim, D., Someya, T., and Sekino, M. 2013. Sensitivity of MRI for directly detecting neuronal electrical activities in rat brain slices. *Proceedings of the Annual International Conference of the IEEE Engineering in Medicine and Biology Society* 1370–1373.

42. Kamei, H., Iramina, K., Yoshikawa, K., and Ueno, S. 1999. Neuronal current distribution imaging using magnetic resonance. *IEEE Transactions on Magnetics* **35**: 4109–4111.

43. Xiong, J., Fox, P. T., and Gao, J. H. 2003. Directly mapping magnetic field effects of neuronal activity by magnetic resonance imaging. *Human Brain Mapping* **20**: 41–49.

44. Truong, T. K., and Song, A. W. 2006. Finding neuroelectric activity under magnetic-field oscillations (NAMO) with magnetic resonance imaging in vivo. *Proceedings of the National Academy of Sciences of the United States of America* **103**: 12,598–12,601.

Magnetic Control of Biological Cell Growth

Shoogo Ueno and Sachiko Yamaguchi-Sekino

CONTENTS

7.1 INTRODUCTION

Cell proliferation and cell death are fundamental phenomena that help in maintaining a human body's condition. The number of cells in the body increases by cell proliferation and decreases by cell death. Because the regulation of cell proliferation and death is strongly connected to various disorders (e.g., cancer or nerve degeneration), these phenomena are strictly regulated by genes or proteins. The basic principle utilized in the magnetic control of cell growth is to control the cells' physiological status using physical stimuli from static or time-varying (and pulsed) magnetic fields. Application of these stimuli might affect the cells' morphological and functional (including genetic and protein expression) features. The greatest advantage of this approach is its noninvasiveness, thus avoiding patient surgery. Another advantage is that this method subjects the patient to nonionizing radiation, thereby minimizing risks to operators and patients from harmful ionizing radiation. Furthermore, the simplicity-of-handling feature of this method implies that the installation of large facilities such as radiation therapy is not required. For these reasons, attempts at controlling the cell state by magnetic fields have been the focus of active research in this decade (Ueno 2012).

Sources of exposure for the magnetic control of cell growth can be static or time-varying (pulsed) magnetic fields. Magneto-mechanical effects and magnetically induced electric currents are the basic mechanisms of physical stimuli in the magnetic control of cell growth. A static magnetic field generates magneto-mechanical effects known as "magnetic orientation" and "magnetic force," although this chapter only focuses on medical applications of the former effect. Historically, the first instance of the magnetic control of cell growth by a static magnetic field was the polymerization of fibrin gel in strong magnetic fields, performed by Torbet et al. (1981) and Torbet and Ronziere (1984). Torbet el al. demonstrated the polymerization of fibrin fibers in an 11 T static magnetic field and reported that the fibers aligned in a particular direction with respect to the magnetic field (known as the "magnetic orientation") when the polymerization was carried out slowly (Torbet et al. 1981). Magnetic orientation has been observed in various biological molecules such as collagen (Murthy 1984; Torbet and Ronziere 1984; Kotani et al. 2000), and cells such as osteoblast cells (Kotani et al. 2000, 2002), Schwann cells (Eguch et al. 2003; Eguchi and Ueno 2005), vascular endothelial cells, and smooth muscle cells (Umeno et al. 2001; Umeno and Ueno 2003; Iwasaka and Ueno 2003a,b; Iwasaka et al. 2003). Owing to these experiments, this phenomenon is now a scientifically established effect of static magnetic fields (WHO EHC 232 2006; ICNIRP 2009). Medical applications of the magnetic orientation have been demonstrated in various targets (Ueno and Sekino 2006).

Apart from static magnetic fields, studies on the medical applications of time-varying magnetic fields (pulsed magnetic fields) are also underway. Magnetic stimulation, established by Barkar et al. (1985) and developed by Ueno et al. (1988), is a noninvasive electrical stimulation technique. Because the promising biological effects of this method toward excitable tissues (e.g., brain or neurons) have been shown, various medical applications of magnetic stimulation have been developed in the fields of brain science or neurology (Ueno 2012). However, applying magnetically induced force or currents to cancers has only recently been studied and the noninvasiveness of this method has been found to be a great advantage for cancer therapy.

We first review the basic principles for biology-magnetism interaction (named biomagnetic effects) in Section 7.2 and introduce the physical background of biological effects caused by exposure to static and time-varying magnetic fields and their multiplicative use with other forms of energy. In the latter half of this chapter, attempts on the magnetic control of tissues for disorders (Section 7.4, bone growth acceleration; Section 7.5, nerve regeneration; and Sections 7.6 and 7.7, cancers) will be introduced.

7.2 BASIC PRINCIPLES FOR BIOMAGNETIC EFFECTS

7.2.1 Effects of Static Magnetic Fields on Biological Systems

Scientific studies on the short-term effects of static magnetic fields on tissues have already been authorized and published by several organizations such as the World Health Organization (WHO) (WHO EHC 232 2006) and the International Commission on Non-Ionizing Radiation Protection (ICNIRP) (ICNIRP 2004, 2009, 2014). A number of books and review articles also describe the effects and experimental results related to static magnetic fields exposure. Chapter 9 of this textbook introduces the effects and mechanisms of static and time-varying magnetic fields on tissues.

Figure 7.1 illustrates an overview of the biological effects of a static magnetic field. Biological effects of a static magnetic field can be classified depending on whether the exposure source is a nonuniform (inhomogeneous) or uniform (homogeneous) magnetic field: (1) A physical force/voltage named "Lorentz force/flow potential" and (2) a rotational force named "magnetic torque" are established effects of a homogeneous magnetic field. The former is a phenomenon that is observed in an electrically conductive fluid flowing perpendicular to a magnetic field that produces

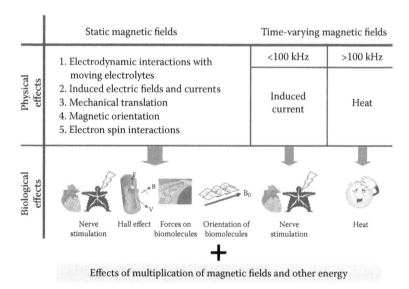

FIGURE 7.1   Basic principles for biomagnetic effects.

a force (Lorentz force) and voltage (flow potential) at right angles to each other. This effect becomes visible as an artifact in ECGs (electrocardiograms), which are recorded under strong static magnetic fields. The latter effect is a phenomenon that the molecules are aligned in the certain direction in which molecules are magnetically stable with dependent on their magnetic anisotropy. Various types of molecules and cells are known to respond to a static magnetic field as well as exhibit morphological changes.

In an inhomogeneous magnetic field, two major properties affect biological organisms; these are (1) the "magnetic force" and (2) magnetically induced currents named "motion-induced currents," which occur because of motion in an inhomogeneous field. Magnetic forces are frequently observed when a paramagnetic material is placed in a magnetic field gradient. This projectile effect is a major concern among magnetic resonance imaging (MRI) operators from the viewpoint of safe handling of the equipment. A diamagnetic material such as water also experiences a repulsive force when exposed to an inhomogeneous magnetic field. A strong magnetic field with a high field gradient, on the order of 50–100 T/m, can generate a partition in water (Ueno and Iwasaka 1994a,b). The latter phenomenon is observable through temporary symptoms, such as dizziness and headache, near a strong inhomogeneous magnetic field such as that produced in the vicinity of an MRI system. Proper management and education are required to protect MRI operators from these symptoms.

The effect of a static magnetic field on a chemical reaction has been described as the "radical pair effect" (Ritz et al. 2000; Okano 2008; ICNIRP 2009). The static magnetic field alters the yield of chemical substances by affecting the intersystem crossing of radicals produced as intermediates in a photochemical reaction.

7.2.1.1 Magnetic Force Acting on a Biological System in a Homogeneous Magnetic Field

7.2.1.1.1 Magnetic Induction 1 (Lorentz Force/Flow Potential)   Magnetic induction, especially magnetohydrodynamics (MHD), describes the electrodynamic interaction of a magnetic field with conductive fluids such as blood flow. The basic principle of MHD is that magnetic fields can induce currents in a moving conductive fluid, which, in turn, creates forces on the fluid and changes the magnetic field itself. The Lorentz force is defined as the vector product of the charge velocity and magnetic flux density and consequently is perpendicular to the direction of electric charge flow (flow potential) (Kangarlu and Robitaille 2000; ICNIRP 2009). It has been a concern that blood or the body itself may be subject to Lorentz force and a flow potential because of magnetic field exposure, because these tissues are electrically conductive. Indeed, it is well established that when major arteries of the circulatory system are placed in a magnetic field, an electric voltage is induced in the blood flowing through them (Kangarlu and Robitaille 2000).

Figure 7.2a shows examples of magnetic forces acting on biological systems in homogeneous magnetic fields. The electrical potential (E) that develops when blood (velocity v) flows in a vessel of cylindrical shape with a diameter D placed at an angle θ with respect to a static magnetic field B can be represented by

$$E = vBD \sin\theta$$

Kangarlu and Robitaille estimated that in order to maintain flow in the aorta against the opposing magnetically induced force, blood pressure may rise by as much as 28 percent at 10 T (Kangarlu and Robitaille 2000). However, a hemodynamics study in dogs at 8 T revealed no elevation of blood pressure during 3 h of exposure (Kangarlu et al. 1999). The authors concluded from calculations that the change between the total hydrodynamic vascular pressure to the total MHD-induced vascular pressure is less than 0.2 percent and that there is no notable pressure effect on the human circulatory system for fields of up to 10 T (Kangarlu and Robitaille 2000). Shiga et al. examined the effect of an external inhomogeneous magnetic

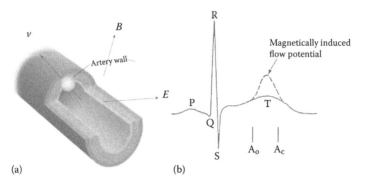

FIGURE 7.2 Magnetic force acting on a biological system in homogeneous magnetic fields. (a) Hall effect by the movement of charged particle in a magnetic field. (b) An artifact in an electrocardiogram due to magnetically induced flow potential. (From World Health Organization, Environmental Health Criteria 69: Magnetic Fields, Geneva, Switzerland, p. 56, 1987.)

field on the flow of erythrocytes containing paramagnetic hemoglobin (Shiga et al. 1993). Red blood cells are known to turn paramagnetic from diamagnetic when the hemoglobin in red blood cells combines with oxygen (Shiga et al. 1993). The authors demonstrated that the paramagnetic attraction takes place with venous blood and that the effect depended on the product of the field strength and its spatial gradient, degree of deoxygenation, flow velocity, and the hematocrit (Shiga et al. 1993).

Figure 7.2b shows a change in ECG caused by magnetically induced flow potential (WHO EHC 69 1987). This magnetically induced flow potential modifies the ECG's T-wave amplitude (Togawa et al. 1967). This effect has been established as an artifact caused by a static magnetic field. Indeed, no hazardous effects were observed in volunteers at 8 T MRI (Kangarlu et al. 1999).

7.2.1.2 Magnetic Torque on Biological Cells in a Homogeneous Magnetic Field

7.2.1.2.1 Magneto-Orientation (Magnetic Torque)   Magneto-mechanical translation (magnetic torque) generates a couple to rotate materials in a stable direction determined by the anisotropy of a material's magnetic susceptibility in a spatially homogenous magnetic field (WHO EHC 232 2006; ICNIRP 2009; Ueno 2012). The torque (magnetic torque) acting on an object is represented as

$$T = -\frac{1}{2}\mu_0 \cdot B^2 \Delta\chi \sin 2\theta$$

where B and μ_0 are defined above, $\Delta\chi$ is the anisotropy in the material's magnetic susceptibility, and θ is the angle between the direction of the magnetic field and the long axis of the material. Magnetic torque aligns various biological molecules and cells where the directions of the orientation depend on the targets. Collagen (Murthy 1984; Torbet and Ronziere 1984; Kotani et al. 2000) and cells such as osteoblast cells (Kotani et al. 2000, 2002), Schwann cells (Eguch et al. 2003; Eguchi and Ueno 2005), vascular endothelial cells, and smooth muscle cells (Umeno et al. 2001; Iwasaka and Ueno 2003a,b; Iwasaka et al. 2003; Umeno and Ueno 2003) have been reported as magnetic torque sensitive. However, Schenck reported that this force is too small to affect biological material *in vivo* owing to the very small (-10^{-5}) values of magnetic susceptibility (Schenck 2000). Details of this effect will also be described in Section 7.3.

7.2.1.3 Magnetic Force Acting on a Biological System in an Inhomogeneous Magnetic Field

7.2.1.3.1 Magneto-Mechanical Translation (Magnetic Force)   A spatially inhomogeneous magnetic field induces magneto-mechanical translation (magnetic force) (ICNIRP 2004, 2009; WHO EHC 232 2006). Materials tend to move along the direction of the steepest field gradient when exposed to an inhomogeneous field. The magnetic force that acts on the material is proportional to the magnetic flux density (B), the gradient of the magnetic flux density B (grad B), and the magnetic susceptibility (χ) of the material. It is represented as

$$F = [\chi B \, (\text{grad } B)]/\mu_0$$

where μ_0 is the magnetic permeability of a vacuum. The direction of the magnetic force depends on the magnetic properties (ferromagnetic, paramagnetic, and diamagnetic) of the materials. This force acts to attract materials to the magnetic source when ferromagnetic materials are exposed to a static magnetic field and could be of substantial magnitude depending on the size and susceptibility of the object. One well-known effect of this force is the "missile effect" or "projectile effect" that occurs around MRI equipment (Shellock and Kanal 1994; Kangarlu and Robitaille 2000; Yamaguchi-Sekino et al. 2011). The projectile effect or the magnetic attraction of ferromagnetic objects within a strong magnetic field might cause injuries to subjects in their proximity (Kangarlu and Robitaille 2000).

Kangarlu et al. calculated the magnetic force acting on a stainless steel wrench of mass 100 g within the magnetic field of an 8 T system (Kangarlu and Robitaille 2000). The magnetic field gradient in their equipment was 7.915 T/m, the maximum force at that position was 42.95 T²/m, and magnitude of the force was calculated as 4.3 N with the χ of an object of density 8 g/cm³ and a volume of 12.5 cm³ is taken to be 0.01. This will impart an acceleration of 43 m/s² on the object.

Exposure to a magnetic field generates a repulsive force on diamagnetic materials. Ueno et al. observed a so-called "Moses effect" phenomenon, in which water is parted when exposed to an 8 T magnetic field with a gradient of 50 T/m (Ueno and Iwasaka 1994a,b). The estimated magnetic force acting on 100 mL of water is 0.288 N, that is, 1/3 of Earth's gravity when the product of the magnetic field and the field gradient is 400 T²/m. Water, a diamagnetic material, is pushed back by the magnetic force to form two "water walls," and the dry "river bed" or dry bottom of the water chamber is seen in between

FIGURE 7.3   Magnetic force acting on biological system in inhomogeneous magnetic fields.

the two water walls, similar to Moses parting the Red Sea as described in the Bible (Ueno 2012). Figure 7.3 is a photograph of the Moses effect. The Moses effect has been a concerning issue in MRI users; however, the degree of the acceleration was suppressed to 1 percent of that of gravity when the force was recalculated for the situation of a whole-body in a 3-T magnet setup. Schenck calculated that the magnetic force at 10 T leads to a change in pressure between the inside of the magnet to the outside of less than 40 mm of H_2O, suggesting that the effect on blood flow is minor (Schenck 2005).

7.2.1.3.2 Magnetic Induction 2 (Motion-Induced Current)   Human body motion in an inhomogeneous static magnetic field causes time-varying magnetic fields inside the body and results in the induction of currents (Schenck et al. 2000; ICNIRP 2004, 2009, 2014; Chakeres and de Vocht 2005; Crozier and Liu 2005; de Vocht et al. 2006; WHO EHC 232 2006; Yamaguchi-Sekino et al. 2011). This effect is due to electromagnetic induction in the body caused by the linear or rotational motion in the magnetic field gradient. In particular, movement along a field gradient generates changes in the flux linkage that induce an electric current, whereas a linear motion within a uniform static field does not. For linear movement in a field gradient, the magnitude of the induced current and the associated electric field increase with the velocity of the movement and the amplitude of the gradient (ICNIRP 2004, 2009, 2014). Rotational motion in a uniform field or in a field gradient causes electromagnetic induction and results in a motion-induced current in a body.

MRI operators are frequently exposed to high magnetic fields from active-shielded MRI scanners. Motion-induced currents due to the stray field from a scanner cause temporal sensations such as vertigo, nausea, magnetophosphenes, and some changes in cognitive functions as described above (de Vocht et al. 2006; Glover et al. 2007). From de Vocht's report, vertigo, metallic taste, and concentration problems were seen to be more prevalent among workers of MRI-fabrication than in reference departments (de Vocht et al. 2006). Occupational exposure to magnetic fields has been the subject of several studies (Hudson 2006; Bradley et al. 2007; Capstick et al. 2008; Wilén and de Vocht 2010; Karpowicz et al. 2011; McRobbie 2012) and the actual exposure levels from 3-T MRI systems exceed 1 T (Yamaguchi-Sekino et al. 2014). Exposure guidelines and a safety survey in MRI applications are introduced in Chapter 9.

7.2.2 Effects of Time-Varying Magnetic Fields on Biological Systems

When a human body is exposed to an electromagnetic field, the biological changes occur in accordance with the intensity and frequency of the field (WHO EHC 238 2007; ICNIRP 2010). The human body responds to electromagnetic fields owing to its conductivity and dielectric constants that are frequency dependent (WHO EHC 238 2007; ICNIRP 2010).

Information is transmitted by nerves in the human body in the form of electrical signals. The coupling mechanism between a low-frequency electromagnetic field and the exposed body is via an induced electric field by Faraday's law that leads to pseudo-nerve information. For example, the excitation threshold of peripheral nerves is 4 V/m at 3 kHz or less (Reilly 1998, 2002; So et al. 2004; ICNIRP 2010). However, sensory changes such as pain or discomfort occur if the induced electric field in the body exceeds this value. For this reason, the exposure guideline of low-frequency electromagnetic field aims at protecting nerves from stimulation effects (ICNIRP 2010).

Exposure to an electromagnetic field at a frequency greater than ~100 kHz can cause a considerable absorption of energy and a subsequent temperature increase. Therefore, protection from heat effects is the purpose of exposure guidelines for high frequency electromagnetic fields (ICNIRP 1998).

7.2.2.1 Low-Frequency Magnetic Fields and Pulsed Magnetic Fields

The physical interaction of a time-varying magnetic field with a human body is well documented (Kato 2006; WHO EHC 238 2007; ICNIRP 2010). Low-frequency magnetic fields induce electric fields and circulating electric currents within the body based on Faraday's law. The magnetic permeability of tissues is the same as that of air, so the field inside a tissue is the same as the external field and therefore human and animal bodies do not significantly perturb a field (ICNIRP 2010). The magnitudes of the induced field strength and current density are proportional to the radius of the loop, the electrical conductivity of the tissue, and the rate of change in the magnitude of the magnetic flux density (ICNIRP 2010). As described in the ICNIRP guideline of 2010, the responsiveness of electrically excitable nerve and muscle tissues to electric stimuli, including those induced by exposure to low-frequency electromagnetic fields, has been well established (e.g., Reilly 2002; Saunders and Jefferys 2007). On the basis of a theoretical calculation using a nerve model, the threshold value of myelinated nerve fibers in human peripheral nervous system has been estimated as 6 V_{peak}/m (Reilly 1998, 2002; ICNIRP 2010). The most widely established effect of magnetically induced electric currents is magnetic phosphenes: the perception of a faint flickering light in the periphery of the visual field occurring in the retinas of volunteers exposed to low-frequency magnetic fields (ICNIRP 2010). As mentioned in the ICNIRP guideline of 2010, the minimum threshold flux density of 5 mT occurs at 20 Hz. The phosphenes are thought to be a result of the interaction of the induced electric field with electrically excitable cells in the retina.

Transcranial magnetic stimulation (TMS) is an example of a medical application of pulsed magnetic fields. Various applications of TMS include neurological diseases, functional mapping of the brain, and cancer therapies (Ueno 2012). A pulsed magnetic field generated by a coil attached to the scalp is applied to the brain during TMS. The applied magnetic field typically has an intensity of 1 T and a pulse duration in the submillisecond range. It induces an electric current in the brain aimed at nerve excitation. Single- or double-pulse stimulation is normally used for diagnosis and functional mapping of the brain.

7.2.2.2 High-Frequency Magnetic Fields and Microwaves

There is a possibility of a temperature increase in the body on exposure to electromagnetic fields of frequency higher than 100 kHz (ICNIRP 1998; Kato 2006). The irreversible structural changes in proteins due to heat cause dysfunction of cell mechanisms and may lead to cell death. Though excessive heat absorption is harmful to the human body, cancer treatments that actively utilize a temperature increase by exposure to high-frequency electromagnetic fields exist: a case in point being "hyperthermia" (Overgaard 1989; Yoo et al. 2006; Japan Society for Thermal Medicine). In general, exposure to a uniform (plane-wave) high-frequency electromagnetic field results in a highly nonuniform deposition of energy within a body that must be assessed by dosimetric measurement and computer calculation.

The specific absorption rate should be minimized to the extent that blood flow and other bodily heat-transfer mechanisms can dissipate the heat. The magnitude of the specific absorption rate is a critical problem, particularly in the case of high-field MRI systems (IEC 2010).

7.2.3 Effects of Multiplicity of Magnetic Fields and Other Forms of Energy

It has been considered that the contribution of the static magnetic field to the chemical reaction might not occur because the induction of energy change by the field exposure is very small. However, several groups have reported on the interactions of a magnetic field on radical pair reactions (Schulten 1982; Grissom 1995; Nagakura et al. 1998; Ritz et al. 2000, 2004; ICNIRP 2009). For instance, a static magnetic field alters the yield of chemical substances by affecting the intersystem crossing of radicals produced as intermediates in a photochemical reaction (Nagakura et al. 1998; ICNIRP 2009). For radicals generated in a living body, the magnetic field effects have been reported in an *in vitro* B12 coenzyme reaction system (Harkins and Grissom 1994).

7.2.3.1 Radical Pair Model and Singlet-Triplet Intersystem Crossing

7.2.3.1.1 Radical Pair Effect and Photochemical Reactions in a Magnetic Field   A change in radical pair recombination rates is one of the few mechanisms by which a magnetic field can interact with a biological system (Harkins and Grissom 1994; ICNIRP 2009). The spin-correlated radical pairs recombine to form reaction products; an applied magnetic field affects the rate and the extent to which the radical pair converts to a triplet state (parallel spins) in which recombination is no longer possible (Figure 7.4) (ICNIRP 2009).

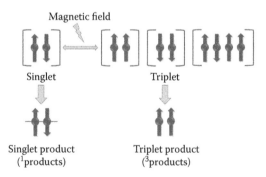

FIGURE 7.4   Radical pair model caused by static magnetic fields.

Using these differences in the radical pair products, Nagakura et al. reported the magnetic field effect in the photolysis reaction of benzoyl peroxide solution at room temperature. As a result of the presence of the magnetic field–favored triplet radicals over the singlet status, yields of [1]products are markedly higher than [3]products (Nagakura et al. 1998).

7.3 MAGNETIC ORIENTATION OF ADHERENT CELLS

7.3.1 Polymerization Processes of Fibrin Fibers under Strong Static Magnetic Fields

As briefly described in Section 7.2.1.2, magnetic torque generates a couple to rotate materials in a stable direction determined by the anisotropy of a material's magnetic susceptibility in a spatially homogeneous magnetic field (WHO EHC 232 2006; ICNIRP 2009; Ueno 2012). Torbet et al. first observed the magnetic orientation of fibrin (Torbet et al. 1981). Fibrin is a fibrous, nonglobular protein involved in the clotting of blood. It is formed by the action of the protease thrombin on fibrinogen which causes the latter to polymerize. Torbet et al. reported that highly oriented fibrin gels are formed when polymerization takes place slowly in a strong magnetic field (Figure 7.5) (Torbet et al. 1981). Scanning electron micrographs indicated that an alignment of fibrin fibers along to the magnetic field. Torbet and Ronziere also reported the magnetic orientation at collagen fibers (Torbet and Ronziere 1984). They showed that the collagen gels formed a high degree of uniaxial alignment.

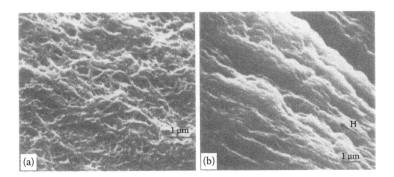

FIGURE 7.5   Polymerization of fibrin fibers under strong static magnetic fields. Scanning electron micrographs of fibrin prepared without (a) or with (b) magnetic field. H represents the direction of the field. The diameter of the fibrin fibers is −1000 Å. (From Torbet, J. et al., *Nature* 289: 91–93, 1981.)

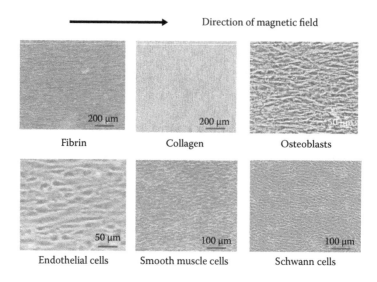

Direction of magnetic field

Fibrin

Collagen

Osteoblasts

Endothelial cells

Smooth muscle cells

Schwann cells

FIGURE 7.6   Magnetic orientation of fibrous proteins and adherent cells.

7.3.2 Fibrin, Collagen, Osteoblast Cells, Endothelial Cells, Smooth Muscle Cells, and Schwann Cells

As well as the fibrin and collagen, when diamagnetic materials such as adherent cells are exposed to strong static magnetic fields of several Tesla, they align either parallel or perpendicular to the direction of the magnetic field because of the anisotropy of the magnetic susceptibility of the materials. For instance, osteoblasts, endothelial cells, and smooth muscle cells exhibit magnetic orientation under the strong static magnetic fields (Figure 7.6) (Ueno and Sekino 2006). Several Japanese researchers observed the magnetic orientation of adherent cells such as osteoblasts (Kotani et al. 2000, 2002), vascular endothelial cells, smooth muscle cells (Umeno et al. 2001; Iwasaka and Ueno 2003a,b; Iwasaka et al. 2003; Umeno and Ueno 2003), human kidney cells (Ogiue-Ikeda and Ueno 2004), and Schwann cells (Eguchi et al. 2003; Eguchi and Ueno 2005) under 8- or 14-T magnetic field exposures. Direction of the orientation depended on the targets.

7.4 BONE GROWTH ACCELERATION BY STATIC MAGNETIC FIELDS

7.4.1 Background

In the orthopedic context, trauma to the musculoskeletal system is the most frequently encountered condition in clinical examinations. Approximately 33 million people in United States are diagnosed with trauma to the musculoskeletal system every year, and fracture accounts for approximately 6.2 million cases (Praemer et al. 1992). The fundamental treatments for fracture are fixation and reset, and recent advancements in the fixing material and orthopedic surgical methods enable satisfactory recovery. However, complications such as pseudarthrosis or delayed recovery still exist in about 5 to 10 percent of the cases (Nakajima et al. 2007).

In the treatment of fractures, bi-side controls (i.e., directional control and the growth control of bones) are indispensable. So far, several cytokines (the bioreactive proteins released from the cells that react as the intercellular signaling molecules) and growth factors (the modulators that promote the proliferation or differentiation to specific cells) are known to exhibit osteogenic activity. In particular, among the growth factors intrinsic to skeletal tissues, bone morphogenetic proteins (BMPs) are known to promote osteogenic activity the most, and are therefore the most clinically used factors (Cook 1999; Schmitt et al. 1999). BMPs exhibit osteogenicity, and they are the only cytokine known to effect ectopic bone formation by itself. BMPs are also responsible for physiological bone formations such as skeletal formation or fracture healing; especially, BMP-2 has been investigated in detail because of its strong osteoinductive activity (Sampath et al. 1990; Marie et al. 2002). In addition to the use of chemicals, the effectiveness of pulsed electromagnetic fields (PEMFs) on fracture healing, spinal fusion, bone defects, bone ingrowth into ceramics, or bone grafts has been reported (Miller et al. 1984; Bruce et al. 1987; Takano-Yamamoto et al. 1992; Glazer et al. 1997). However, neither chemical agents, including BMPs, nor electrical stimuli such as PEMFs can control the direction of bone formation. Thus, directional control in the bone growth during treatment is a matter requiring attention in clinical applications.

Accumulated evidence has revealed that strong magnetic fields of the order of a Tesla are capable of regulating the orientation of matrix proteins and cells. As described in Section 7.3, this phenomenon refers to the tendency of materials to rotate in a stable direction determined by the anisotropies in their magnetic susceptibility. So far, extracellular matrix proteins such as collagen fibers (Torbet and Ronziere 1984; Murthy 1984), fibrin fibers (Torbet et al. 1981; Ueno et al. 1993; Iwasaka et al. 1994), and several types of cells such as erythrocytes and platelets have been reported to orient in regular patterns under static magnetic fields (Vassilev et al. 1982; Yamagishi et al. 1992; Higashi et al. 1993a,b, 1995). Therefore, magnetic control of bone formation from both the perspectives of growth and directional control has been proposed as a new approach to fracture treatment.

7.4.2 Magnetic Control of Bone Growth Acceleration

Kotani et al. have demonstrated magnetic control of bone growth acceleration by using static magnetic fields of 8 T (Kotani et al. 2002). Figures 7.7a through 7.7c illustrate the magnet and experimental setup used in their study. The intensity of the magnetic field was estimated as 7.84–7.91 T (Figure 7.7c). To determine the effect of a magnetic field exposure on the differentiation of cultured mice osteoblast cells (MC3T3-E1 cells), Kotani et al. examined the alkaline phosphatase (ALP) activity of cultured MC3T3-E1 cells. After 14 days in culture post a magnetic field exposure for 60 h, cells showed greater ALP expression, compared to unexposed cells (Figure 7.7d). Matrix nodules were identified by alizarin red staining. They also showed that the orientation of ALP⁺ cells and alizarin red–positive matrix could be maintained parallel to the direction of the magnetic field

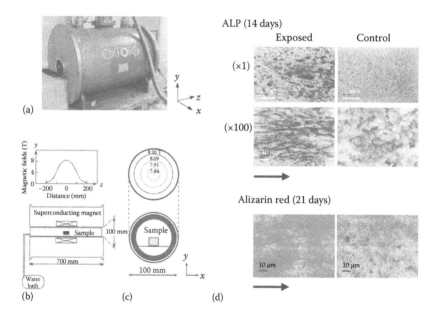

FIGURE 7.7 Bone growth acceleration by magnetic fields (*in vitro* experiments). (a) Photograph of the horizontal cylindrical-type superconducting magnet. (b) The longitudinal section of the bore and the distribution of the magnetic field along the *z* axis. (c) The cross section of the bore and the magnetic field at the center of the *z* axis (*x*-*y* plane). (d) Effects of the SMF on the differentiation and matrix synthesis of cultured MC3T3-E1 cells. Fourteen days and 21 days in culture after 8-T SMF exposure for 60 h; cells were stained with ALP and alizarin red, respectively. Arrows indicate the direction of the magnetic fields. (From Kotani, H. et al., *Journal of Bone and Mineral Research* 17: 1814–1821, 2002.)

for 14 and 21 days, respectively, after an exposure for 60 h, although there was no significant difference between the growth curves of the cultured MC3T3-E1 cells between the exposed and unexposed groups.

Through *in vivo* experiments, Kotani et al. also investigated the effect of a strong static magnetic field on bone formation by using an ectopic bone formation model in and around subcutaneously implanted BMP-2/collagen pellets in mice. Twenty-one days after 60 h of 8-T magnetic field exposure, the BMP-2/collagen pellets were harvested and radiological and histological analyses were performed. The newly formed bones were found to extend parallel to the direction of the magnetic field in the exposed group, and only small spherical-shaped ossicles were seen in the unexposed group. The bone mineral content of the exposed group was approximately 4 times greater than that of the unexposed group. Histological examinations revealed that the pellets were replaced by newly formed bone tissue, including bone marrow, in both exposed and unexposed groups (Figure 7.8a).

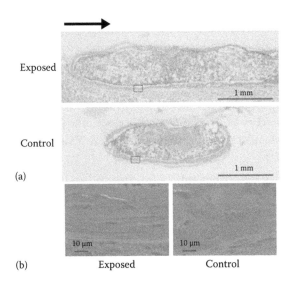

FIGURE 7.8 Bone growth acceleration by magnetic fields (*in vivo* experiments). Histological analyses of the SMF effect on bone formation in and around the BMP-2/collagen pellets implanted subcutaneously in mice. Twenty-one days after BMP-2/collagen pellet implantation; samples were obtained and embedded in paraffin and 5-μm-thick paraffin sections were stained with H&E. (a) Light microscopic and (b) differential interference contrast microscopic findings of a representative pellet from each group. The squares in panel a represent respective areas in panel b. The arrow indicates the direction of the magnetic fields. (From Kotani, H. et al., *Journal of Bone and Mineral Research* 17: 1814–1821, 2002.)

Differential interference contrast microscopic analysis showed that the periphery of the tissues consisted of lamellar bones in the exposed group, whereas woven bones were partially observed in the unexposed tissue (Figure 7.8b). The authors thereby suggest that a strong static magnetic field exposure may increase not only the bone mass but also bone maturation *in vivo*.

7.5 CONTROL OF DIRECTION AND GROWTH OF NERVE AXONS DURING NERVE REGENERATION BY STATIC MAGNETIC FIELDS

7.5.1 Background

Nerve regeneration is an issue for which a medical resolution is awaited. Unlike in the central nervous system, the peripheral nervous system (PNS) has some capability for nerve regeneration (Yiu and Zhigang 2006). However, there is no solution for complete PNS nerve regeneration in severely damaged cases. In those cases, scaffolding neurons using biomaterials or biocompatible materials would be useful in nerve regeneration if these materials play the roles of endoneurial tubes to guide regrowing axons.

Figure 7.9 illustrates the mechanisms of nerve regeneration in PNS. Schwann cells provide a supportive role in the PNS. These cells form a layer or myelin sheath along single segments of an axon. However, when axonotmesis occurs, a series of degenerative processes take place before regeneration of nerve fibers. This phenomenon is called "Wallerian degeneration"

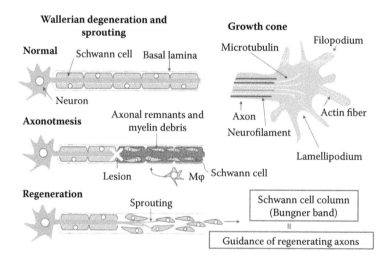

FIGURE 7.9   Mechanisms of nerve regeneration in PNS.

(Ide 1996). After significant injury, a nerve begins to degrade in an antero-grade fashion. The axon and the surrounding myelin break down during this process. Round mast cells as well as phagocytic macrophages that inter-act with Schwann cells to remove the injured tissue debris can be seen. As the degradation of the distal nerve segment continues, its connection with the target muscle is lost, leading to muscle atrophy and fibrosis. Once the degenerative events are complete, all that remains is a column of collapsed Schwann cells (bands of Büngner). Axon sprouts with fingerlike growth cones advance using the Schwann cells as guides. After reinnervation, the newly connected axon matures and the pre-injury cytoarchitecture and function are restored. In this phase, Schwann cells play essential roles in neuronal regeneration of the PNS by guiding the regrowth of axons. In other words, Schwann cells can serve as scaffolds for nerve regeneration. Therefore, as in the case of directional control of bone extension mentioned in Section 7.4, directional control of the Schwann cells in the PNS regenera-tion is also an important issue in peripheral nerve regeneration.

As mentioned in Sections 7.3 and 7.4, an alignment effect on cells and biological molecules occurs in static magnetic fields of the order of a tesla. Controlling the orientation of Schwann cells may be useful in medical and tissue engineering for promoting PNS regeneration. Section 7.5.2 intro-duces the application of a strong static magnetic field to Schwann cells aimed at controlling PNS regeneration.

7.5.2 Magnetic Control of Direction and Growth of Nerve Cells

Eguchi et al. (2003) investigated the magnetic orientation of Schwann cells by exposure to an 8-T magnetic field. The authors also observed the growth of Schwann cells in a mixture containing Schwann cells and col-lagen after the field exposure.

In their study, a magnetic field exposure of 8 T was first given only to Schwann cells cultured from the sciatic nerves of newborn rats. The authors used a horizontal-type superconducting magnet that produced 8 T at its center the same as that reported by Kotani et al. (2002). The flasks with the Schwann cells and the culture medium were positioned at the center of the magnet and exposed to an 8-T magnetic field for 60 h at 37°C. The magnetic orientation of Schwann cells was then observed microscopically. A simi-lar experiment was carried out for collagen gel alone, and collagen fibers were found to align perpendicular to the magnetic field after 2-h exposure. Subsequently, the Schwann cells harvested on the collagen gel were incu-bated at 37°C in the absence of a magnetic field. Morphological analysis

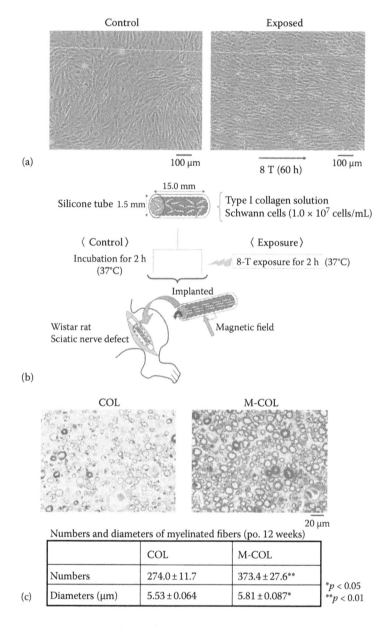

FIGURE 7.10 Magnetic field effects on Schwann cells and nerve regeneration. (a) Light micrographs of a mixed culture of Schwann cells and collagen on the second day in culture after 2 h with and without 8-T magnetic field exposure. Scale bars: 50 mm. (b) Set up for *in vivo* experiment. (c) Morphological examination with toluidine blue staining. The number and the diameter of regenerated axons increased significantly in the M-COL group compared with the COL group by histological evaluation. (b and c, From Eguchi, Y. et al., *Bioelectromagnetics* 36: 233–243, 2015.)

of the Schwann cells was performed microscopically after 60 h for the two groups, namely, the exposed group (magnetically aligned collagen) and the control group (unaligned collagen). As shown in Figure 7.10a, light micrographs of a mixed culture of Schwann cells and collagen indicated that cells cultured with magnetically oriented collagen aligned along the collagen fibers oriented perpendicular to the magnetic field.

The same author also investigated the effectiveness of using magnetically aligned collagen (after exposure to a maximum 8-T magnetic field) for nerve regeneration in both an *in vitro* and *in vivo* model (Eguchi et al. 2015). Eguchi et al. investigated the orientation of Schwann cells embedded in collagen gel within a silicone tube as an artificial nerve guide to bridge sciatic nerve defects. A collagen-cell suspension by mixing collagen type I solution with culture medium containing Schwann cells were prepared (Figure 7.10b). As shown in Figure 7.10b, the collagen-cell suspensions were seeded into a silicone tube (length, 15.0 mm; inner diameter, 1.5 mm) and the tube was exposed to an 8-T magnetic field for 2 h with the axis perpendicular to the magnetic field. The sciatic nerve trunk was cut in the central part of the thigh and the silicone tubes were affixed to both stumps of the nerve (Figure 7.10b). Four groups were examined: collagen gel only (COL), magnetically aligned collagen gel (M-COL), collagen gel mixed with Schwann cells (S-COL), and magnetically aligned collagen gel mixed with Schwann cells (M-S-COL). The ratio of infiltrating regenerated nerves was higher in the M-COL group compared to the COL group at 8 weeks postoperation, and there were no significant differences between the two groups with and without Schwann cells. The number and diameter of regenerated axons increased significantly in the M-COL compared with the COL group at 12 weeks post-operation (Figure 7.10c).

The authors demonstrated that magnetically oriented collagen promoted nerve regeneration using both an *in vitro* and *in vivo* model and also suggest that magnetically aligned collagen acts as a scaffold for Schwann cells and in combination with aligned Schwann cells can also be used for biohybrid nerve guide implants to bridge nerve lesions.

7.6 DISTRACTION OF LEUKEMIC CELLS BY MAGNETIZABLE BEADS AND PULSED MAGNETIC FORCE

7.6.1 Background

Treatment modalities for eradicating cancer cells using heat from electromagnetic waves, particularly involving "hyperthermia," have widely been investigated (Overgaard 1989; Yoo et al. 2006; Japan Society for Thermal Medicine). Hyperthermia is a type of cancer treatment in which a body tissue is exposed to high temperatures (42°C–43°C) for 30–60 min (Japan Society for Thermal Medicine). The idea of using heat to treat cancer is based on the fact that temperatures higher than 42.5°C lead to the death of somatic cells. Research reports indicate that hyperthermia may render some cancer cells more sensitive to radiation or chemotherapy, in addition to a killing effect by heating itself. Numerous clinical trials of hyperthermia for a variety of cancers such as breast cancer, melanoma (malignant melanoma), cervical cancer, colorectal cancer, bladder cancer, and cervical lymph node metastasis have been demonstrated (Yoo et al. 2006; Japan Society for Thermal Medicine).

However, irradiation of only the targeted tumors noninvasively remains an issue. For example, to accomplish sufficient localization of heat requires the implantation of an electrode to the body, instead of using a whole-body electrode. In addition, whole-body electrodes are ineffective for deep-seated tumors, implying that only invasive methods using implanted antennas are available for deep-site tumors.

Alternatively, hyperthermia using harmless magnetizable particles as a method to treat targeted tumors has been reported (Yanase et al. 1997; Hafeli and Pauer 1999; Wada et al. 2001; Ogiue-Ikeda et al. 2003, 2004). The most important properties of magnetic particles for clinical diagnostics and medical therapies are nontoxicity, biocompatibility, injectability, and high-level accumulation in the target tissue or specific organ, i.e., being strictly and spatially confined to the planned region of the internal body. However, the possibility that the drugs might damage normal tissues as they circulate throughout the body before reaching their target remains. Further, although magnetized particles are accumulated only in the cancer tissue, application of a whole-body electrode for hyperthermia may induce heat inside the normal tissue as well as the cancer because of poor localization.

Instead, pulsed magnetic stimulation has recently been proposed as a source of electromagnetic field for cancer therapy using magnetizable particles (Ogiue-Ikeda et al. 2003, 2004). As introduced earlier, pulsed magnetic stimulation is a method for stimulating biomedical tissues noninvasively and with good localization. The next section introduces a new method to destruct targeted cells by using magnetizable beads and pulsed magnetic force.

7.6.2 Magnetic Destruction of Leukemia Cells by Physical Force

Ogiue-Ikeda investigated a new method to destruct targeted cells by using magnetizable particles and pulsed magnetic force (Ogiue-Ikeda et al. 2003, 2004). The cells were combined with the beads by an antigen-antibody reaction (cell/bead/antibody complex), aggregated by a magnet, and subsequently stimulated by a magnetic stimulator. The first step of this idea is to allow cancer cells and magnetized beads to react via an antigen-antibody reaction and then collect them using a permanent magnet. Pulsed magnetic stimulation was then applied to the cells and the induced magnetic force was expected to destroy cells by physical force.

The authors used TCC-S leukemic cells (expressing CD33 on the surface) (Kano et al. 2001; Ogiue-Ikeda et al. 2001) extracted from a patient with chronic myelogenous leukemia and Dynabeads Pan Mouse IgG (Dynal, Oslo, Norway) as magnetizable beads. Dynabeads are monosized superparamagnetic macroporous particles containing narrow pores in which magnetizable materials may be distributed in a volume-filling manner. The beads were coated with a monoclonal antibody specific for the Fc region of mouse immunoglobulin G (IgG). TCC-S cells and anti-CD33 mouse IgG antibodies (Cytotech, Hellebaek, Denmark) were then mixed gently. After the binding of cells and beads by an antigen-antibody reaction, the cell/bead/antibody complexes were isolated using a magnetic particle concentrator (MPC-2, Dynal, Oslo, Norway) and then aggregated by a magnet (diameter, 10 mm; thickness, 12 mm; 450 mT) placed under the tube. Monophasic pulses of 150 μs and a maximum of 2.4 T produced at the center of the coil of a circular-shaped magnetic stimulator (inner diameter, 15 mm; outer diameter, 75 mm) (Nihon Kohden, Tokyo, Japan) were applied to the cells. The magnetic forces acting on each of the beads and the aggregated cell/bead/antibody complexes were expected to be approximately 1.8×10^{-9} N and 0.036 N, respectively. The samples were stimulated 10 times at 5-sec intervals.

Ogiue-Ikeda et al. performed two types of experiments (Ogiue-Ikeda et al. 2003, 2004). In the first, the authors established a procedure to destruct targeted cells by using magnetizable particles and a pulsed magnetic force by the methodology shown in Figure 7.11 (Ogiue-Ikeda et al. 2003). The cells were classified into four groups: (1) cells, (2) cell/bead (diameter 4.5 ± 0.2 mm) mixture, (3) cell/antibody complex, and (4) cell/bead/antibody complex. Each group consisted of four subgroups: (a) without magnet, nonstimulated; (b) without magnet, stimulated; (c) with magnet, non-stimulated; and (d) with magnet, stimulated (Figure 7.11). The viability of

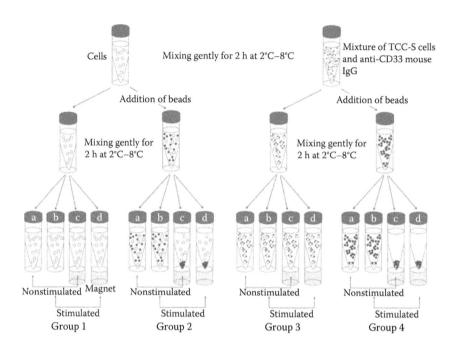

FIGURE 7.11   Methodology of physical destruction of leukemia cells by pulsed magnetic field. (From Ogiue-Ikeda, M. et al., *IEEE Trans Nanobioscience* 2: 262–265, 2003.)

the aggregated and stimulated cell/bead/antibody complexes was significantly decreased and the cells were observed to have been destructed by the penetration into or rupturing of the cells by the beads.

The authors then examined how bead size affected the results (Ogiue-Ikeda et al. 2004). Two types of magnetizable beads (diameter, 4.5 ± 0.2 mm and 2.8 ± 0.2 mm) were used in the second experiment. After stimulation, cell destruction was observed (Figure 7.12a). The viabilities of the stimulated groups were significantly lower than those of the control groups, for both 4.5- and 2.8-mm beads (Figure 7.12b). In addition, the viabilities of the stimulated groups containing the 4.5-mm beads were lower than those containing the 2.8-mm beads. Furthermore, the viabilities of the stimulated groups with the 4.5-mm beads decreased as the number of beads increased (Figure 7.12b).

From these two studies, the authors propose that instantaneous pulsed magnetic forces cause the aggregated beads to forcefully penetrate or rupture the target cells and conclude that the use of a large number of magnetizable beads of a large diameter can effectively destroy cells targeted by an antigen-antibody reaction.

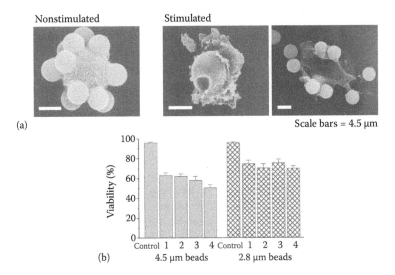

(a)

Scale bars = 4.5 μm

(b)

FIGURE 7.12   Physical destruction of leukemia cells by pulsed magnetic fields. (a) A scanning electron micrograph of a TCC-S cell combined with beads by an antigen-antibody reaction. (b) Viabilities of the cell/bead/antibody complexes with and without magnetic stimulation. (b, From Ogiue-Ikeda, M. et al., *IEEE Trans Magnetics* 40: 3018–3020, 2004.)

7.7 INHIBITION OF TUMOR GROWTH BY PULSED MAGNETIC STIMULATION

7.7.1 Background

Instead of a thermal effect such as hyperthermia, attempts at cancer therapy using direct electrical effects have been carried out since 1950s. Table 7.1 summarizes electrical stimulation and cancer therapy efforts. Electrical stimulation, using, for example, direct current (DC), is known to induce various biological responses including antitumor effects (Humphrey and Seal 1959; David et al. 1985; Nordenstrom 1989) and/or immunomodulatory effects; a relationship between the antitumor effects has been suggested (Sersa et al. 1992; Chou et al. 1997; Miklavcic et al. 1997; Cabrales et al. 2001). Mechanisms underlying these biological effects by DC involve not only pH changes but also inhibition of cell division and/or protein synthesis, activation of lysosome, and influence on immune systems (Harguindey 1982; Sersa et al. 1992; Chou et al. 1997; Cabrales et al. 2001). However, as is the case with hyperthermia, deep-site stimulation is not possible by electrodes placed on the body surface. Furthermore, an implantation of electrodes is also required for the enhancement of anticancer effects by DC.

TABLE 7.1   Electric Stimulation and Cancer Therapy

Method	Condition	Application	Mechanism
Direct current (DC)	0.5–5 mA 2 hours/day etc…	Kidney cancer Epidermoid cancer Rectal cancer	Heat pH change Inhibition of cell division Activation of lysosome Inhibition of protein synthesis Activation of immune system
Electrochemotherapy (ECT)	1000–2000 V/cm	Epithelial cancer Bone cancer	Enhancement of anticancer agents
Hyperthermia	Radio wave Microwave	Recurrence breast cancer Malignant melanoma Cervical cancer Rectal cancer Cervical lymph node metastasis	Heat Activation of immune system

Attempts at noninvasive anticancer therapy using noninvasive methods instead, such as with electric currents from magnetic stimulation, have been investigated (Yamaguchi et al. 2004, 2005, 2006a,b). Magnetic stimulation induces eddy currents in the body, and the resultant biological responses have been reported in regard to neurological tissues such as from the hippocampus (Ueno 2012). However, the effects of magnetically induced eddy currents on tumors and immune systems have not been studied in detail. Additionally, the mechanisms underlying antitumor effects of magnetic stimulation have not been classified sufficiently. In the next section we introduce a report on the anticancer effects of magnetic stimulation and the possible mechanisms.

Another approach for cancer therapy by electrical stimulation is known as electrochemotherapy (ECT). ECT uses electrical pulses, which have an enhancing effect on anticancer agents (Sersa et al. 2008). The reduction of adverse events in chemotherapy has been a key factor in the improvement of the quality of life. However, implantation of an electrode is required for deep-site stimulation and for enhancing the efficacy of anticancer treatment. Thus, novel approaches are required to develop a noninvasive therapy that can be used to target cancer while enhancing the effectiveness of the treatment. In the next section, investigations on the combinational use of anticancer agents and magnetic stimulation are described.

7.7.2 Magnetic Control of Tumor Growth Inhibition

Yamaguchi et al. investigated the effects of pulsed magnetic stimulation on tumor development processes and immune functions in mice (Yamaguchi et al. 2004, 2005, 2006a,b). A circular coil (inner diameter, 15 mm; outer diameter, 75 mm) was used in the experiments (Figure 7.13a and b). Stimulus conditions were pulse width, 238 µs; peak magnetic field, 0.25 T (at the center of the coil); frequency, 25 pulses/sec; dosage, 1000 pulses/sample/day; and magnetically induced eddy currents in mice, 0.79–1.54 A/m^2. In an animal study, B16-BL6 melanoma model mice were exposed to pulsed magnetic stimulation for 16 days from the day of injection of cancer cells (Figure 7.13b). A study of the tumor then revealed a significant tumor weight decrease in the stimulated group (54 percent of the nonexposed group) (Figure 7.13c and d). In a cellular study, B16-BL6 cells were also exposed to the magnetic field (1000 pulses/sample and eddy currents at the bottom of the dish equal to 2.36–2.90 A/m^2); however, the magnetically induced eddy currents were found to have no effect on cell viabilities. Cytokine production in mice spleens was measured to analyze the immunomodulatory effect after pulsed magnetic stimulation. Tumor necrosis factor (TNF-α) production in mice spleens was significantly activated after an exposure to

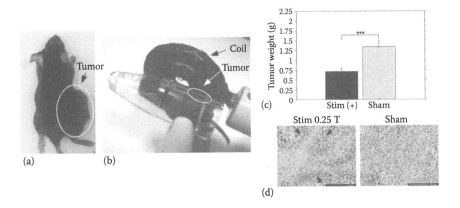

FIGURE 7.13 Inhibition of tumor growth using pulsed magnetic stimulation. (a) Example of tumor mouse on day 17, induced by injection of B16-BL6 cells. (b) Magnetic stimulation coil (75 mm outer diameter, 25 mm inner diameter) was placed on the right side of mice. (c) The tumor weights of each group on day 17 were expressed as mean ± SE. ***$P < 0.001$; $n = 14$. (d) Specimen treated by (left) magnetic stimulation and that of (right) sham group. The arrow shows where tissue necrosis occurred. Magnification 400×. (From Yamaguchi, S. et al., *Bioelectromagnetics* 27: 64–72, 2006a.)

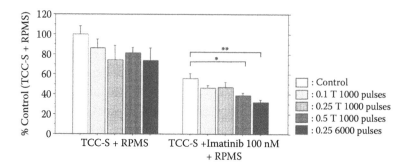

FIGURE 7.14 Combination effects of magnetic stimulation (RPMS) and an anticancer agent (imatinib). (From Yamaguchi, S. et al., *IEEE Transactions on Magnetics* 42: 3581–3583, 2006b.)

the stimulus condition described above. The authors concluded that these results showed the first evidence of antitumor and immunomodulatory effects brought about by the application of repetitive magnetic stimulation and also suggested the possible relationship between antitumor effects and the increase of TNF-α levels caused by pulsed magnetic stimulation.

Next, the combined effect of magnetic stimulation and an anticancer agent was examined on human chronic myelogenous leukemia-derived cell line TCC-S using molecular target drug (selective tyrosine kinase inhibitor) imatinib mesylate (imatinib). The stimulus conditions were determined as follows: 0.1, 0.25, and 0.5 T; 25 pulses/sec; 1000, 3000, and 6000 pulses/day. TCC-S cells were cultured with 100 nM imatinib and exposed to RPMS at 1, 12, 24, 36, 48, and 56 h after drug treatment. The combined effect of magnetic stimulation and imatinib depended on the stimulus intensity and the pulse dose (Figure 7.14). To further clarify the effect of magnetic stimulation on human normal lymphocytes, they too were exposed to magnetic stimulation with or without imatinib. Magnetic stimulation was found to have no effect on the viability of normal lymphocytes. These results indicate that magnetic stimulation possibly improves the effectiveness of anticancer agents.

REFERENCES

Barkar, A. T, Jalinous, R. I., and Freestone, I. L. 1985. Non-invasive magnetic stimulation of human motor cortex. *Lancet* 1: 1106–1107.

Bradley, J. K., Nyekiova, M., Price, D. L., Lopez, L. D., and Crawley, T. 2007. Occupational exposure to static and time-varying gradient magnetic fields in MR units. *Journal of Magnetic Resonance Imaging* 26: 1204–1209.

Bruce, G. K., Howlen, C. A., and Huckstep, R. L. 1987. Effect of a static magnetic field on fracture healing in a rabbit radius. *Clinical Orthopaedics* 222: 300–306.

Cabrales, L. B., Ciria, H. C., Bruzon, R. P., Quevedo, M. S., Aldana, R. H., De Oca, L. M., Salas, M. F., and Pena, O. G. 2001. Electrochemical treatment of mouse Ehrlich tumor with direct electric current. *Bioelectromagnetics* 22(5): 316–322.

Capstick, M., McRobbie, D., Hand, J., Christ, A., Kuhn, S., Hansson Mild, K., Cabot, E., Li, Y., Melzer, A., Papadaki, A., Prüssmann, K., Quest, R., Rea, M., Ryf, S., Oberle, M., and Kuster, N. 2008. An investigation into occupational exposure to electro-magnetic fields for personnel working with and around medical magnetic resonance imaging equipment. Report on Project VT/2007/017 of the European Commission Employment, Social Affairs and Equal Opportunities DG. 2008. Available from http://www.myesr.org/html /img/pool/VT2007017FinalReportv04.pdf (accessed August 15, 2013).

Chakeres, D. W., and de Vocht, F. 2005. Static magnetic field effects on human subjects related to magnetic resonance imaging systems. *Progress in Biophysics and Molecular Biology* 87: 255–265.

Chou, C.-K., McDougall, J. A., Ahn, C., and Vora, N. 1997. Electrochemical treatment of mouse and rat fibrosarcomas with direct current. *Bioelectromagnetics* 18: 14–24.

Cook, S. D. 1999. Preclinical and clinical evaluation of osteogenic protein-1 (BMP-7) in bony sites. *Orthopedics* 22: 669–671.

Crozier, S., and Liu, F. 2005. Numerical evaluation of the fields induced by body motion in or near high-field MRI scanners. *Progress in Biophysics and Molecular Biology* 87: 267–278.

David, S. L., Absolom, D. R., Smith, C. R., Gams, J., and Herbert, M. A. 1985. Effect of low level direct current on *in vivo* tumor growth in hamsters. *Cancer Research* 45(11 Pt 2): 5625–5631.

de Vocht, F., van Drooge, H., Engels, H., and Kromhout, H. 2006. Exposure, health complaints and cognitive performance among employees of an MRI scanners manufacturing department. *Journal of Magnetic Resonance Imaging* 23: 197–204.

Eguchi, Y., and Ueno, S. 2005. Stress fiber contributes to rat Schwann cell orientation under magnetic field. *IEEE Transactions on Magnetics* 41: 4146–4148.

Eguchi, Y., Ogiue-Ikeda, M., and Ueno, S. 2003. Control of orientation of rat Schwann cells using an 8-T static magnetic field. *Neuroscience Letters* 351: 130–132.

Eguchi, Y., Ohtori, S., and Ueno, S. 2015. The effectiveness of magnetically aligned collagen for neural regeneration *in vitro* and *in vivo*. *Bioelectromagnetics* 36: 233–243.

Glazer, P. A., Heilmann, M. R., Lotz, J. C., and Bradford, D. S. 1997. Use of electromagnetic fields in a spinal fusion. A rabbit model. *Spine* 22: 2351–2356.

Glover, P. M., Cavin, I., Qian, W., Bowtell, R., and Gowland, P. A. 2007. Magnetic-field-induced vertigo: A theoretical and experimental investigation. *Bioelectromagnetics* 28: 349–361.

Grissom, C. B. 1995. Magnetic field effects in biology—A survey of possible mechanisms with emphasis on radical-pair recombination. *Chemical Reviews* 95: 3–24.

Hafeli, U. O., and Pauer, G. J. 1999. In vitro and in vivo toxicity of magnetic microspheres. *Journal of Magnetism and Magnetic Materials* **194**: 76–82.

Harguindey, S. 1982. Hydrogen ion dynamics and cancer: An appraisal. *Medical and Pediatric Oncology* **10**(3): 217–236.

Harkins, T. T., and Grissom, C. B. 1994. Magnetic field effects on B12 ethanolamine ammonia lyase: Evidence for a radical mechanism. *Science* **263**(5149): 958–960.

Higashi, T., Yamagishi, A., Takeuchi, T., Kawaguchi, N., Sagawa, S., Onishi, S., and Date, M. 1993a. Orientation of erythrocytes in a strong static magnetic field. *Blood* **82**: 1328–1334.

Higashi, T., Sagawa, S., Kawaguchi, N., and Yamagishi, A. 1993b. Effects of a strong static magnetic field on blood platelets. *Platelets* **4**: 341–342.

Higashi, T., Yamagishi, A., Takeuchi, T., and Date, M. 1995. Effects of static magnetic fields on erythrocyte rheology. *Bioelectrochemistry and Bioenergetics* **36**: 101–108.

Hudson, M. 2006. The EU Physical Agents (EMF) Directive and its impact on MRI imaging in animal experiments: A submission by FRAME to the HSE. *Alternatives to Laboratory Animals* **34**: 343–347.

Humphrey, C. E., and Seal, E. H. 1959. Biophysical approach toward tumor regression in mice. *Science* **130**: 388–389.

Ide, C. 1996. Peripheral nerve regeneration. *Neuroscience Research* **25**: 101–121.

International Commission on Non-Ionizing Radiation Protection. 1998. Guidelines for limiting exposure to time-varying electric, magnetic, and electromagnetic fields (up to 300 GHz). *Health Physics* **74**: 494–522.

International Commission on Non-Ionizing Radiation Protection. 2004. Medical magnetic resonance (MR) procedures: Protection of patients. *Health Physics* **87**: 197–216.

International Commission on Non-Ionizing Radiation Protection. 2009. Guidelines on limiting exposure to static magnetic fields. *Health Physics* **96**: 504–514.

International Commission on Non-Ionizing Radiation Protection. 2010. ICNIRP guidelines for limiting exposure to time-varying electric and magnetic fields (1 Hz to 100 kHz). *Health Physics* **99**: 818–836.

International Commission on Non-Ionizing Radiation Protection. 2014. Guidelines for limiting exposure to electric fields induced by movement of the human body in a static magnetic field and by time-varying magnetic fields below 1 Hz. *Health Physics* **106**(3): 418–425.

International Electrotechnical Commission. 2010. Medical electrical equipment—Part 2Y33: Particular requirements for the basic safety and essential performance of magnetic resonance equipment for medical diagnosis. Geneva: IEC; IEC 60601-2-33 ed 3.0.

Iwasaka, M., and Ueno, S. 2003a. Polarized light transmission of smooth muscle cells during magnetic field exposures. *Journal of Applied Physics* **93**: 6701–6703.

Iwasaka, M., and Ueno, S. 2003b. Detection of intracellular macromolecule behavior under strong magnetic fields by linearly polarized light. *Bioelectromagnetics* **24**: 564–570.

Iwasaka, M., Ueno, S., and Tsuda, H. 1994. Effects of magnetic fields on fibrinolysis. *Journal of Applied Physics* 75: 105–107.

Iwasaka, M., Miyakoshi, J., and Ueno, S. 2003. Magnetic field effects on assembly pattern of smooth muscle cells. *In Vitro Cellular & Developmental Biology—Animal* 39: 120–123.

Japan Society for Thermal Medicine. http://www.jsho.jp.

Kangarlu, A., and Robitaille, P. M. L. 2000. Biological effects and health implications in magnetic resonance imaging. *Concepts in Magnetic Resonance*, 12: 321–359.

Kangarlu, A., Burgess, R. E., Zhu, H., Nakayama, T., Hamlin, R. I., Abduljalh, A. M., and Robitaille, P. M. 1999. Cognitive, cardiac, and physiological safety studies in ultrahigh field magnetic resonance imaging. *Magnetic Resonance Imaging* 17: 1407–1416.

Kano, Y., Akutsu, M., Tsunoda, S., Mano, H., Sato, Y., Honma, Y., and Furukawa, Y. 2001. In vitro cytotoxic effects of a tyrosine kinase inhibitor STI571 in combination with commonly used antileukemic agents. *Blood* 97: 1999–2007.

Karpowicz, J., Gryz, K., Politański, P., and Zmyślony, M. 2011. Exposure to static magnetic field and health hazards during the operation of magnetic resonance scanners (in Polish). *Medycyna Pracy* 62: 309–321.

Kato, M. (Ed.) 2006. *Electromagnetics in Biology*. Springer.

Kotani, H., Iwasaka, M., Ueno, S., and Curtis, A. 2000. Magnetic orientation of collagen and bone mixture. *Journal of Applied Physics* 87: 6191–6193.

Kotani, H., Kawaguchi, H., Shimoaka, T., Iwasaka, M., Ueno, S., Ozawa, H., Nakamura, K., and Hoshi, K. 2002. Strong static magnetic field stimulates bone formation to a definite orientation in vivo and in vitro. *Journal of Bone and Mineral Research* 17: 1814–1821.

Marie, P. J., Debiais, F., and Haÿ, E. 2002. Regulation of human cranial osteoblast phenotype by FGF-2, FGFR-2 and BMP-2 signaling. *Histology and Histopathology* 17(3): 877–885.

McRobbie, D. W. 2012. Occupational exposure in MRI. *British Journal of Radiology* 85: 293–312.

Miklavcic, D., An, D., Belehradek, Jr., J., and Mir, L. M. 1997. Host's immune response in electrotherapy of murine tumors by direct current. *European Cytokine Network* 8(3): 275–279.

Miller, G. J., Burchardt, H., Enneking, W. F., and Tylkowski, C. M. 1984. Electromagnetic stimulation of canine bone grafts. *Journal of Bone and Joint Surgery, American* 66: 693–698.

Murthy, N. S. 1984. Liquid crystallinity in collagen solutions and magnetic orientation of collagen fibrils. *Biopolymers* 23: 1261–1267.

Nagakura, S., Hayashi, H., and Azumi, T. 1998. *Dynamic Spin Chemistry: Magnetic Controls and Spin Dynamics of Chemical Reactions*. Kodansha, Tokyo.

Nakajima, F., Nakajima, A., Ogasawara, A., Nakajima, A., Goto, K., Shimizu, S., Moriya, H., and Einhorn, T. A. 2007. Effects of a single percutaneous injection of basic fibroblast growth factor on the healing of a closed femoral shaft fracture in the rat. *Calcified Tissue International* 81: 132–138.

Nordenstrom, B. E. 1989. Electrochemical treatment of cancer. I: Variable response to anodic and cathodic fields. *American Journal of Clinical Oncology* 12(6): 530–536.

Ogiue-Ikeda, M., and Ueno, S. 2004. Magnetic cell orientation depending on cell type and cell density. *IEEE Transactions on Magnetics* 40: 3024–3026.

Ogiue-Ikeda, M., Kotani, H., Iwasaka, M., Sato, Y., and Ueno, S. 2001. Inhibition of leukemic cell growth under magnetic fields of up to 8 T. *IEEE Transactions on Magnetics* 37: 2912–2914.

Ogiue-Ikeda, M., Sato, Y., and Ueno, S. 2003. A new method to destruct targeted cells using magnetizable beads and pulsed magnetic force. *IEEE Trans Nanobioscience* 2: 262–265.

Ogiue-Ikeda, M., Sato, Y., and Ueno, S. 2004. Destruction of targeted cancer cells using magnetizable beads and pulsed magnetic forces. *IEEE Tansactions on Magnetics* 40: 3018–3020.

Okano, H. 2008. Effects of static magnetic fields in biology: Role of free radicals. *Frontiers in Bioscience* 1(13): 6106–6125.

Overgaard, J. 1989. Sensitization of hypoxic tumour cells—clinical experience. *International Journal of Radiation Biology* 56: 801.

Praemer, A., Furner, S., and Price, O. P. 1992. Musculoskeletal conditions in the United States. Rosemont, IL. *Journal of the American Academy of Orthopaedic Surgery* 85–91.

Reilly, J. 1998. *Applied Bioelectricity: From Electrical Stimulation to Electropathology.* Springer-Verlag, New York.

Reilly, J. P. 2002. Neuroelectric mechanisms applied to low frequency electric and magnetic field exposure guidelines—Part I: Sinusoidal waveforms. *Health Physics* 83: 341–355.

Ritz, T., Adem, S., and Schulten, K. 2000. A model for photoreceptor-based magnetoreception in birds. *Biophysics Journal* 78: 707–718.

Ritz, T., Thalau, P., Phillips, J. B., Wiltschko, R., and Wiltschko, W. 2004. Resonance effects indicate a radical-pair mechanism for avian magnetic compass. *Nature* 429: 177–180.

Sampath, T. K., Coughlin, J. E., Whetstone, R. M., Banach, D., Corbett, C., Ridge, R. J., Ozkaynak, E., Oppermann, H., and Rueger, D. C. 1990. Bovine osteogenic protein is composed of dimers of OP-1 and BMP-2A, two members of the transforming growth factor-beta superfamily. *Journal of Biological Chemistry* 265(22): 13,198–13,205.

Saunders, R. D., and Jefferys, J. G. 2007. A neurobiological basis for ELF guidelines. *Health Physics* 92: 596–603.

Schenck, J. F. 2000. Safety of strong, static magnetic fields. *Journal of Magnetic Resonance Imaging* 12(1): 2–19.

Schenck, J. F. 2005. Physical interactions of static magnetic fields with living tissues. *Progress in Biophysics and Molecular Biology* 87: 185–204.

Schmitt, J. M., Hwang, K., Winn, S. R., and Hollinger, J. O. 1999. Bone morphogenetic proteins: An update on basic biology and clinical relevance. *Journal of Orthopaedic Research* 17: 269–278.

Schulten, K. 1982. Magnetic field effects in chemistry and biology. In *Festkrperprobleme*, vol. 22. Treusch, J., ed. Vieweg, Braunschweig: 61–83.

Sersa, G., Miklavcic, D., Batista, U., Novakovic, S., Bobanovic, F., and Vodovnik, L. 1992. Anti-tumor effect of electrotherapy alone or in combination with interleukin-2 in mice with sarcoma and melanoma tumors. *Anti-Cancer Drugs* 3(3): 253–260.

Sersa, G., Jarm, T., Kotnik, T., Coer, A., Podkrajsek, M., Sentjurc, M., Miklavcic, D., Kadivec, M., Kranjc, S., Secerov, A., and Cemazar, M. 2008. Vascular disrupting action of electroporation and electrochemotherapy with bleomycin in murine sarcoma. *British Journal of Cancer* 98(2): 388–398.

Shellock, F. G., and Kanal, E. 1994. *Magnetic Resonance: Bioeffects, Safety, and Patient Management*. Raven Press, New York.

Shiga, T., Okazaki, M., Seiyama, A., and Maeda, N. 1993. Paramagnetic attraction of erythrocyte flow due to an inhomogeneous magnetic field. *Bioelectrochemistry and Bioenergetics* 30: 181–188.

So, P. P. M., Stuchly, M. A., and Nyenhuis, J. A. 2004. Peripheral nerve stimulation by gradient switching fields in magnetic resonance imaging. *IEEE Transactions on Biomedical Engineering* 51: 1907–1914.

Takano-Yamamoto, T., Kawakami, M., and Sakuda, M. 1992. Effect of a pulsing electromagnetic field on demineralized bone-matrix-induced bone formation in a bony defect in the premaxilla of rats. *Journal of Dental Research* 71: 1920–1925.

Togawa, T., Okai, O., and Oshima, M. 1967. Observation of blood flow E.M.F. in externally applied strong magnetic field by surface electrodes. *Medical and Biological Engineering* 5(2): 169–170.

Torbet, J., and Ronziere, M. 1984. Magnetic alignment of collagen during self-assembly. *Biochemical Journal* 219: 1057–1059.

Torbet, J., Freyssinet, J. M., and Hudry-Clergeon, G. 1981. Oriented fibrin gels formed by polymerization in strong magnetic fields. *Nature* 289: 91–93.

Ueno, S. 2012. Studies on magnetism and bioelectromagnetics for 45 years: From magnetic analog memory to human brain stimulation and imaging. *Bioelectromagnetics* 33(1): 3–22.

Ueno, S., and Iwasaka, M. 1994a. Properties of diamagnetic fluid in high gradient magnetic fields. *Journal of Applied Physics* 75: 7177–7179.

Ueno, S., and Iwasaka, M. 1994b. Parting of water by magnetic fields. *IEEE Transactions on Magnetics* 30: 4698–4700.

Ueno, S., and Sekino, M. 2006. Biomagnetics and bioimaging for medical applications. *Journal of Magnetism and Magnetic Materials* 304: 122–127.

Ueno, S., Tashiro, T., and Harada, K. 1988. Localized stimulation of neuronal tissues in the brain by means of paired configuration of time-varying magnetic fields. *Journal of Applied Physics* 64: 5862–5864.

Ueno, S., Iwasaka, M., and Tsuda, H. 1993. Effects of magnetic fields on fibrin polymerization and fibrinolysis. *IEEE Transactions on Magnetics* 29: 3352–3354.

Umeno, A., and Ueno, S. 2003. Quantitative analysis of adherent cell orientation influenced by strong magnetic fields. *IEEE Transactions on Nanobioscience* 2: 26–28.

Umeno, A., Kotani, H., Iwasaka, M., and Ueno, S. 2001. Quantification of adherent cell orientation and morphology under strong magnetic fields. *IEEE Transactions on Magnetics* 37: 2909–2911.

Vassilev, P. M., Dronzine, R. T., Vassileva, M. P., and Georgiev, G. A. 1982 Parallel assays of microtubules formed in electric and magnetic fields. *Bioscience Reports* **2**: 1025–1029.

Wada, S., Yue, L., Tazawa, K., Furuta, I., Nagae, H., Takemori, S., and Minamimura, T. 2001. New local hyperthermia using dextran magnetic complex (DM) for oral cavity: Experimental study in normal hamster tongue. *Oral Diseases* **7**(3): 192–195.

Wilén, J., and de Vocht, F. 2010. Health complaints among nurses working near MRI scanners—A descriptive pilot study. *European Journal of Radiology* **80**: 510–513.

World Health Organization. 1987. Environmental Health Criteria 69: Magnetic Fields. Geneva, Switzerland.

World Health Organization. 2006. Environmental Health Criteria 232: Static Fields. Geneva, Switzerland.

World Health Organization. 2007. Environmental Health Criteria 238: Extremely Low Frequency (ELF) Fields. Geneva, Switzerland.

Yamagishi, A., Takeuchi, T., Higashi, T., and Date, M. 1992. Diamagnetic orientation of blood cells in high magnetic field. *Physica B* **177**: 523–526.

Yamaguchi, S., Ogiue-Ikeda, M., Sekino, M., and Ueno, S. 2004. The effect of repetitive magnetic stimulation on the tumor development. *IEEE Transactions on Magnetics* **40**: 3021–3023.

Yamaguchi, S., Ogiue-Ikeda, M., Sekino, M., and Ueno, S. 2005. Effects of magnetic stimulation on tumors and immune function. *IEEE Transactions on Magnetics* **41**: 4182–4184.

Yamaguchi, S., Ogiue-Ikeda, M., Sekino, M., and Ueno, S. 2006a. The effect of repetitive magnetic stimulation on tumor and immune functions in mice. *Bioelectromagnetics* **27**: 64–72.

Yamaguchi, S., Sato, Y., Sekino, M., and Ueno, S. 2006b. Combination effects of the repetitive pulsed magnetic stimulation and the anticancer agent imatinib on human leukemia cell line TCC-S. *IEEE Transactions on Magnetics* **42**: 3581–3583.

Yamaguchi-Sekino, S., Sekino, M., and Ueno, S. 2011. Biological effects of electromagnetic fields and recently updated safety guidelines for strong static magnetic fields. *Magnetic Resonance in Medical Sciences* **10**: 1–10.

Yamaguchi-Sekino, S., Nakai, T., Imai, S., Izawa, S., and Okuno, T. 2014. Occupational exposure levels of static magnetic field during routine MRI examination in 3T MR system. *Bioelectromagnetics* **35**: 70–75.

Yanase, M., Shinkai, M., Honda, H., Wakabayashi, T., Yoshida, J., and Kobayashi, T. 1997. Intracellular hyperthermia for cancer using magnetite cationic liposomes: Ex vivo study. *Japanese Journal of Cancer Research* **88**: 630–632.

Yiu, G., and Zhigang, H. 2006. Glial inhibition of CNS axon regeneration. *Nature Reviews Neuroscience* **7**: 617–627.

Yoo, J. L., Kim, H. R., and Lee, Y. J. 2006. Hyperthermia enhances tumour necrosis factor-related apoptosis-inducing ligand (TRAIL)-induced apoptosis in human cancer cells. *International Journal of Hyperthermia* **22**(8): 713–728.

Effects of Radio Frequency Magnetic Fields on Iron Release and Uptake from and into Cage Proteins

Oscar Cespedes and Shoogo Ueno

CONTENTS

8.1 INTRODUCTION

8.1.1 Interactions of Radio Frequency Magnetic Fields with Iron Release and Uptake

Most biological components are diamagnetic or paramagnetic and only interact weakly with magnetic fields. For this reason, the effect of nonionizing radio frequency (RF) radiation in biology is commonly described in terms of heating.[1] However, iron is an essential element to most living organisms, including, of course, humans. A healthy adult needs between 10 (males) and 18 mg (females) of iron per day, the highest concentration of any metallic element.[2] Although it has a crucial role in biochemical mechanisms such as oxygen transport and redox processes, iron is also a potential toxic agent to cells via the Fenton reaction. Therefore, it requires a complex set of regulatory mechanisms to meet the demands of the body while avoiding accumulation or release in the wrong environment.[3]

A key constituent in the chemical control of iron is ferritin. This cage protein is the main form of storage for mineral iron in biology, to the point that the clinical assessment of iron content in blood is done via ferritin concentration. The role of the protein is to oxidize and store iron ions in the form of a ferrihydrite nanoparticle. Ferritin proteins have also been associated with the function of the mitochondria.[4,5] This emphasizes the role of the protein in protecting against oxidative damage, given that most of the iron must pass through the organelle to participate in biological processes.[4-6] The core nanoparticle can contain up to 4500 iron ions, which results in a high magnetic moment and a superparamagnetic magnetization at room temperature. The only components found in biology with a higher magnetic moment than ferritin are mineral biogenic magnetite nanocrystals. These have also been considered as a possible transducer for field effects,[7,8] but no effect was observed in the mortality rates of magnetic bacteria when the nanocrystals were exposed to RF fields.[9,10]

Being the organic bioelement with the highest magnetic moment, it is perhaps not unexpected that ferritin shows strong effects even at weak magnetic fields and relatively low frequencies in the RF range. The inner superparamagnetic ferrihydrite nanoparticle increases its internal energy when exposed to alternating magnetic fields via Néel absorption/relaxation. This energy is irradiated to the surrounding proteic cage, altering the protein ability to uptake and release iron. Proteins exposed during several hours have iron intake and chelation rates up to a factor of 3 smaller than control samples although for concentrations not observed in physiological conditions. These results are examined more closely in the following sections, as they open new paths of research for biological effects of alternating magnetic fields.

8.1.2 Possible Medical Consequences and Applications of RF Magnetic Fields on Iron Biochemistry

In recent studies, we have shown that exposure to an RF magnetic field affects the ability of horse spleen ferritin to release and uptake iron. However, in ferritin without an inner superparamagnetic nanoparticle, RF magnetic fields have no effect. The results were attributed to the energy released by the inner superparamagnetic nanoparticle due to the magnetization lag.[11-13] If this hypothesis is correct, the effect of the RF magnetic field will depend not only on the applied field but also on the magnetic properties of the ferrihydrite nanoparticle inside ferritin. Proteins from different animal species, individuals, and organs have different magnetic properties.[14] In particular, although the susceptibility will be roughly proportional to the amount of iron in the body, the relaxation times and frequency dependence of power dissipation will vary greatly with the type of ferritin. From estimates of the power released, it is easy to see that the energy loss by the nanoparticle inside ferritin is unlikely to lead to a thermal effect at the concentrations usually found in the body, even at relatively large magnetic fields and frequencies. This is due to the relatively low susceptibility of ferrihydrite (as compared to, for example, magnetite nanoparticles) and to the low concentration of ferritin in healthy organisms. It is also difficult to estimate how much power is transmitted to the organic cage or how well the peptides may dissipate the energy into the surrounding medium. If the effects were to be significant for *in vivo* experiments, there are two possible considerations: Will the RF exposure result in impaired protein function? And, can we use this exposure to regulate iron chemistry?

As for the first question, any damage to the role of ferritin in biochemistry may have serious consequences to the health of an organism. For example, it could result in iron release in the brain, which would then lead to oxidative-stress related illnesses, e.g., some forms of dementia. However, *a priori* it seems that the frequency and amplitude of the fields that would be needed for these severe effects would have to be very large, unlikely to be found in any natural environment. These large RF fields would probably generate effects via heating of the bio-environment.

On the other hand, it is possible to generate a controlled exposure to high-frequency/amplitude magnetic fields without generating dipolar electric heat, see the numerous experiments in hyperthermia.[15–17] Similarly, it might be possible to expose certain regions in the body to alternating magnetic fields so that the iron release/uptake is done more slowly, which could be beneficial in case of chelating agents being present in the body, genetic misregulations, etc. Further research could also help to distinguish the mechanisms affecting iron release from those in iron uptake, in which case the magnetic field exposure could be used to stimulate one above the other.

8.2 IRON CAGE PROTEINS

8.2.1 Introduction to Ferritins and Their Physiological Role

The iron cage protein ferritin is present in living organisms from bacteria to humans. In order to make use of iron chemistry without generating radicals that can lead to oxidative stress, organisms need to capture the harmful Fe (II) ions, oxidize, and store them in a safe manner. This is done inside the peptidic cage of ferritin as a relatively harmless ferrihydrite nanoparticle ($9H_2O \cdot 5Fe_2O_3$) with small concentrations of P, Cs, Cd, Zn, Cu, and S. Ferritin will later deliver the iron ions under certain chemical signals or agents when it is needed for metabolic purposes. The proteic cage (apoferritin) is formed by 24 subunits about 180 peptides each and arranged in a 4-3-2 symmetry to form a roughly spherical cage of 440 kDa in mass and 12 nm in diameter. The inner cavity is roughly 8 nm in size, so it can store particles with up to 4500 iron ions. A typical ferritin protein contains of the order of 1000–1500 ions.[18] Usually, ferritin is found mostly in the liver and the spleen, the main places for iron storage, but in some pathologies, ferritin can also be found in high concentrations in other organs, such as the heart and the brain. In order to ensure that the iron is protected in the different biological environments, ferritin has evolved into a very stable and robust configuration. The peptidic arrangement only dissociates at extreme conditions with temperatures above 80°C or at pH extremes of 2.8 and 11.2, unlikely to be found in environments that host living organisms (see Figure 8.1).

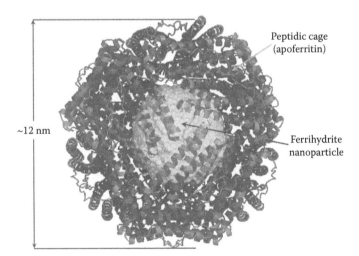

Peptidic cage (apoferritin)

~12 nm

Ferrihydrite nanoparticle

FIGURE 8.1 Ferritin is a protein for iron storage. It is composed of an outer organic cage (apoferritin) formed by 24 subunits of about 130 peptides each arranged in a 4-3-2 symmetry to form a rough sphere some 12 nm in size. Iron is stored in an inner ferrihydrite nanoparticle (ferric hydroxide: $5Fe_2O_3 \cdot 9H_2O$) up to some 8 nm in size.

8.2.2 Iron Release and Absorption

8.2.2.1 Spectroscopic Techniques in the Determination of Iron Contents

The methods to detect iron release from (or absorption into) ferritin are commonly based on optical titration experiments. During titration, the optical density (OD) of a ferritin solution with the desired chemicals is measured. In the case of iron release, an iron chelator should be added to the ferritin solution in order to extract the metal ions. This chelator can be ferrozine, 6-Hydroxidopamine (6-OHDA) or meso-2,3-dimercaptosuccinic acid (DMSA), for example. To test the effects of RF magnetic fields in ferritin, we measure the iron release rates using ferrozine, a molecule with a very strong affinity for Fe (II). When ferrozine is added to a ferritin solution, the molecules bind one iron ion per every three ferrozine molecules. Once it combines with the iron ion released by the protein, ferrozine acquires a very strong molar absorption coefficient at 562 nm of $\varepsilon = 29{,}700$ $M^{-1}cm^{-1}$. This gives a characteristic purplish color to the ferritin solution (originally red-yellow). By measuring the OD at that wavelength, we can determine the amount of Fe-ferrozine generated down to the nM scale, and hence the number of iron ions released. This reaction is usually slow because ferritin is naturally protected against strong chelators.[19] Release rates are typically of order 0.5 ions per ferritin and hour in our experiments.

Iron levels in all the chemicals employed are nominally below 0.1–1 ppm. Mixture of ferrozine with the solution without ferritin does not give rise to any measurable titration at the concentrations used in our study. The change in OD at 562 nm after 1 hour of adding the ferrozine is below our limit of detection 10^{-3} over 1 cm. This implies that the amount of trace or contaminant iron is $\lesssim 10$ nM. Addition of reducing agents (e.g., 6-OHDA) does not change this value. On the other hand, the iron concentration due to the ferrihydrite nanoparticles is of order 1–10 mM, at least 5 to 6 orders of magnitude more abundant than any iron impurity present in the solution. When ferritin is added, the typical change in OD at 562 nm we measure after 1 hour is of the order of 0.01 (1 μM) without reducing agents and 0.1–1 ($\gtrsim 100$ μM) with reducing agents, so trace iron cannot account for it.

The reducing agents 6-OHDA and DMSA can also be used to survey iron release from ferritin. They can be used in combination with ferrozine as optical marker or measured by themselves at wavelengths of 480 and 562 nm, respectively. However, the results we obtained using these chelators on their own were noisier and had worse reproducibility than using ferrozine. Owing to changes in the solution turbidity while the chemical are still not well mixed, we only considered the iron release and change in OD during the equilibrium state, a few minutes after adding the chelators.

Iron intake, on the other hand, does not need a chemical chelator or optical agent to be monitored. It can be followed by measuring the OD of a ferritin solution with iron (II) ions at 310 or 420 nm, with ε of 2745 and 550 $(mol \cdot cm)^{-1}$, respectively. In our experiments, the data is averaged over 14 to 20 samples and the error bars are the standard error of the mean. The use of buffers is generally avoided in order not to increase the solution conductivity and the subsequent heating effects during magnetic field exposure. The changes in OD during iron uptake are some 10 times smaller than when measuring iron release using ferrozine. For a typical measurement, this implies an increase of OD of 5×10^{-3} (10^{-3}) at 310 nm (420 nm) when all ferritin molecules incorporate one iron ion.

8.2.2.2 Function of the Symmetry Points in Iron Release and Uptake

Although the research is still ongoing, there is a more or less established consensus that the iron release in ferritins takes place through the three-fold symmetry point of the molecule. At these (eight) points, three subunits form the edges of a parallelepiped with a small pore in between them. This is where the weakest subunit interactions occur, with the three interacting subunits that surround the pore linked by just three side chain-main chain hydrogen bonds (lysine-glutamine).[20] This interaction is weaker than the pair or quadruple interactions in the two- and fourfold points, and much weaker than the bonds in each subunit, let alone the covalent bonds within each peptide and the atomic bonds within the ferrihydrite nanoparticle. Therefore, when considering the effects of RF magnetic fields, these threefold symmetry points are where the power released from the nanoparticle may be expected to result in more significant changes (see Figure 8.2).

Some iron chelators (e.g., 6-OHDA) can easily penetrate the protein through the empty space between the three protein subunits in order to reduce and release the ions. Ferrozine, owing to its relatively large size and negative charge in neutral pH, takes much longer to infiltrate inside the cage if at all. When considering aggregation and precipitation of the protein in solution, the threefold symmetry point plays as well an essential role: They act as hydrophilic terminals that ensure the solubility of the protein. This is due to the strong negative charge density at the pore, which

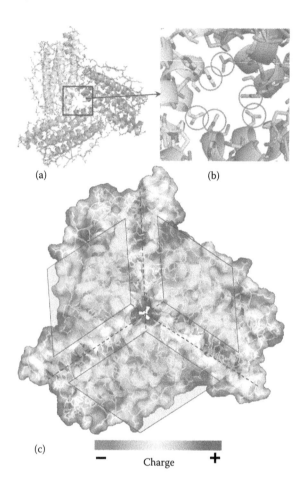

(a)   (b)

(c)

−   Charge   **+**

FIGURE 8.2   (a) Molecular representation of the polar threefold symmetry point, exit point for Fe (III) ions, and the protein aggregation center. (b) Zoom on the six carboxyl groups from the glutamic acids (COO⁻) that form the hydrophilic terminals at the threefold symmetry point. (c) Charge distribution at the symmetry point.

originates from the six carboxyl groups of aspartic and glutamic acids at the terminals. If the threefold point is blocked, for example, because of a rapid release and aggregation of iron at the pore, the proteins may aggregate and precipitate. The fourfold symmetry point is nonpolar and therefore does not contribute to molecular aggregation. Whereas iron release seems to take place predominantly or exclusively at the threefold symmetry point, iron uptake occurs probably at the two- and fourfold symmetry points.[21,22]

The process of ferroxidation by which iron ions are oxidized and later incorporated into the core ferrihydrite nanoparticle is similar to that of Fenton chemistry (see Section 8.2.2.3). Ferroxidase enzyme molecules in the proteic cage act as catalysts and binding agents to facilitate the oxidation and incorporation of iron (II). These ferroxidase centers are commonly found at the protein symmetry point with four helical bundles, although they can also be present in the two subunit links. For some bacterial ferritins, these centers may also be found at all 24 subunits. The role of the ferroxidase is to act as catalyst in the reactions:[23]

$$2Fe\ (II) + O_2 + 4H_2O \rightarrow 2Fe(O)OH_{nanoparticle} + H_2O_2 + 4H^+ \quad (8.1)$$

$$2Fe\ (II) + O_2 + H_2O_2 + 2H_2O \rightarrow 2Fe(O)OH_{nanoparticle} + 4H^+ \quad (8.2)$$

Ferritins without a ferroxidase center can still incorporate iron and form a mineral core, but the process is slower and these proteins can probably function only in environments low in iron and/or where it is not essential.

8.2.2.3 Role of Iron in Oxidative Stress and Dementia

Iron is needed for the good functioning of the body thanks to its polyvalent role in oxidation and reduction of electrons. However, it is a strong generator of oxidative stress via free radicals generated in the Fenton chemistry.[24–26] The simplest representation of the Fenton reaction is:[24]

$$Fe(II) + H_2O_2 \rightarrow Fe(III) + \left[H_2O_2^- \right] \rightarrow OH + OH^- \quad (8.3)$$

However, this mechanism is thermodynamically unfavorable and therefore unlikely to occur without the mediation of a ligand (L) such as EDTA, phosphate, or a biological chelator:

$$L\text{-}Fe\ (II) + H_2O_2 \rightarrow L\text{-}Fe(H_2O_2)^{2+} \rightarrow L\text{-}Fe\ (III) + OH + OH^- \quad (8.4)$$

or through the generation of an iron (IV) species:

$$\rightarrow L\text{-}Fe^{4+} + 2OH^- \quad (8.5)$$

or by oxidizing a substrate (R):

$$+R \rightarrow L\text{-}Fe\ (III) + OH^- + ROH \quad (8.6)$$

Given its relationship to oxidative stress and the formation of free radicals, it has been speculated that the presence of iron in the brain could be associated to neurodegenerative disorders. Recently, biomagnetite nanocrystals[27] and ferrihydrite[28] have been associated to Alzheimer's disease. For example, the superior temporal gyrus brain tissue from 11 patients with Alzheimer's disease and 11 age-matched control subjects shows an exponential correlation between the concentrations of the Fe (II)-ion-containing iron oxide, magnetite (Fe_3O_4), and the fraction of those particles that are smaller than 20 nm in diameter. There was also some potential evidence of particles being originated from ferritin proteins. A correlation between ferritin levels and onset of dementia has been studied,[29] although other factors, such as gender and age, make it difficult to find a straight correlation. The relationship between iron and dementia may simply be due to the role of iron in lipid peroxidation, e.g., during the breaking of the cell membrane by β-amyloid in Alzheimer's disease.[30]

Conditions such as neuroferritinopathy and Friedreich ataxia are associated with mutations in genes that encode proteins that are involved in iron metabolism, and as the brain ages, iron accumulates in regions that are affected by Alzheimer's and Parkinson's disease. High concentrations of reactive Fe (II) can increase oxidative stress-induced neuronal

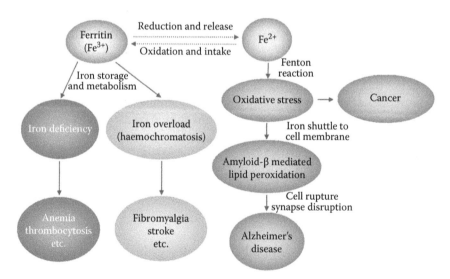

FIGURE 8.3 Potential physiological pathways for iron-related illnesses and the role of ferritin function.

vulnerability, and iron accumulation might increase the toxicity of the environment in neurodegenerative disorders.[31] To determine precisely the role of ferritin in these diseases, further methods to study and image the iron content *in vivo* would likely be needed, and also to separate iron in form of ferrihydrite within ferritin with free particles and/or stored in malfunctioning proteins with different oxidation states, i.e., as Fe (II) ions (see Figure 8.3). Studying the accumulation and cellular distribution of iron during ageing must be correlated also with protein and genomic deficiencies.

From the point of view of RF effects, we have seen no evidence that, at least at the amplitudes and frequencies we have used, the functioning of ferritin could result in an increased release of harmful iron ions. On the contrary, it would seem that iron release is slowed by the magnetic fields more significantly that iron absorption, which would result in a net iron deficiency for the experimental conditions and chemicals used. Even then, the concentrations used for our study are orders of magnitude above those found in any healthy organism. However, studies into the form of the iron stored (i.e., the oxidation state), the size, and the composition of the particles and exposures under different exposure conditions would be needed. Furthermore, therapies to regulate iron biochemistry using RF magnetic fields could be considered.

8.3 MAGNETIC PROPERTIES OF FERRIHYDRITE BIONANOPARTICLES

8.3.1 Nanoscale Magnetism

Here, we give a brief description of magnetic phenomena in iron oxide magnetic nanoparticles—and, in particular, the characteristics of the ferrihydrite nanoparticle inside ferritin. The physics of these materials have been studied in depth, in particular, since the discovery of their application for hyperthermia, drug delivery, and other biochemical processes.[15,32–34] Ferritins, or rather the inner ferrihydrite nanoparticles, are conventionally characterized through superconducting quantum interference device (SQUID) magnetometers and Mössbauer spectroscopy. The reason to use these techniques is partly due to the very small moment of the samples used, that requires a highly sensitive magnetometry tool, and the fact that iron is the magnetic ion in ferrihydrite. Here, magnetization and AC susceptibility measurements are taken from liquid-frozen samples using a SQUID.

8.3.1.1 Superparamagnetism

In order to minimize their internal energy, ferro-, ferri-, and anti-ferromagnetic materials form domains with aligned magnetization. These domains can have sizes from ~0.1 to 100 μm, depending on the exchange energy, temperature, and magnetic anisotropy of the material. If the magnetic material is nanostructured below the single domain size, its magnetic behavior is different from that of conventional ferromagnets or paramagnets. The particle behaves then as a single macrospin whose direction fluctuates owing to the thermal energy but can also be aligned in a magnetic field. Below a certain temperature, the magnetization direction at the measuring speed is blocked, and the hysteresis loop has a net remanence and coercivity different from zero. Above the blocking temperature, the nanomaterial has no coercivity or remanence but still has a very large susceptibility well above that of conventional paramagnets, hence the name of superparamagnetic particles. Different from paramagnets with a constant susceptibility, superparamagnets above the blocking temperature display two different susceptibilities for low and high magnetic fields, the characteristic sigma shape.

One of the most common methods to study superparamagnetic nanoparticles is to measure the difference between zero field cooled and field cooled low field susceptibility. If the particles are cooled with no field applied below the blocking temperature, only a small percentage of the particles will align once a small field is applied. As the temperature increases, more particles can surmount the energy barrier with the aligned state and the magnetization increases with increasing temperature. Once the system reaches the blocking temperature, the magnetization reaches a maximum and then decays as the magnetic energy becomes weaker compared to the temperature. However, if the sample is cooled in a field, particles that are aligned with the field at high temperatures are frozen in the high magnetization state as we cross the blocking temperature, and, as we cool down the sample, more particles go into a high magnetization state. In the case of the ferrihydrite nanoparticles in ferritin, this blocking temperature is of the order of 10–20 K (see Figure 8.4).

8.3.1.2 Ferrimagnetism in Iron Oxides and Surface Effects

Most iron oxides and hydroxides are antiferromagnetic or ferrimagnetic materials. The highest magnetization is found in magnetite (Fe_3O_4), where

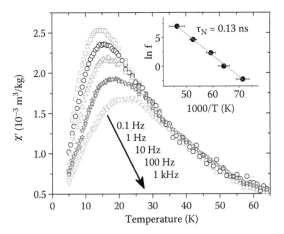

FIGURE 8.4 Real component of the susceptibility χ' and the dependence of the blocking temperature (T_B) with the measuring frequency for superparamagnetic ferrihydrite nanoparticles (ferritin solution). The dependence using Equation 8.11 leads to a Néel relaxation time τ_N of 0.13 ns ($\tau_0 = 3$ psec). Here, the AC field was set to 0.5 mT and DC field to zero.

uncompensated Fe (II) spins give rise to a magnetization of some 470 emu/cc above the Verwey transition $(T_V \sim 120$ K). Maghemite $(\gamma\text{-Fe}_2\text{O}_3)$ also has a high magnetization of ~400 emu/cc. Other iron oxides and hydroxides such as ferrihydrite in particular have a large saturation field and small magnetization arising from frustrated magnetic interactions and partly uncompensated spins.

In ferri/antiferromagnetic particle, the surface spins will not be matched by a layer with opposite magnetization. Surface canting of the spins, or misalignment due to shape and magnetocrystalline aniso-tropy, can take place. These uncompensated spins are not significant in bulk owing to the very small fraction of surface atoms. However, in nanometer-size particles such as the ferrihydrite found in ferritin, cant-ing and uncompensated spins contribute to a net magnetization of some 500–1500 μ_B at low temperatures and fields of several Tesla.[35] This is small compared to the magnetization of a magnetite particle of similar size (~10,000 μ_B at 10–100 mT), but it is enough to give rise to a measur-able susceptibility even at low fields of the order of mT.[14] At room tem-perature, the samples behave as weak superparamagnets, with a small but positive susceptibility.

8.3.2 Power Absorption and Dissipation

When a magnetic material is exposed to an alternating magnetic field, it absorbs and dissipates power. The amount of energy the magnet can store is equal to the area of its hysteresis loop, which is therefore called its energy product:

$$\Delta U = -\mu_0 \oint M \, dH \tag{8.7}$$

where μ_0 is the magnetic permeability in vacuum, H is the magnetic field intensity, M is the magnetization, and ΔU is the energy absorbed/dissipated per cycle.

In the case of nanoparticles, the physical description of the power absorption can be more complicated. Having no coercivity, the area of the hysteresis loop of a superparamagnetic nanoparticle is nominally zero and therefore should dissipate no energy. However, for sufficiently fast alternating magnetic fields, the nanoparticles' magnetization lags the external applied field. In that case, the imaginary component of the susceptibility is different from zero, and this results in an energy gain by the nanoparticle of:

$$\Delta U = 2\mu_0 H_0^2 \chi'' \int_0^{2\pi/\omega} \sin^2 \omega t \, dt \tag{8.8}$$

where χ'' is the imaginary component of the susceptibility and ω is the angular frequency of the applied magnetic field H_0.

As discussed in Section 8.3.1, the protein is a roughly spherical peptidic cage formed by 24 subunits of some 180 peptides each, arranged in a 4-3-2 symmetry and with an inner ferrihydrite nanoparticle of up to 4500 iron ions and 4 nm in radius inside. We will show in the following sections that RF magnetic fields have an effect on protein aggregation and iron release from ferritin at high concentrations. We hypothesize that a novel mechanism of interaction between RF magnetic fields and biomolecules is responsible for these changes. In this mechanism, the ferrihydrite nanoparticles dissipate power under an RF field, and, within a few hours after exposure, the released energy affects the protein functioning. Although the mechanism is based on the same power dissipation from nanoparticles as in hyperthermia experiments, the power loss is small and has not macroscopic (thermal) effect due to the small protein concentrations, which is of order of µM in our study and nM in the organism.

The power loss of a superparamagnetic nanoparticle in an alternating magnetic field depends on the frequency and amplitude of the applied field and on the magnetic susceptibility and relaxation time of the nanoparticle.[17,36] In the following sections we will show the magnetic properties of the ferritin samples used in our study and give an estimate for their dissipated power. This power is far from the values required to denaturalize or alter the structure of the protein, but here it might be able to promote higher dynamical states, i.e., low-energy vibrations, in the protein. Further evidence for protein alterations and the role of the inner nanoparticle is found from comparisons of iron uptake and release rates between control and RF exposed proteins.

8.3.2.1 Néel and Brownian Relaxation

The magnetization of a nanoparticle may follow the direction of an external AC field using one of two mechanisms: the so-called Néel and Brownian relaxations. In the first case, the magnetization rotates within the particle, whereas in the second one the particle itself rotates physically when following the applied magnetic field. Obviously, the particle can only do so if it is free to move, i.e., if it is in a solution. In the Néel mechanism, the particle does not move, but the spins rotate within it trying to align themselves with the AC field. The characteristic time that determines how fast the magnetization of the nanoparticle can rotate depends on the mechanism of relaxation of the nanoparticle and is expressed as $(\tau)^{-1} = (\tau_N)^{-1} + (\tau_B)^{-1}$, with τ_N and τ_B the Néel and Brownian relaxation times, respectively. The particles will always relax through the fastest mechanism, which depends on the size of the particle, its magnetic anisotropy and the viscosity of the medium.

The Néel relaxation time τ_N can be written as:[14,36,37]

$$\tau_N = \tau_0 \exp\left(\frac{KV}{k_B T}\right) \tag{8.9}$$

with τ_0 the characteristic attempt time, K the magnetic anisotropic energy, V the particle volume, k_B Boltzmann's constant, and T the temperature.

τ_0 and K can be found from the AC magnetic susceptibility measurements using the Néel-Arrhenius relation, which in our case can be expressed as:

$$T_B = \frac{K}{k_B \ln\left(\dfrac{f}{f_0}\right)} \tag{8.10}$$

with T_B the Blocking temperature, f_0 the attempt frequency ($1/\tau_0$), and f the frequency of the applied magnetic field. By plotting $\ln(f)$ versus $1/T_B$, K/k_B is the slope, and $\ln(f_0)$ the crossing with the abscise. In ferritin samples we obtain $\tau_0 \sim 3 \times 10^{-11}$ s and $K \sim 370$ K, which gives $\tau_N \sim 10^{-10}$ s (Figure 8.4).

On the other hand, the Brownian relaxation time τ_B is:

$$\tau_B = \frac{3\eta V_H}{k_B T} \tag{8.11}$$

with η the medium viscosity (~ 1 mPa $\cdot$ s^{-1}) and V_H the hydrodynamic particle volume ($\sim 10^{-25}$–10^{-24} m^3 for ferritin). The result is that $\tau_B \sim 10^{-7}$–10^{-8} s $\gg \tau_N \Rightarrow \tau_N \cong \tau$, so the relaxation mechanism for ferrihydrite nanoparticles in ferritin is of Néel type, i.e., the particle, and the protein around it, will not move will the magnetic field.

8.3.2.2 AC Susceptibility

If the magnetic field varies faster that the characteristic time for the magnetization (τ_N or τ_B), the particle finds itself in the "wrong" magnetic state after each cycle; that is, it is not aligned with the magnetic field. This represents an energy gain that is then dissipated in the form of spin waves and/or heat. From Equation 8.8, it can be derived that the power loss of a superparamagnetic nanoparticle in a magnetic field is:[36]

$$P = \pi\mu_0\chi_0 H_0^2 f \times \frac{2\pi f \tau}{1+(2\pi f \tau)^2} \tag{8.12}$$

where χ_0 is the DC magnetic susceptibility, f is the frequency of the applied field, and τ the characteristic time of the nanoparticle as seen in the previous section. The magnetic susceptibility of ferritin χ_0 at room temperature and low fields is of order 2.2×10^{-6} m^3 kg^{-1} (Figures 8.4 and 8.5).

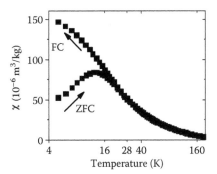

FIGURE 8.5   Zero field cooled – field cooled (ZFC – FC) magnetic susceptibility χ for a saturated ferritin solution. The curve is characteristic of superparamagnetic nanoparticles with a blocking temperature of ~15 K (note the log x-scale).

In order to calculate the dissipated power, the defining experimental parameter is the $\omega \cdot B$ product. For the experiments we performed, $\omega \cdot B$ is about 190 Ts^{-1}. They are operated at frequencies below 10 MHz (period of magnetic field $\gg \tau_N$), which gives a constant power of some 5–10 μWkg^{-1}. However, the magnetic susceptibility we use to calculate the values of the dissipated power is that measured in hysteresis fields of up to 10 mT, whereas the AC fields used here do not exceed 120 μT. There is as well a large uncertainty in the characteristic attempt time of the nanoparticle ($\tau_0 \sim 10^{-11}$–10^{-10}), which may also lead to lower relaxation times.

We must emphasize that the power calculated using Equation 8.12 is the power dissipated by the inner ferrihydrite nanoparticle and not the power absorbed by the medium or the protein. The power density is several orders of magnitude below the power usually generated in hyperthermia experiments with magnetite or maghemite because of the much smaller susceptibility of the ferrihydrite nanoparticle. It is roughly equivalent to the power dissipated by a nanoparticle with a magnetic moment of 630 μ_B,[17] in agreement with previous estimates for the low-coercivity magnetic moment of ferritin.[35,37]

8.4 MAGNETIC FIELD EFFECTS ON IRON RELEASE

8.4.1 Experimental Setup and Titration Measurements

For our experiments we use ferritin from equine spleen, which is the most readily available. The RF magnetic fields were generated via a power source connected to a set of Helmholtz coils. They are calibrated to produce 2.8838 mT per Ampere, with a deviation of less than 2 percent at the sample space. The fields vary from up to 120 μT at 250 kHz to 15 μT at 2 MHz. The fields are smaller at higher frequencies because the coils have higher impedance while the applied RF power is constant. The field magnitude was chosen to be high enough to generate an effect but well below the heating limit. The period of the magnetic field at those frequencies is below the lifetime of the radicals involved in Fenton reaction[38] but above the relaxation time of the ferrihydrite nanoparticle, which should give an invariant effect for constant frequency-field product. Other experimental details can be found.[11–13]

The electric field (E) in the ferritin solution is measured with an oscilloscope MS-5100A from Iwatsu Electric Co. (Tokyo, Japan) connected to electrodes dipped in the ferritin solution and spaced by 1 cm. It varies from 0.1 V/m (250 kHz) to 0.4 V/m (2 MHz). This variation may be due to changes in coil impedance and reflected power. Solution conductivity s is of order 10^{-3} S/m. The resultant specific absorption rate (SAR) is calculated as $\sigma/\rho\, E^2$, with ρ the medium density[39] and varies roughly from 10^{-8} to 10^{-7} W/kg. No heating effect is expected. The control and exposed samples' temperature remains equal within our 0.1°C limit of detection during the exposure and the titration measurements (no field exposure during the measurements). Samples are kept in the same light conditions. The coils are not around the sample but above and below. Background radiation (50 Hz to 10 MHz) is measured to be the same inside and outside the coils.

Optical titration measurements were carried out with 1 nm and 10^{-4} wavelength and OD resolution, respectively. Only the release in the equilibrium state, after the initial burst, is considered, as discussed in Section 8.4.2. Each titration measurement was carried out in 14 to 20 samples, half of which were control and half exposed. Error bars are the added standard error of the mean of the control and the exposed samples. Large deviations from the mean ($\gtrsim 40$ percent) are due to an error in adding the chelating agent to some samples and therefore discarded. All measurements reported hereafter were done after (not during) field exposure.

8.4.2 Changes in Iron Release

Here, we consider the iron chelation speed (iron release) when using ferrozine as only a reducing and chelating agent. Iron release measurements frequently use ferrozine as a color marker in combination with other agents such as 6-OHDA to facilitate the chelation of iron from the protein.[19] However, the use of several chemicals would at first complicate the data analysis. When ferrozine is added to a solution of ferritin proteins, the molecule may penetrate the proteic cage through the threefold symmetry points. The protein is then well protected against strong, negatively charged, iron chelators such as ferrozine. When using ferrozine as chelating agent, we found that samples previously exposed to magnetic fields of 1 MHz and 30 μT released up to 40 to 50 percent less iron than control samples.[11] We define the magnetic field effect on the change of iron released as $\Delta Fe_{released} \equiv (Fe_{released}|_{control} - Fe_{released}|_{exposed})/Fe_{released}|_{control}$, where $Fe_{released}|_{control}$ and $Fe_{released}|_{exposed}$ are the total iron released 1 hour after adding the chelating agent discounting the initial burst in control and exposed samples, respectively. This effect was dependent on several parameters, including the magnetic field amplitude and frequency, the exposure time, the ferritin and ferrozine concentration, and the pH of the solution.

The variation of the magnetic field effect with chemical concentrations of ferrozine and ferritin (Figure 8.6) can be explained as due to pH variations and changes in the intermolecular interactions. The pH of the ferritin solution decreases when the acidic ferrozine is added, which will result in iron reduction and increased rates of release. On the other hand, the protein itself and the NaCl solution in which it is dissolved act as a buffer to the pH change. Higher rates of iron release due lower pH will affect both control and exposed samples, decreasing the magnetic field effect. The reason for smaller effects at lower ferritin concentrations (1 μM) may

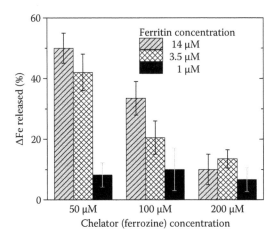

FIGURE 8.6 Changes in iron release after exposure to magnetic field as a function of the chemical concentrations of ferritin and ferrozine (chelating agent) used.

also be due to the buffer effect of the proteins or to a decrease in proteic interactions, which being energetically weaker, on the order of 8 kJM⁻¹ instead of 300 kJM⁻¹ for protein subunit interaction,[40,41] would be expected to be more affected by the irradiated energy.

To test this hypothesis, we studied the effect of a pH 7 buffer. We used solutions with 3.5 µM ferritin and 50 µM ferrozine concentrations. The magnetic field reduced the iron released by 53 ± 5% when 10 percent of pH 7 buffer was added after the magnetic field exposure, 27 ± 6% when the buffer was added before the magnetic field exposure and 42 ± 6% when no buffer was added. Adding the pH buffer after RF exposure rather than before ensures that the solution conductivity remains low. The effect of a temperature increase is to induce faster iron release, not slower as seen after exposure to RF magnetic fields, with rates 4 times higher in samples cultured at 45°C than at 25°C. Therefore, a temperature change induced by the RF fields would have the opposite effect to the one reported here. Adding buffer before exposure is actually when the effect is smallest. We hypothesize that this is because the pH buffer acts as a protecting agent against the effect of the magnetic field. On the other hand, the effect is higher when the pH buffer is added after the measurement; the pH buffer ensures that the iron release is not overly fast and avoids iron reduction due to low pH.

Because all these measurements are done after the RF magnetic field exposure, it is expected that the protein will eventually relax to its original state. In our measurements of ΔFe versus time, the maximum effect occurs 5 to 30 min after the addition of the ferrozine and afterward the effect decays slowly over several hours. As expected, the effect depends on the exposure time of the proteins to the RF magnetic fields. However, even though the effect initially increases from 17 percent after 1 hour exposure to 40 percent after 3 hours, it eventually decreases once the exposure exceeds 5 hours, dropping to 22 percent after being exposed for nine hours. This is due to a degradation of the protein and/or the ferrihydrite core when the solutions have been kept for long time at room temperature without physiological conditions. The degradation can be observed by comparing the amount of iron chelated in proteins freshly prepared with that of older solutions: about 4 times more iron is chelated from proteins left at in solution at room temperature after nine hours than when the solution was fresh.

The effect has a power dependence on the applied magnetic field: ΔFe $\propto$ $B^{1/2}$. Still, it is invariant at constant $\omega \times B$, with ω the angular frequency of the field. On the other hand, if the effect depended on the electrical field, such as due to induced polarization changes, it would be expected to increase at higher frequencies because the measured electrical field generated by the coils at 15 µT and 2 MHz is 4 times higher than at 120 µT and 250 kHz (see Figure 8.7). The effect should also in that case vanish shortly after stopping the applied field, in a time scale similar to the dipolar relaxation, i.e., the relaxation of the electrical dipoles in the protein cage. In fact, the equivalent effect obtained for constant products is in agreement with a mechanism mediated by the RF power dissipation of superparamagnetic nanoparticles as calculated in Equation 8.12.

Assuming that the energy irradiated by the nanoparticle is responsible for the effects in iron chelation, it remains the question of what are the changes induced by the irradiated power that lead to the slower release

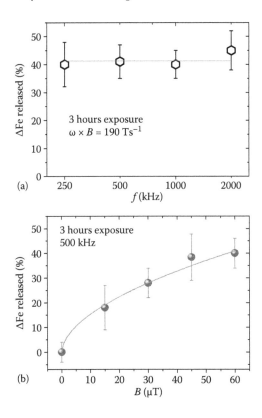

FIGURE 8.7 Dependence of the change in chelated iron after 3 hours of RF magnetic field exposure as a function of (a) the field frequency maintaining the frequency × amplitude product constant and (b) with the magnetic field amplitude.

rates. This release rate by ferrozine is determined by the chelation of iron and the dynamics of the ferrozine molecules going in and out of the proteic cage; i.e., the rate is probably decided by how fast ferrozine can penetrate in the proteic cage and reduce iron atoms from the ferrihydrite nanoparticle or by how many iron ions it can obtain from the proteic cage itself without penetrating. Therefore, the effect must originate either in the nanoparticle or in the entry-exit process (see Figure 8.8).

We then consider the potential effect that the RF magnetic field may have on the eight pores at the threefold symmetry axis of the protein, the entry and exit point for ferrozine and discussed in Section 8.2.2.2. In Fourier-transformed infrared measurements (FTIR) of ferritin solutions exposed to RF fields, we do not observe any significant change in the protein structure. This was expected, as the energy released by the nanoparticle should not be sufficient to break any proteic link. It is possible, however, that the effect of the power release affects not the structure but the charge distribution in or around the protein pores. Because ferrozine is a negatively charged molecule, a change in the charge distribution due to the RF field would result in an effect in the ferrozine chelation rate.

The effect of RF magnetic fields on the charge state of the threefold symmetry points is supported by aggregation experiments. The carboxyl groups at the threefold points form hydrophilic terminals that increase

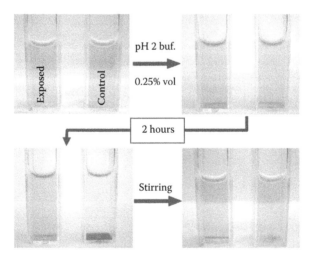

FIGURE 8.8 The threefold point act as hydrophilic terminals essential to protein solubility. Surface protonation and a sudden release of iron via pH reduction and iron reducing agents lead to blocking of the negative terminals and protein precipitation. This effect is quenched in solutions exposed to RF magnetic fields.

protein solubility. When the pH of a ferritin solution is lowered to a value of 3 to 4 in the presence of a high concentration of iron chelators or other iron reducing agents, surface protonation, and the fast release of Fe (II) block the negative carboxyl groups, which in turn lead to the aggregation and precipitation of the proteins in the solution.[42,43]

In ferritin solutions exposed to the magnetic fields of 1 MHz and 30 μT, the blocking of the hydrophilic terminals in conditions of low pH and high chelators concentration takes longer or does not happen at all unless the pH is further lowered. To generate aggregation and precipitation, exposed samples require 22 percent volume of pH 2 buffer in comparison with 4 percent for control samples (1 mM ferrozine in both cases). The effect depends on the exposure time and is accumulative.

8.4.3 Effects of Reducing Agents

So far, we have discussed only the changes in the iron release rate after adding ferrozine as chelating agent. However, the action of different chelators may be affected by the RF fields differently. The 6-Hydroxidopamine (6-OHDA), in particular, is an interesting iron reducing agent given its role in models of Parkinson's disease.[44–46] In contrast with ferrozine, 6-OHDA is a small, positively charged molecule that chelates iron by entering into the proteic cage with relative ease.[19] If an iron reducing agent such as 6-OHDA is added to a ferritin-ferrozine mixture, the Fe (III) in the ferrihydrite nanoparticle is reduced to Fe (II) and leaves the protein. It can then be easily bound, without the ferrozine going inside the peptidic cage. Therefore, the chelation rates are much faster and depend on the entry of the reducing agent and the reduction rates (see Figure 8.9).

Differently from using ferrozine on its own, the RF magnetic field does not change the iron release rate induced by 6-OHDA.[12] Because ferrozine is negatively charged, whereas 6-OHDA is positive, the difference in RF magnetic field effect may point in the direction of a mechanism mediated by the charge distribution at the protein. When testing another negatively charged iron chelator (Meso 2,3-dimercaptosuccinic acid or DMSA), we observe again that iron release is again slower for proteins exposed to RF magnetic fields, although the effect is weaker in DMSA. The effect of the RF magnetic fields cannot be in the chelating agents directly, because these are added after the exposure. Therefore, we may conclude that the RF magnetic field effect can indeed be related to the charge state of the protein and, in particular, to the threefold symmetry pore. Chelators with positive or neutral charge that can gain easy access to the inner nanoparticle such

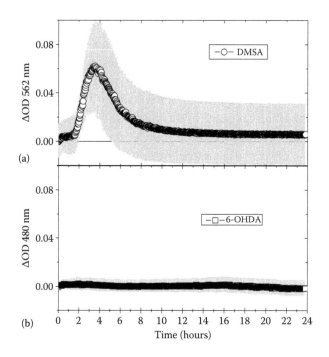

FIGURE 8.9   Changes in the optical absorption after the addition of (a) meso 2,3-dimercaptosuccinic acid (DMSA) and (b) 6-hydroxydopamine (6-OHDA) between control and samples exposed for 5 hours to fields of 1 MHz and 30 µT.

as 6-OHDA are not affected by exposure to RF magnetic fields; at least for the frequencies and amplitudes employed in our experiments.

8.5 MAGNETIC FIELD EFFECTS ON IRON UPTAKE

8.5.1 Raman Spectroscopy during RF Exposure

The horse spleen ferritin samples have a magnetic susceptibility at room temperature of 2.2×10^{-6} m^3kg^{-1} and their Néel relaxation time is of 0.1 nsec, with a characteristic attempt time of 3×10^{-11} sec. Using the expression for released power by a magnetic nanoparticle in Equation 8.12 gives a total energy dissipation of some 20 JM^{-1} after a typical exposure of three hours in fields of 1 MHz and 30 µT. Because this energy is much smaller than the intermolecular and subunit cohesion energies,[40,41] it is not expected that the magnetic fields will damage the structure of the protein, and no changes are observed in the IR spectra at 1000–4000 cm^{-1} of ferritins exposed to RF fields. However, this energy could be enough to increment the population of low-energy vibrational states. It has been shown that, given spatiotemporal symmetry, an increase in the population of a vibrational mode may result in a growth or broadening in the anti-Stokes region of the Raman spectrum associated with the excited state.[47] In order to look for changes in the protein structure and dynamics, we took Raman measurements of ferritin solutions before, during, and after an RF magnetic field was applied.[13] The samples had concentrations of 30–50 mg/mL. Although intensity filters were used to reduce the laser intensity and avoid sample heating, samples were left for three hours exposed to the laser for thermalization before the measurement. Temperature in the chamber was constant during the experiment. We observe such an asymmetric increase in the low-energy region of the anti-Stokes spectrum measured during the application of a magnetic field (1 MHz and 30 μT). The change is of up to 13 percent of the initial spectrum at low frequencies. On the other hand, the Stokes spectrum shows a smaller and less defined growth at frequencies below 80 cm^{-1}. For a rough calculation of the energy absorbed by the proteic cage, the low-frequency anti-Stokes Raman spectrum appears displaced by about 0.6 cm^{-1}, compared to some 0.03 cm^{-1} for the Stokes spectrum. This is equivalent to an energy $E = h\nu \cdot 6 \times 10^{23} \approx 7$ J/M. Our estimate for the dissipated power would be of the order of 1–2 J/M for a 10-min exposure (Raman measurement time). The disagreement may be due to an underestimation of the relaxation time, which could lead to larger power loss.

The vibrational frequencies with increased population could be characteristic of breathing or backbone vibration modes involving many atoms. These low-frequency modes in molecular dynamics are still about 6 orders of magnitude faster than the applied fields, 0.1–1 THz vibrations compared with 0.1–1 MHz fields. The temperature during the experiment remains constant and the Raman intensity goes back to its original value once the field is removed. As the exposure time is increased, the energy dispersed through vibrations and/or Brownian motion can lead to changes in protein function that remain once the field is removed. However, the Raman spectrum of the ferritin solution was dominated by a fluorescence signal and very sensitive to the experimental conditions, so further work is needed in order to confirm a change in the protein dynamics.

8.5.2 Changes in Iron Uptake

There is also an effect of RF magnetic fields on the ability of ferritin to uptake Fe (II) ions, although it is weaker than for iron chelation. In these experiments, Fe (II) ions were added to ferritin/apoferritin solutions as ammonium iron sulfate 6-hydrate (Mohr's salt). When iron ions added to a ferritin solution are uptaken by the protein, there is an increase in the optical absorption of the solution. The changes in OD at 310 and 420 nm have been calibrated as ε = 2745 and 550 per mol of iron uptaken and centimeter of solution, respectively.[48,49] The temperature of the samples, measured during and after the field exposure and during the titration measurements, is once again equal for control and exposed samples within 0.1°C. Measurements are done after (not during) field exposure.

As it was the case for iron release (Section 8.4.2), we define the magnetic field effect on the change of iron uptaken by the protein as $\Delta Fe_{uptake} \equiv$ $(Fe_{up}|_{control} - Fe_{up}|_{exposed})/Fe_{up}|_{control}$, where $Fe_{up}|_{control}$ and $Fe_{up}|_{exposed}$ are the concentrations of iron uptaken 1 hour after adding the Fe (II) ions in control and exposed samples, respectively. We found that the ability of ferritin to oxidize and store iron is reduced after being exposed to RF magnetic fields, i.e., $\Delta Fe_{uptake} > 0$. The change is a function of the molecular concentrations, exposure time and $\omega \cdot B$ product (Figures 8.10 and 8.11). The iron absorbed in 3.5 μM ferritin solutions exposed to field of 30 μT (H = 25 Am^{-1}) at 1 MHz for five hours (total energy released ~ 35 JM^{-1}), is 20 ± 10% smaller than for control samples for added iron concentrations between 0.25 and 1 mM. However, for the same conditions in apoferritin (protein without inner ferrihydrite nanoparticle), the effect is not significant: ΔFe_{uptake} (apoferritin) = 1 ± 4%. The effect depends on the relative ferritin-iron concentrations, and it has not statistical significance for large Fe (II) concentrations (~1 mM). The fact that there is no effect in apoferritin

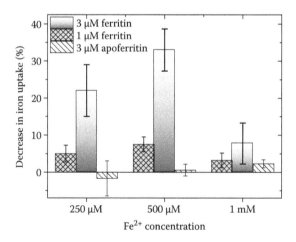

FIGURE 8.10 Changes in iron uptake after exposure to magnetic field as a function of the chemical concentrations of iron cations and ferritin.

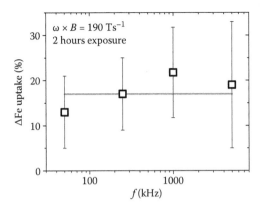

FIGURE 8.11 Changes in iron uptake 2 hours exposure to RF magnetic fields as a function of the field frequency maintaining the frequency × amplitude product constant.

reinforces our hypothesis that the mechanism of interaction is mediated by the nanoparticle: in proteins without a ferrihydrite core, there is no effect.

In agreement with our hypothesis of a mechanism governed by Equation 8.1, the effect remains roughly the same if the product $\omega \cdot B$ is constant, even over large frequency ranges: $\Delta Fe_{uptake} \propto \omega \times B$ when the frequencies are such that $\omega^{-1} \ll \tau_N$. The relatively weak effect measured at low ferritin and/or high iron concentrations may be due to the rapid decrease of the effect after the iron salt is added from an initial 50 to 17 percent after one hour. For iron release, the decrease over one hour was much smaller, going from 55 to 42 percent. The rapid decrease of the effect in iron uptake results in a comparatively smaller effect after one hour at high field-frequency products and a nonsignificant effect at field-frequency products below 190 Ts^{-1}. However, the results shortly after adding the iron are not reliable owing to the small signal measured in the first minutes and the initial changes to the solution optical density when mixing Mohr's salt. The larger difference in the early stages of the experiment may be due to a loss power effect over the proteic region responsible for the early formation of the diferric peroxo complex[50] rather than an effect over the incorporation of the iron ions to the nanoparticle.

8.6 ROLE OF THE THREEFOLD SYMMETRY POINT AND FUTURE CONSIDERATIONS

8.6.1 Effects of Cationization at the Threefold Symmetry Point

Our hypothesis to explain the effect of the RF fields on the Fe (III) release, Fe (II) uptake and protein solubility is related to the charge distribution at the threefold symmetry point, As we saw in the previous section, the pore at the center of the three subunits has a net negative charge around it because of the presence of six carboxyl groups in the peptides at the union, each subunit contributing with one glutamic and one aspartic acid. Owing to this negative charge, it also has the characteristic of being able to trap cations, so in a ferritin sample, this negative charge is, at least partially, screened by the solution and/or compensated by cations, so if the irradiated power has altered the charge distribution around the pore, the entrance of charged molecules through it will also be affected. The pore is involved in iron exchange, and, in some cases, it is the entry point for chelating agents, so it has an essential role in the protein function as studied here.

Ferrozine is a large, acidic molecule that would find difficult to penetrate the proteic cage, although diffusion over long periods of time may be possible.[19] Another possibility is that other acidic molecules, say, from the pH buffer employed, act as reducing agents, liberating the Fe (II) ions that are then bound by three ferrozined molecules. Directly after adding the ferrozine, there is a sudden increase in the OD of the ferritin solution. This is probably due in part to a general increase of the absorbance of the solution with the added chemicals and partly to the quick combination of ferrozine with Fe (II) ions present in the solution. These ions may have been previously reduced by the acidic molecules in the buffer. As mentioned in Section 8.4.2, we do not take into account this initial burst in our results because it is too fast to measure accurately; this takes place in a few seconds. However, we find that this initial ratio of formation of the iron-ferrozine complex is roughly the same for control and exposed samples. As the chelator reaches equilibrium with the ferritin, i.e., when the ferrozine may penetrate inside the cage, the release in exposed samples falls behind. The effect reaches a maximum after some 20 min and then slowly decays over several hours. It is therefore possible that the magnetic field reduces the screening effect of the solvent or removes the attached metallic ions, increasing the effective negative charge at the threefold symmetry axis and making more difficult for ferrozine and other acidic molecules to penetrate through the pore, delaying iron release in particular at low pH.

However, small changes in the charge distribution around the pore, or in the solvent ionic screening, can greatly modify the entry times of charged molecules such as ferrozine.

In cationized ferritin,[12,51] the six carboxyl groups from the glutamic acids that form the hydrophilic terminals at the threefold symmetry point are neutralized with N,N dimethyl-l,3propanediamine (NNPD). The NNPD could be functionalized to other carboxyl or negative molecules at the protein surface, but only at the threefold symmetry pore are there six free COO− groups. Because the charge distribution after cationization is different, changed from negative to neutral, an effect of the RF magnetic fields on the charge distribution around the threefold symmetry point should now give a different result. The entry pore will now also be smaller owing to the attached molecules that reduce the pore frame. This results in slower chelation rates by up to a factor 2 to 3 (see Figure 8.12).

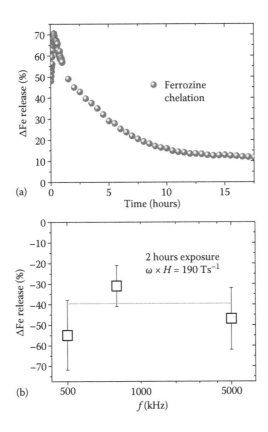

FIGURE 8.12 (a) The effect has a maximum just 20 min after the addition of the iron chelator (note that the magnetic field exposure is stopped previous to the measurement) and then decays slowly over several hours. (b) Effect of RF magnetic fields in the iron release from cationized ferritin when using ferrozine as chelating agent. The change has the opposite sign than for natural ferritin with negatively charged threefold symmetry points.

We find that the exposure to the same fields that resulted in a factor 2 slower release rate in ferritin (500 kHz to 2MHz; 60–15 µT), now result in faster release rates from exposed ferritin. Furthermore, the effect now extends over longer periods of time, with a roughly constant increase of 30 to 50 percent one day after the magnetic field exposure has been stopped and the ferrozine added. Following our initial argument, we would conclude that the charge redistribution induced by the magnetic effect has now the effect of making it easier for the negative ferrozine to penetrate the proteic cage across the now neutralized carboxyl groups.

8.6.2 Fluorescence Determination of Iron Contents

The ferrihydrite nanoparticle inside ferritin is not fluorescent, but it absorbs the photons emitted by the organic fluorophores at the peptide cage (see Table 8.1 and Figure 8.13), quenching the photoemission. This can be used to determine the iron content inside the protein. Usually, physiological iron levels are measured in function of ferritin concentration. To test whether the changes in the rates of iron chelation are due to changes in the oxidation states of the iron ions or in the amount of iron inside the nanoparticle, we have used fluorescence measurements. The proteic cage apoferritin shows fluorescence because of almost equal contributions from tryptophan and tyrosine residues, with strong absorbance around 275 nm and emission at 325 nm. We have found that it is also possible to determine the average amount of iron inside the protein by measuring the fluorescence polarization. A reduction in the nanoparticles size induced by the RF magnetic fields would have as consequence an increase in the fluorescence of the proteins. However, measurements show no change in fluorescence emission or polarization after RF exposure. Therefore, we can conclude that the RF magnetic fields do not affect directly the size or iron content of the inner ferrihydrite nanoparticle.

TABLE 8.1   Fluorescent Characteristics of Peptides[a]

	Absorbance	Emission	Quantum Yield
Tryptophan 93 (peptide position)	280 nm (5600 relative absorbance)	300–348 nm	0.2 f
Tyrosine 12,18,22,24,30,168	274 nm (1400)	303 nm	0.14 f
Phenylalanine 39,41,54,82,132,137,170	257 nm (200)	282 nm	0.04 f

[a] After Gabor Mocz in http://dwb.unl.edu/Teacher/NSF/C08/C08Links/pps99.cryst.bbk .ac.uk/projects/gmocz/fluor.htm.

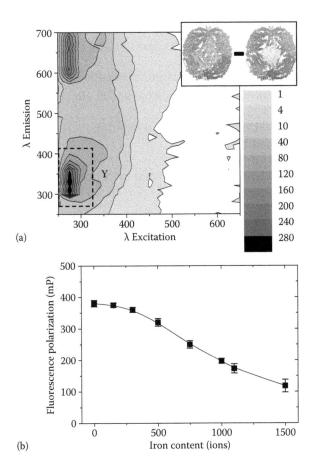

(a)

(b)

FIGURE 8.13 (a) Change in fluorescence between apoferritin and ferritin. (b) The fluorescence polarization as a function of the iron content in ferritin.

8.6.3 Summary of the Effects, Future Considerations to RF Magnetic Fields Exposure, and Their Effects on Iron Cage Proteins

The effects due to the suggested new mechanism of interaction between ferritin with RF fields can be resumed as follows:

- Reduction in the iron chelation rate by negatively charged molecules such as ferrozine and DMSA. No changes when the molecule used was 6-OHDA (positively charged).

- Stronger effects when the ferritin concentrations were high and/or the chelator concentrations were low.

- Increase in the iron chelation rate for cationized ferritin when using the same molecules; opposite effect to standard protein.

- Effect is increased when a pH buffer is added previous to the exposure and reduced when the buffer is added after the exposure.

- An increased protein solubility in exposed proteins when acidic pH buffers were added.

- A reduced rate in the oxidation and incorporation of Fe (II) ions for exposed proteins.

The effect does not seem to be thermal or electric, as the temperature in the exposed and control samples remain constant, and the effect does not depend on the electrical field or sample conductivity. Furthermore, changes due to an increased temperature would be opposite to those measured, e.g., increased iron released rates rather than decreased. The relationships with the pH of the solution and, in particular, the inverse effect observed in cationized ferritin indicate a mechanism mediated by the charge distribution in the protein and around the threefold symmetry point. At the same time, the constant effect for equivalent frequency-amplitude products point to an interaction mediated by the power dissipation of the superparamagnetic ferrihydrite nanoparticle.

Although the described effects only happen at large ferritin concentrations not found in healthy organisms, individuals with high levels of iron (hematomacrosis) or exposed to large fields may be affected. Given the role of Fe (II) in oxidative processes via the Fenton reaction, a protein malfunction would have serious consequences. This could be at the origin of some RF magnetic field effects in oxidative-stress related disorders.

Another hypothesis for a relationship between oxidative stress and magnetic fields arises from the well-established band splitting on free radicals in a magnetic field. However, reproducible experimental effects due to band splitting on free radicals have been measured only for fields of the order of, or above the hyperfine interaction, i.e., above some 1 mT (and generally fields of the order of 10–100 mT are employed). Some of the best experiments are performed at low temperatures to reduce the effect of the thermal energy (see, for example, the work of van Dijk and coworkers[52]). This kind of experiment with controlled conditions, only one chemical species, reproducible results and a clear dependence on magnetic field is, in our opinion, a better indication of the fields involved than experiments done for example on comet analysis for DNA strands, which are not a direct measurement of radical lifetime, have little reproducibility and many variables with a complex procedure.

Another consideration when using RF, rather than DC magnetic fields, is that once the period of the magnetic field is comparable with the life-time of the free radical, the band split effect decreases as the frequency increases. The lifetime of free radicals is for example 4 μsec ($\rightarrow$ 40 kHz) for a free radical involved in Fenton chemistry.[38,53] If our effect was due to changes in the Fenton reaction, as we increase the frequency from 250 kHz to 2 MHz, the effect should quickly decrease because of (1) the reduced averaged field over the lifetime of the radical and (2) the fields used at 250 kHz are 4 times higher than those at 2 MHz: from 60 to 15 μT. In fact, at frequencies of the order of 1 MHz, the averaged field over the radical lifetime would be zero, and no effect would be expected. However, we do measure an effect, and it remains constant over the frequency range. We think that if the effect was due to changes in the Fenton chemistry, the use of pH buffers would surely reduce it because the buffer would control the oxidation state of iron. We see in our measurements that it is not the case. Finally, the addition of chelating agents is done after the exposure.

Our measurements are performed at frequencies of 0.25 to 2 MHz, but ferrihydrite nanoparticles can have small relaxation times down to 10 psec.[14] In theory, the described mechanism could vary with ω·B up to much higher frequencies, with a dependence only on B once the period of the field is smaller than the relaxation time of the nanoparticle.[36] However, measurements at higher frequencies would be necessary to asses this hypothesis and the effects at higher frequencies. In the case of physiologi-cal experiments, the iron/ferritin levels together with the frequency and amplitude of the magnetic (not only electric) field should be considered. From this calculation, it is easy to see that the energy loss by the nanopar-ticle inside ferritin is unlikely to lead to a thermal effect at the concentra-tions usually found in the body, even at relatively large magnetic fields and frequencies. This calculation is nevertheless independent of the effects that the loss power may have on the proteic cage itself or on whether our suggested mechanism is indeed the correct explanation for the observed effects (see Figure 8.14).

We do not know what the implications could be. For example, our estimates calculate that to denaturalize human spleen ferritin we would need fields of the order of 100 mT at frequencies of 100 kHz operating for about 15 min, but these fields are one order of magnitude higher than the

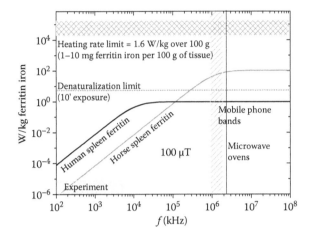

FIGURE 8.14 Differences in the effect of iron releasing agent using the time scale for a ferrihydrite nanoparticle with $\tau_N \sim 0.13$ nsec. The denaturalization limit assumes that all the power dissipated is absorbed by the protein, which is not likely and would need to be probed experimentally.

commonly employed in hyperthermia and the estimation depends on the magnetic characteristics of the nanoparticle. It would depend on the body region the relaxation time of the ferrihydrite nanoparticle varies with the organ, the amount of iron, etc. More studies would also be needed to determine how the protein dissipates the energy to the medium. We do not think it is likely that the effect could cause macroscopic heating in addition to the heat generated by the artificial nanoparticles, due to the relatively low ferritin iron concentrations; in the brain, the maximum is of some 200 ng/g of tissue, found in the globus pallidus.

Extrapolating to higher-frequency fields, such as those used by mobile phones, microwave ovens, and other applications, is just speculation. However, if the RF magnetic field effect described in the previous sections is confirmed and if its origin is indeed in the power released by the superparamagnetic nanoparticle, there is no physical reason why the mechanism we describe could not maintain the frequency field dependence up to or near to the GHz range. The period of the magnetic fields we employed in these research is below the lifetime of the radicals involved in Fenton reaction, but above the relaxation time of the ferrihydrite nanoparticle. This relaxation time is usually 10 to 100 psec for horse spleen ferritin but can go up to 10 nsec (~16 MHz) in thalassemic human spleen ferritin.[14] At periods above the relaxation time, the frequency field relation is maintained, but below the relaxation time of the nanoparticle only a field dependence would be expected, limiting the experiments we could do to confirm the mechanism. Given the characteristic time of the ferrihydrite nanoparticle, which can be as low as 10 to 100 psec in some ferritins but depends on many factors, even at higher frequencies the power dissipation would theoretically remain constant as long as the field amplitude was maintained. That is, in principle, the effect at 1 μT and 1 GHz would be approximately equivalent to also 1 μT at 10 GHz, but it could be similar to 10 mT at 100 kHz. Of course, at those frequencies other factors, such as molecular vibrations, would also play a role and may change the dissipation and absorption mechanisms. Furthermore, experiments *in vitro* with high ferritin concentrations may differ very significantly from tests *in vivo* and biological effects due to proteic relaxation, chemical buffer, and other environmental considerations.

ACKNOWLEDGMENTS

We thank *Bioelectromagnetics* for the permission to reproduce ideas and results as reported in References 11,13.

REFERENCES

1. Bavrnes, F. S. 2005. Mechanisms for electric and magnetic fields effects on biological cells. *IEEE Transactions on Magnetics* **41**: 4219–4224.
2. Mertz, W. 1981. The essential trace-elements. *Science* **213**: 1332–1338.
3. Beard, J. L. 2001. Iron biology in immune function, muscle metabolism and neuronal functioning. *Journal of Nutrition* **131**: 568S–579S.
4. Levi, S., Corsi, B., Bosisio, M., Invernizzi, R., Volz, A., Sanford, D., Arosio, P., and Drysdale, J. 2001. A human mitochondrial ferritin encoded by an intronless gene. *Journal of Biological Chemistry* **276**: 24,437–24,440.
5. Arosio, P., and Levi, S. 2002. Ferritin, iron homeostasis, and oxidative damage. *Free Radical Biology and Medicine* **33**: 457–463.
6. Carrondo, M. A. 2003. Ferritins, iron uptake and storage from the bacterioferritin viewpoint. *Embo Journal* **22**: 1959–1968.
7. Kirschvink, J. L. 1996. Microwave absorption by magnetite: A possible mechanism for coupling nonthermal levels of radiation to biological systems. *Bioelectromagnetics* **17**: 187–194.
8. Kirschvink, J. L., Kobayashikirschvink, A., and Woodford, B. J. 1992. Magnetite Biomineralization in the Human Brain. *Proceedings of the National Academy of Sciences of the United States of America* **89**: 7683–7687.
9. Cranfield, C. G., Weiser, H. G., and Dobson, J. 2003. Exposure of magnetic bacteria to simulated mobile phone-type RF radiation has no impact on mortality. *IEEE Transactions on Nanobioscience* **2**: 146–149.

10. Cranfield, C., Wieser, H. G., Al Madan, J., and Dobson, J. 2003. Preliminary evaluation of nanoscale biogenic magnetite-based ferromagnetic transduction mechanisms for mobile phone bioeffects. *IEEE Transactions on Nanobioscience* **2**: 40–43.

11. Cespedes, O., and Ueno, S. 2009. Effects of radio frequency magnetic fields on iron release from cage proteins. *Bioelectromagnetics* **30**: 336–342.

12. Cespedes, O., Inomoto, O., Kai, S., and Ueno, S. 2009. Effects of cationization and 6-hydroxydopamine on the reduced iron release rates from ferritin by radio-frequency magnetic fields. *IEEE Transactions on Magnetics* **45**: 4865–4868.

13. Cespedes, O., Inomoto, O., Kai, S., Nibu, Y., Yamaguchi, T., Sakamoto, N., Akune, T., Inoue, M., Kiss, T., and Ueno, S. 2010. Radio frequency magnetic field effects on molecular dynamics and iron uptake in cage proteins. *Bioelectromagnetics* **31**: 311–317.

14. Allen, P. D., St Pierre, T. G., Chua-anusorn, W., Strom, V., and Rao, K. V. 2000. Low-frequency low-field magnetic susceptibility of ferritin and hemosiderin. *Biochimica Et Biophysica Acta-Molecular Basis of Disease* **1500**: 186–196.

15. Pankhurst, Q. A., Connolly, J., Jones, S. K., and Dobson, J. 2003. Applications of magnetic nanoparticles in biomedicine. *Journal of Physics D-Applied Physics* **36**: R167–R181.

16. Jordan, A., Scholz, R., Wust, P., Fahling, H., and Felix, R. 1999. Magnetic fluid hyperthermia (MFH): Cancer treatment with AC magnetic field induced excitation of biocompatible superparamagnetic nanoparticles. *Journal of Magnetism and Magnetic Materials* **201**: 413–419.

17. Hergt, R., Andra, W., d'Ambly, C.G., Hilger, I., Kaiser, W.A., Richter, U., and Schmidt, H. G. 1998. Physical limits of hyperthermia using magnetite fine particles. *IEEE Transactions on Magnetics* **34**: 3745–3754.

18. Theil, E. C. 2011. Ferritin protein nanocages use ion channels, catalytic sites, and nucleation channels to manage iron/oxygen chemistry. *Current Opinion in Chemical Biology* **15**: 304–311.

19. Jameson, G. N. L., Jameson, R. F., and Linert, W. 2004. New insights into iron release from ferritin: direct observation of the neurotoxin 6-hydroxydopamine entering ferritin and reaching redox equilibrium with the iron core. *Organic & Biomolecular Chemistry* **2**: 2346–2351.

20. Gallois, B., d'Estaintot, B. L., Michaux, M. A., Dautant, A., Granier, T., Precigoux, G., Soruco, J. A., Roland, F., Chavas Alba, O., Herbas, A., and Crichton, R. R. 1997. X-ray structure of recombinant horse L-chain apoferritin at 2.0 angstrom resolution: Implications for stability and function. *Journal of Biological Inorganic Chemistry* **2**: 360–367.

21. Levi, S., Luzzago, A., Cesareni, G., Cozzi, A., Franceschinelli, F., Albertini, A., and Arosio, P. 1988. Mechanism of ferritin iron uptake-activity of the H-chain and deletion mapping of the ferro-oxidase site—A study of iron uptake and ferro-oxidase activity of human-liver, recombinant H-chain ferritins, and of 2 H-chain deletion mutants. *Journal of Biological Chemistry* **263**: 18,086–18,092.

22. Harrison, P. M., and Arosio, P. 1996. Ferritins: Molecular properties, iron storage function and cellular regulation. *Biochimica Et Biophysica Acta-Bioenergetics* **1275**: 161–203.
23. Arosio, P., Ingrassia, R., and Cavadini, P. 2009. Ferritins: A family of molecules for iron storage, antioxidation and more. *Biochimica Et Biophysica Acta-General Subjects* **1790**: 589–599.
24. Winterbourn, C. C. 1995. Toxicity of iron and hydrogen peroxide: The Fenton reaction. *Toxicology Letters* **82-3**: 969–974.
25. Valko, M., Morris, H., and Cronin, M. T. D. 2005. Metals, toxicity and oxidative stress. *Current Medicinal Chemistry* **12**: 1161–1208.
26. Stohs, S. J., and Bagchi, D. 1995. Oxidative mechanisms in the toxicity of metal-ions. *Free Radical Biology and Medicine* **18**: 321–336.
27. Pankhurst, Q., Hautot, D., Khan, N., and Dobson, J. 2008. Increased levels of magnetic iron compounds in Alzheimer's disease. *Journal of Alzheimers Disease* **13**: 49–52.
28. Everett, J., Cespedes E., Shelford, L. R., Exley, C., Collingwood, J. F., Dobson, J., van der Laan, G., Jenkins, C. A., Arenholz, E., and Telling, N. D. 2014. Evidence of redox-active iron formation following aggregation of ferrihydrite and the Alzheimer's disease peptide beta-amyloid. *Inorganic Chemistry* **53**: 2803–2809.
29. Bartzokis, G., Tishler, T. A., Shin, I. S., Lu, P. H., and Cummings, J. L. 2004. *Redox-Active Metals in Neurological Disorders*, vol. 1012, *Annals of the New York Academy of Sciences*. LeVine, S. M., Connor, J. R., and & Schipper, H. M. Academy of Sciences, New York: 224–236.
30. Rottkamp, C. A., et al. 2001. Redox-active iron mediates amyloid-beta toxicity. *Free Radical Biology and Medicine* **30**: 447–450.
31. Zecca, L., Youdim, M. B. H., Riederer, P., Connor, J. R, and Crichton, R. R. 2004. Iron, brain ageing and neurodegenerative disorders. *Nature Reviews Neuroscience* **5**: 863–873.
32. Lee, J.-H., Huh, Y.-M., Jun, Y.-W., Seo, J.-W., Jang, J.-T., Song, H.-T., Kim, S., Cho, E.-J., Yoon, H.-G., Suh, J.-S., and Cheon, J. 2007. Artificially engineered magnetic nanoparticles for ultra-sensitive molecular imaging. *Nature Medicine* **13**: 95–99.
33. Zhang, Y., Pilapong, C., Guo, Y., Zhenlian, L., Cespedes, O., Quirke, P., and Zhou, D. 2013. Sensitive, simultaneous quantitation of two unlabeled DNA targets using a magnetic nanoparticle-enzyme sandwich assay. *Analytical Chemistry* **85**: 9238–9244.
34. Neuberger, T., Schopf, B., Hofmann, H., Hofmann, M., and von Rechenberg, B. 2005. Superparamagnetic nanoparticles for biomedical applications: Possibilities and limitations of a new drug delivery system. *Journal of Magnetism and Magnetic Materials* **293**: 483–496.
35. Brem, F., Stamm, G., and Hirt, A. M. 2006. Modeling the magnetic behavior of horse spleen ferritin with a two-phase core structure. *Journal of Applied Physics* **99**: 123906.
36. Rosensweig, R. E. 2002. Heating magnetic fluid with alternating magnetic field. *Journal of Magnetism and Magnetic Materials* **252**: 370–374.

37. Gilles, C., Bonville, P., Wong, K. K. W., and Mann, S. 2000. Non-Langevin behaviour of the uncompensated magnetization in nanoparticles of artificial ferritin. *European Physical Journal B* **17**: 417–427.

38. Cheeseman, K. H., and Slater, T. F. 1993. An introduction to free-radical biochemistry. *British Medical Bulletin* **49**: 481–493.

39. Chou, C. K., Bassen, H., Osepchuk, J., Balzano, Q., Petersen, R., Meltz, M., Cleveland, R., Lin, J. C., and Heynick, L. 1996. Radio frequency electromagnetic exposure: Tutorial review on experimental dosimetry. *Bioelectromagnetics* **17**: 195–208.

40. Stefanini, S., Cavallo, S., Wang, C. Q., Tataseo, P., Vecchini, P., Giartosio, A., and Chiancone, E. 1996. Thermal stability of horse spleen apoferritin and human recombinant H apoferritin. *Archives of Biochemistry and Biophysics* **325**: 58–64.

41. Yau, S. T., Petsev, D. N., Thomas, B. R., and Vekilov, P. G. 2000. Molecular-level thermodynamic and kinetic parameters for the self-assembly of apoferritin molecules into crystals. *Journal of Molecular Biology* **303**: 667–678.

42. Funk, F., Lenders, J. P., Crichton, R. R., and Schneider, W. 1985. Reductive mobilization of ferritin iron. *European Journal of Biochemistry* **152**: 167–172.

43. Santambrogio, P., Levi, S., Cozzi, A., Corsi, B., and Arosio, P. 1996. Evidence that the specificity of iron incorporation into homopolymers of human ferritin L- and H-chains is conferred by the nucleation and ferroxidase centres. *Biochemical Journal* **314**: 139–144.

44. Fahn, S., and Cohen, G. 1992. The oxidant stress hypothesis in Parkinsons-disease—Evidence supporting it. *Annals of Neurology* **32**: 804–812.

45. Blum, D., Torch, S., Lamberg, N., Nissou, M. F., Benabid, A. L., Sadoul, R., and Verna, J. M. 2001. Molecular pathways involved in the neurotoxicity of 6-OHDA, dopamine and MPTP: Contribution to the apoptotic theory in Parkinson's disease. *Progress in Neurobiology* **65**: 135–172.

46. Bove, J., Prou, D., Perier, C., and Przedborski, S. 2005. Toxin-induced models of Parkinson's disease. *Journal of the American Society for Experimental NeuroTherapeutics* **2**: 484–494.

47. Li, P., Sage, J. T., and Champion, P. M. 1992. Probing picosecond processes with nanosecond lasers—Electronic and vibrational-relaxation dynamics of heme-proteins. *Journal of Chemical Physics* **97**: 3214–3227.

48. Macara, I. G., Hoy, T. G., and Harrison, P. M. 1973. Formation of ferritin from apoferritin—Inhibition and metal ion-binding studies. *Biochemical Journal* **135**: 785–789.

49. Macara, I. G., Hoy, T. G., and Harrison, P. M. 1973. Formation of ferritin from apoferritin—Catalytic action of apoferritin. *Biochemical Journal* **135**: 343–348.

50. Theil, E. C., Takagi, H., Small, G.W., He, L., Tipton, A. R., and Danger, D. 2000. The ferritin iron entry and exit problem. *Inorganica Chimica Acta* **297**: 242–251.

51. Danon, D., Goldstein, L., Marikovs, Y., and Skutelsk, E. 1972. Use of cationized ferritin as a label of negative charges on cell surfaces. *Journal of Ultrastructure Research* **38**: 500–510.

52. vanDijk, B., Gast, P., and Hoff, A. J. 1996. Control of radical pair lifetime by a switched magnetic field. *Physical Review Letters* **77**: 4478–4481.

53. Nedoloujko, A., and Kiwi, J. 1997. Transient intermediate species active during the Fenton-mediated degradation of quinoline in oxidative media: Pulsed laser spectroscopy. *Journal of Photochemistry and Photobiology A: Chemistry* **110**: 141–148.

CHAPTER 9

Safety Aspects of Magnetic and Electromagnetic Fields

Sachiko Yamaguchi-Sekino,

Tsukasa Shigemitsu, and Shoogo Ueno

CONTENTS

9.1 INTRODUCTION

The magnetic resonance imaging (MRI) was introduced into the diagnostic imaging technology in medicine in the early 1980s. It originally comes from the technique of nuclear magnetic resonance (NMR). In 1970s, Damadian observed the differences of the relaxation times between tumor and normal tissues and proposed the use of NMR as imaging for the detection of cancer (Damadian 1971). In Japan, during the 1970s, using NMR, Abe et al. proposed and measured noninvasively detection of the biological image information by magnetic focusing method (Abe et al. 1974).

The basic principle and the technical development of imaging formation using NMR were proposed by Lauterbur, who developed the MRI (Lauterbur 1973). The Nobel Prize in Medicine and Physiology was awarded in 2003 to Lauterbur and Mansfield for the fundamental application of a static magnetic field in combination with a gradient (time-varying) magnetic field and the development of the image acquisition and processing (Mansfield and Maudsley 1977). In this way, the root of MRI is known as nuclear magnetic resonance. To avoid alarming the public and medical experts, the word "nuclear" was deleted from the term magnetic resonance, although it has nothing to do with radioactivity.

The operation of MRI basically utilizes three different types of electromagnetic field, the strong static magnetic field, the rapidly changing gradient (time-varying) magnetic field, and the radio frequency electromagnetic field. The static magnetic field is the main magnetic field in MRI. It aligns the proton spins and generates a total magnetization in the human body. The static magnetic field is responsible for a measure of the proton density. It is usually generated by a strong superconducting magnet. The increase in the static magnetic field strength goes toward improvements in image resolution. Today, the commercially available MRI system typically has a static magnetic field strength of up to 3 T. The gradient magnetic field is used to localize aligned protons in the image reconstruction process. The exposure to the gradient (time-varying) magnetic field with frequencies in the kHz range is the character to the MRI technique. The radio frequency electromagnetic fields in the range of 10 to 400 MHz are used to excite the protons within the stable magnetic fields.

The objective of this chapter is to describe some of the recent information on the biological effects and safety aspects of electromagnetic fields related to MRI. Section 9.2 is a short description of the plausible mechanisms of electromagnetic fields related to MRI. Section 9.3 introduces the evaluation of the biological effects of static magnetic field, gradient (time-varying) magnetic field, and radio frequency electromagnetic fields as well as the health effects of these three fields as they relate to MRI. Recently published documents are used to provide an overview of the three different electromagnetic fields with the addition of a brief review of more recent experimental observations (ARPANSA 2014; HPA 2008, 2012; IARC 2002, 2013; ICNIRP 2003a, 2009d; NRPB 2003; WHO 2006a,b). Indirect interactions between human and medical devices (or metal implants) are not considered because of the newest research results and evidence-based information on the biological and health effects.

Through a discussion of biological and health effects, Section 9.4 reviews and summarizes safety guidelines that provide a consideration in human risk for both patients and medical staff on limiting the operation to static magnetic fields as well as the other electromagnetic fields for MRI (ICNIRP 1998, 2003b, 2009b, 2010, 2014).

Section 9.5 covers the health risk assessments associated with the environment and during the operation of MRI system, with an emphasis on occupational exposure. In order to understand the hazard identification, measurement, work-environment risk assessment, and health survey of the working staffs, occupational exposure, and the MRI safety policy and procedure should be reviewed. In Europe, there have been attempts to create a Directive on the protection of occupational exposure of workers to the risks from physical agents. The Directive applies to occupational exposure, not patient exposure, to electromagnetic fields. The progress, update, scope, and conditions of the EU electromagnetic fields Directive will be given shortly to meet the request of protection of MRI working staffs and employers. Followed by the evaluation of the biological and health effects, Section 9.5 concludes with a discussion of safety issues related to MRI in clinical diagnosis, with the highlighting of the guidelines and existing MRI standard. Within these frameworks, Section 9.6

discusses the conclusion remarks on the biological and human health issues related to MRI.

9.2 MECHANISMS FOR BIOLOGICAL INTERACTION

Basically, the interaction between the static magnetic field and biological tissues including the human body are proposed and established on the basis of experimental results. The other two, gradient magnetic field and radio frequency electromagnetic field interaction, are primarily based on Faraday's law of induction.

9.2.1 Static Magnetic Field

The physical mechanisms of interactions between static magnetic fields and biological systems are basically well investigated. There are three established mechanisms: (1) magnetic induction, (2) magneto-mechanical interactions, and (3) electronic interactions (ICNIRP 2003a; WHO 2006a).

1. There are two types of interaction for magnetic induction.

    The electrodynamic interactions with moving charged particles can lead to an induced electric field. The change in electrocardiograms (ECG) is a well-known example of this electrodynamic interaction. These induced electrical potentials have been experimentally observed by ECG changes on mammalians (e.g., rat, rabbit, dog, baboon, and monkey) in the presence of a static magnetic field (Gaffey and Tenforde 1981; Tenforde 1983).

    In the presence of a static magnetic field, the electrical potential is induced. This is the result of the Lorenz force exerted on moving charged particles (electrolytes) in the blood. The induced electric potential is given by

$$\phi = v \cdot B \cdot d \cdot \sin \theta$$

    where v is the velocity, d is the diameter of the artery, and θ is the angle between the direction of the blood flow and the magnetic field. Kinouchi et al. carried out the detailed theoretical treatment of the effects of magnetic fields on blood flow by using the Navier-Stokes equation (Kinouchi et al. 1996). In the case of magnetic fields perpendicular to the blood flow, they found a reduction in the flow rate of blood.

In addition to electrodynamic interaction, induced electric fields and currents are created by gradient (time-varying) magnetic fields in MRI technique or by the movement of subject in the static magnetic fields, which are accordance with Faraday's law of induction. The amplitude of the induced electric fields and currents increases with movement speed and with the degree of gradient.

2. There are two types of interactions for magneto-mechanical effects.

The first type is magneto-orientation: This concerns the orientation of paramagnetic molecules in the static magnetic field. This effect is involved in magneto-reception in certain species of animal including fish and birds. The second type of interaction is the magneto-mechanical translation. This occurs in the presence of a field gradient for paramagnetic or diamagnetic materials (Ueno and Iwasaka 1994a,b).

3. The third mechanism is electronic spin interaction.

This interaction can affect the rate of recombination of pairs of free radicals in chemical reaction intermediates. It seems that this mechanism plays a part in the navigation systems of certain birds. Okano gave the excellent review on the role of free radicals in biology (Okano 2008).

9.2.2 Gradient (Time-Varying) Magnetic Field

The gradient magnetic field serves for the spatial localization in the image reconstruction process in MRI technique. Exposure to gradient magnetic field induces time-varying electric fields and currents in human body. It is considered time-varying magnetic field below 1 kHz. Many studies have investigated the biological effects of low-frequency time-varying magnetic fields (up to 100 kHz) (IARC 2002; ICNIRP 2003a; NIEHS 1998; WHO 1987, 2007a).

The coupling between the time-varying magnetic field and the body is summarized below from the document of International Commission on Mon-Ionizing Radiation Protection (ICNIRP 2010). For magnetic fields, the permeability of tissue is the same as that of air so the field in tissues is the same as the external field. Human and animal bodies do not significantly perturb the field. The main interaction of magnetic fields is the Faraday induction of electric fields and associated currents in the tissues. Key features of dosimetry for exposure of humans to low-frequency magnetic fields include

- For a given magnetic field strength and orientation, higher electric fields are induced in the bodies of larger people because the possible conduction loops are larger.

- Induced electric field and current depend on the orientation of the external magnetic field to the body. Generally, induced fields in the body are greatest when the field is aligned from the front of the back of the body, but for some organs the highest values are for different field alignments.

- Weakest electric fields are induced by a magnetic field oriented along the principal body axis.

- Distribution of the induced electric field is affected by the conductivity of the various organs and tissues.

As mentioned above, in the patient during MRI examination, the time-varying magnetic field in this range induces the electric currents in the body by the Faraday's law of induction. The induction stimulates nerves and muscles. The nerve stimulation may cause discomfort. Electric fields may also be induced by movement in static magnetic field.

In the case of sinusoidal magnetic fields with amplitude B_0 and frequency f, the magnitude of the induced current density is given by

$$J = \pi \cdot r \cdot f \cdot \sigma \cdot B_0$$

This means that the induced currents are proportional to the loop radius r and tissue conductivity σ. A well-known biological effect of magnetic fields in this range is the induction of visual sensations called magnetophosphene.

9.2.3 Radio Frequency Electromagnetic Field

Sheppard et al. quantitatively evaluated potential mechanisms of interaction between radio frequency electromagnetic fields and biological systems

(Sheppard et al. 2008). They categorized biophysical mechanisms as established and proposed. The established mechanisms are dielectric relaxation and ohmic loss, which lead to elevated temperature in tissue through heating. On the other hand, the radical pair mechanism is one of the proposed.

Exposure to radio frequency electromagnetic fields can induce heating in biological tissues. So, biological effects caused by radio frequency electromagnetic fields ranging between 100 kHz and 300 GHz can be divided into two categories: thermal effects and nonthermal effects. Thermal effects are due to tissue heating, and nonthermal effects are due to unknown mechanism.

The devices operating with radio frequency electromagnetic fields have been introduced for therapeutic and diagnostic applications. The therapeutic application comprises cancer treatment with hyperthermia and tissue heating. The latter is mainly associated to MRI, which generates detectable MR signals. Before the 1980s, there were no reports concerning the thermophysiologic responses of humans exposed to radio frequency electromagnetic fields during MRI procedures.

Heating is classically given by a quantity of specific absorption rate (SAR) with units of watts per kilogram (W/kg). The SAR is derived from the square of electric field strength in tissue.

$$SAR = \frac{\sigma |E|^2}{2\rho}$$

where σ is the tissue conductivity, ρ is the tissue density, and E is the (instantaneous) electric field within the tissue. For a time-varying electric field with sinusoidal, the factor of 1/2 may be omitted and the root mean square (rms) value of the field substituted. The SAR cannot be measured directly in humans and is usually estimated from computer-based simulation models of the human body.

9.3 BIOLOGICAL EFFECTS RELATED TO MRI

This section is divided into the three types of electromagnetic field utilized MRI systems. It gives an overview of the findings about the biological effects of static magnetic field, the gradient (time-varying) magnetic field, and radio frequency electromagnetic field. In each section, the biological effects related to each field are summarized and evaluated briefly.

In parallel with the brief introduction of recent studies in each section, the evaluation was conducted with the help of authoritative reviews by well-recognized organizations such as World Health Organization (WHO) and the International Commission on Non-Ionizing Radiation Protection (ICNIRP).

9.3.1 Static Magnetic Field

Apart from investigating interaction mechanisms, a large number of studies have been conducted in the last 50 years on the possible biological and health effects of the static magnetic fields ranging from geomagnetic field level (micro Tesla) to several Tesla in magnetic flux densities. *In vivo* and *in vitro* investigations have included experiments on animal cognition and behavior, cell growth, cell proliferation, cell cycle, apoptotic cell death, tissue development, the cardiovascular system, cancer, the reproductive system, teratogenicity, the neuro-endocrine system, circadian rhythms, and haematologic parameters. Studies on human volunteers were conducted to assess the health risk of short-term exposure to high static magnetic fields. Its endpoints included the central nervous system, behavioral and cognitive functions, sensory perception, cardiovascular system, and brain activity. In addition, few epidemiological studies related to static magnetic fields generated from DC supply in industries have been conducted.

There have been many comprehensive reviews on the biological and health effects of the static magnetic field (HPA 2008; ICNIRP 2003a; Ueno and Okano 2012; Ueno and Shigemitsu 2007; WHO 1987, 2006a; Yamaguchi-Sekino et al. 2011).

9.3.1.1 In Vivo *Studies*

Here, three research areas are considered for the evaluation of the biological effects of static magnetic fields: (1) cancer-related endpoints, (2) reproduction and development, and (3) physiological and behavioral response.

In terms of cancer-related issues, *in vivo* studies have been conducted using animals to investigate the potential carcinogenicity of static magnetic fields. These studies were carried out for genotoxic and carcinogenic effects, through direct, initiation, or promotional means. Mevissen et al. investigated the effect of the magnetic field of 15 mT for 13 weeks on the incidence of DMBA-induced tumor in rats (Mevissen et al. 1993). There are no significant effects on the incidence. Compared with control groups, the number of tumors per animal was not affected, although the weight per tumor was significantly increased.

Several studies have been conducted on the possible teratogenic effects of exposure to static magnetic fields. Most studies have investigated possible effects on the embryo and fetus. A German research group investigated and reported, in two associated papers, the effect of repetitive exposure of mice to static magnetic field (Zahedi et al. 2014; Zaun et al. 2014). The group evaluated possible risks of strong static magnetic fields for embryo implantation, gestation, organogenesis, and embryonic development. The mice were exposed daily for 75 min during the entire course of pregnancy at the bore entrance (the position of MRI medical staff) or at the bore isocenter (the position of patients) of a 1.5- and 7-T MRI scanner. At the entrance positions, the magnetic gradient fields were presented, from 0.2 to 1 T for 1.5-T MRI scanner and from 0.6 to 1.4 T for 7-T MRI scanner. Exposures started at day 1.5 of pregnancy for 18 days. After delivery, development of the offspring was monitored. No effects of any static magnetic fields were observed with regard to pregnancy rate, duration of pregnancy, litter size, still births, malformation, sex distribution, or postpartum death of offspring (Zahedi et al. 2014). The effects were a slight delay in weight gain in the groups exposed at both positions in the 1.5-T MRI scanner or at the entrance of the 7-T MRI scanner and a slight but significant delay in the opening of the eyes in all exposed groups. Further, Zaun et al. investigated the effect of daily exposure in utero to MRI-generated static magnetic field during prenatal development on germ cell development and fertility of exposed offspring in adulthood (Zaun et al. 2014). Offspring at 8 weeks were mated with unexposed mice. In the in utero-exposed male mice, no exposure effect was measured on the weight of testes and epididymis or on sperm count, sperm morphology, or fertility. In the in utero-exposed female mice, there was no effect on fertility in terms of pregnancy rate and litter size. However, in the offspring of the exposed female mice in the bore or at the entrance of a 7-T MRI scanner, a reduced placental weight was observed. This correlated with a decrease in embryonic weight only in those mice exposed at the strongest magnetic field. These studies indicate possible effects of repetitive exposure on fertility and development.

The research group of Houpt et al. published a series of studies on the effects of strong static magnetic field on behavior of animals (Cason et al. 2009; Houpt and Houpt 2010; Houpt et al. 2011). In the past, it was reported that vertigo is a side effect of exposure to the high magnetic fields in MRI. So, Cason et al. tested the effect of static magnetic field, 14.1 T on the vestibular apparatus in the inner ear. Chemically, labyrinthecotomized rats (injection of sodium arsanalite, destroying the hair cells) and intact rats (sham-labyrinth-ecotomized with saline injection) were exposed to the 14.1 T for 30 min. Intact rats acquired a profound conditioned taste aversion (CTA), saccharin avoidance, and labyrinth-ecotomized rats did not acquire a CTA and showed a high preference for saccharin, which is similar to sham-exposed rats. Significant increase in c-Fos expression was observed in intact rats, but magnetic field exposure did not elevate c-Fos levels in labyrinth-ecotomized rats. This result demonstrated that an intact inner ear is necessary for all the observed effects of exposure to high static magnetic fields in rats. In another study from the same group, rats were preexposed two times to a 14.1-T static magnetic field for 30 min on two consecutive days (Houpt and Houpt 2010). This result showed that repeated treatment to the 14.1 T causes habituation. Here, the habituation included locomotor circling, rearing, and conditioned taste aversion. Compared to a sham-exposed rat, the preexposed rat showed less locomotor circling and an attenuated CTA. Rearing was suppressed in all magnet-exposed groups regardless of pre-exposure. Further, Houpt et al. tested the effects on the movement of rat though 14.1-T static magnetic field (Houpt et al. 2011). Only momentary passage of the rat into and out of the static magnetic fields was enough to suppress rearing and induce a significant CAT. They concluded that in rats, movement though the steep gradient of a high magnetic field has some behavioral effects, but sustained exposure to the homogeneous center of the magnetic field is required for the full behavioral consequences.

The document of the WHO stated, that "A large number of animal studies on the effects of static magnetic fields have been carried out. Most of those considered relevant to human health have examined the effect of fields considerably larger than the natural geomagnetic field. A number of studies have been carried out of fields in the millitesla region, comparable to relatively high industrial exposure. More recently, with the advent of superconducting magnetic technology and MRI, studies of behavioral, physiological and reproductive effects have been carried out at flux densities around, or exceeding, 1 T. Few studies, however, have examined possible chronic effects of exposure, particularly in relation to carcinogenesis. The most consistent responses seen in neurobehavioral studies suggest that the movement of laboratory rodents in static magnetic fields equal to or greater than 4 T may be unpleasant, inducing aversive responses and conditioned avoidance. Such effects are thought to be consistent with magnetohydrodynamic effects on the endolymph of the vestibular apparatus. The data are otherwise variable. There is good evidence that exposure to fields greater than about 1 T (0.1 T in larger animals) will induce flow potentials around the heart and major blood vessels, but the physiological consequences of this remain unclear. Several hours of exposure to very high flux densities of up to 8 T in the heart region did not result in any cardiovascular effects in pigs. In rabbits, short and long exposures to fields ranging from geomagnetic levels to the millitesla range have been reported to affect the cardiovascular system, although the evidence is not strong" (WHO 2006a, p. 5).

9.3.1.2 In Vitro *Studies*

In vitro studies investigating the effects of static magnetic fields have covered a wide range of biological systems from bacteria to animal and human cells. The endpoints cover various issues, including genotoxic effects, mutation, chromosomal damage, DNA damage, cell viability, proliferation, differentiation, cellular responses, gene expression, and signal transduction.

Ikehata et al. reported that the effects of exposures to static magnetic field of 2 and 5 T for 20 min up to 48 h in a bacterial mutation test using *Salmonella typhimurium* and *Escherichia coli* strains (Ikehata et al. 1999). They observed no effect on mutagenicity or growth rate of bacterial strains.

Sakurai et al. have studied the effect of strong static magnetic fields up to 10 T with gradient field of 0–41.7 T/m on insulin-secreting cells and on myotube orientation of a mouse-derived myoblast cell line (C2C12) (Sakurai et al. 2009, 2012). First, after insulin-secreting cells exposed to above the static magnetic field for 0.5 or 1 h, the insulin secretion, mRNA expression, glucose-stimulated insulin secretion, insulin content, cell proliferation, and cell number were analyzed. This result suggested that the high static magnetic field of MRI systems might not cause cell proliferative or functional damage on insulin-secreting cells. In the experiment, where effects of above static magnetic field on the orientation of myotubes formed from a mouse-derived myoblast cell line, C2C12 are performed. Cell cultures were exposed for 6 days. The results indicated that the formation of oriented myotube is dependent on the magnetic flux density and the gradient magnetic field. The myogenic differentiation and cell number were not affected for any of the experimental conditions.

Zhao et al. exposed human-hamster hybrid cells, mitochondria-deficient cells, and double-strand break repair-deficient cells to 8.5 T (Zhao et al. 2011). Adenosine triphosphate (ATP) content was significantly decreased in human-hamster cells exposed to 8.5 T but not 1 or 4 T for either 3 or 5 h. ATP content significantly decreased in the two deficient cell lines exposed to 8.5 T for 3 h. With further incubation of 12 or 24 h without static magnetic field exposure, ATP content could retrieve to the control level in human-hamster hybrid cells but in the two deficient cell lines. The levels of reactive oxygen species (ROS) in the three cell lines were significantly increased by exposure to 8.5 T for 3 h. They indicated that the cellular ATP content was reduced by 8.5 T for 3 h and was mediated by ROS.

Lee et al. investigated the genotoxic potential of 3-T MRI scans in cultured human lymphocytes *in vitro* by analyzing chromosome aberrations (CA), micronuclei (MN), and single-cell electrophoresis (Lee et al. 2011). They exposed human lymphocytes to the condition of a routine brain examination protocol and observed a significant increase in the frequency of single-strand DNA breaks following exposure to a 3-T MRI. The frequency of both CAs and MN in exposed cells increased in a time-dependent manner. Their results suggest that exposure to 3-T MRI induces genotoxic effects in human lymphocytes.

The document of the WHO stated that "a number of different biological effects of static magnetic fields have been explored *in vitro*. Different levels of organization have been investigated, including cell free systems (employing isolated membranes, enzymes, or biochemical reactions) and various cell models (using both bacteria and mammalian cells). Endpoints studied included cell orientation, cell metabolic activity, cell membrane physiology, gene expression, cell growth and genotoxicity. Positive and negative findings have been reported for all these endpoints. However, most data were not replicated. The observed effects are rather diverse and were found after exposure to a wide range of magnetic flux densities. There is evidence that static magnetic fields can affect several endpoints at intensities lower than 1 T, in the mT range. Thresholds for some of the effects were reported, but other studies indicated nonlinear responses without clear threshold values" (WHO 2006a, p. 4).

9.3.1.3 Human Experimental and Epidemiological Studies

Studies on human volunteers exposed up to 8 T have been investigated to assess the relationship between exposure to high static magnetic fields and human health. Endpoints investigated have included central and peripheral nerve function, brain activity, neurobehavioral and cognitive functions, sensory perception, cardiac function, blood pressure, heart rate, serum proteins and hormone levels, body, and skin temperature (WHO 2006a).

The document of the WHO stated that "The results do not indicate that there are effects of static magnetic field exposure on neurophysiological responses and cognitive functions in stationary volunteers, nor can they rule out such effects. A dose-dependent induction of vertigo and nausea was found in workers, patients and volunteers during movement in static fields greater than about 2 T. One study suggested that eye-hand coordination and near visual contrast sensitivity are reduced in fields adjacent to a 1.5 T MRI unit. Occurrence of these effects is likely to be dependent on the gradient of the field and the movement of the subject. A small change in blood pressure and heart rate was observed in some studies, but were in the range of normal physiological variability. There is no evidence of effects of static magnetic fields on other aspects of cardiovascular physiology, or on serum proteins and hormones. Exposure to static magnetic fields of up to 8 T does not appear to induce temperature changes in humans" (WHO 2006a, p. 7).

In the past, several epidemiological studies have been conducted to examine mortality and cancer incidence among workers exposed to static magnetic fields at aluminum reduction and chloralkali plants. Historically, Marsh et al. presented the occupational exposure study (Marsh et al. 1982). This study with 320 male workers carried out the occupational exposure to static magnetic fields ranging from about 3 to 15 mT generated from large electrolytic cells. The control group was 186 male workers. Although the cohort was small, they found that there are significant increases in heart disease and cancer. Barregard et al. carried out very limited occupational study on workers exposed to static magnetic field from 4 to 29 mT (average 14 mT) (Barregard et al. 1985). They studied mortality and cancer incidence among workers in a Swedish chloralkali plant where about 100 kA direct currents are used to produce chlorine. The exposed group comprised of 157 men employed between 1951 and 1983, and their mortality has been compared with that of Swedish men using calendar-year and age-specific mortality rates. Cancer incidence has been compared with the expected incidences among Swedish men. They reported that there is no increased mortality among the exposed men. No excess incidence of cancer is found.

Regarding the human and epidemiological studies, a document stated that "Increased risks of various cancers, e.g., lung cancer, pancreatic cancer, and haematological malignancies, were reported. But results were not consistent across studies. The few epidemiological studies published to date leave a number of unresolved issues concerning the possibility of increased cancer risk from exposure to static magnetic fields. Assessment of exposure has been poor, the number of participants in some of the studies has been very small, and these studies are thus able to detect only very large risks for such rare diseases. The inability of these studies to provide useful information is confirmed by the lack of clear evidence for other, more established carcinogenic factors present in some of the work environments. Other noncancerous health effects have been considered even more sporadically. Most of these studies are based on very small numbers and have numerous methodological limitations. Other environment with a potential for high fields have not been adequately evaluated, e.g., those for MRI operators. At present, there is inadequate date for a health evaluation" (WHO 2006a, p. 8).

9.3.2 Gradient (Time-Varying) Magnetic Field

After the review and evaluation of low-frequency electric and magnetic fields (IARC 2002; ICNIRP 2003a; NIEHS 1998; WHO 1987, 2007a), the International Agency for Research on Cancer (IARC) classified power frequency magnetic fields as possibly carcinogenic to humans (Group 2B). This classification was strongly influenced by epidemiological studies that have observed increased risks of childhood leukemia at magnetic fields greater than 0.3–0.4 µT (IARC 2002).

Many laboratory studies of the effects of extremely low frequency magnetic field on the cells have shown no induction of genotoxicity and have not yet provided good explanation of the cause of leukemia. Although epidemiological studies in children show an increased risk of childhood leukemia exposed to extremely low frequency magnetic field, there is a lack of supporting evidence for such an effect from animal studies and/or *in vitro* studies. In addition, there is no plausible mechanism.

Owing to the absence of a plausible mechanism and no induction of cancer in animals study, the expert report has not concluded that there is no causal relationship between the magnetic field exposure and the risk of childhood leukemia (SCENIHR 2009, 2013). In the same way, Leitgeb revealed that the assumption of a causal link between magnetic field exposure and childhood leukemia is no longer plausible and hence that the magnetic field's classification as possibly carcinogenic needs revision (Leitgeb 2014).

9.3.3 Radio Frequency Electromagnetic Field

During MRI examinations, patients are exposed to the radio frequency electromagnetic field. In his review article, Shellock discussed the characteristics of radio frequency electromagnetic field–induced heating associated with MRI procedure (Shellock 2000). He emphasized thermal and other physiologic responses in human subjects due to heating.

The possible effects of radio frequency electromagnetic fields have been reviewed (Ahlbom et al. 2004; ARPANSA 2014; HPA 2012; IARC 2013; ICNIRP 2009d; NRPB 2003). These possible effects of radio frequency electromagnetic field exposure have been evaluated with the *in vitro* and *in vivo* studies covering areas including the genotoxicity, carcinogenicity, cell transformation, cell proliferation, apoptosis, gene expression, intracellular signaling, cellular physiology, neurotransmitter, electrical activity, blood-brain barrier, autonomic functions, behavior, endocrine system (melatonin), auditory system, immunology, haematology, reproduction, and development. The studies in humans are covered from neurocognitive cancers (e.g., brain tumor and acoustic neuroma) and noncancers (e.g., reproduction, morbidity, and hypersensitivity) to epidemiological studies.

In May 2011, after the review and evaluation of scientific papers by 30 experts from 14 countries, the IARC assessed the carcinogenicity of radio frequency electromagnetic fields and classified them as a possibly carcinogenic to humans (Group 2B). The IARC monograph on radio frequency electromagnetic field was published in May 2013 (Baan et al. 2011; IARC 2013). The above classification was based on (1) there is limited evidence in humans for the carcinogenicity of radio frequency electromagnetic fields, based on positive association between glioma and acoustic neuroma and exposure to radio frequency electromagnetic fields from wireless phones (epidemiological studies), (2) there is limited evidence in experimental animals for the carcinogenicity of radio frequency electromagnetic fields, and (3) there is "weak mechanistic evidence" relevant to radio frequency electromagnetic field-induced cancer in humans. The evidence for other exposures and outcomes was considered insufficient for any conclusion. There was a minor opinion in the monograph that current evidence in humans was inadequate, therefore permitting no conclusion about a causal association.

Using this evaluation and the classification made by the IARC to revise and update the Environmental Health Criteria (EHC) document on radio frequency fields, the WHO opened a first draft monograph of EHC of radio frequency electromagnetic field. The draft document will be finalized by an expert group and will be published in the series of EHC.

9.3.4 Possible Health Effect Related to MRI Operation

After the publication of the EHC in 2006, there have been several studies on the health effects of exposure to static magnetic fields related to MRI. It is well accepted that the high static magnetic field leads to the mild sensory effect for vertigo, metallic taste, and magnetophosphene. These effects are not considered to be hazardous per se. However, they may result in a loss of working ability during practice. MRI workers experience exposure to static magnetic fields on an almost daily basis. It is important to characterize potential hazards. A number of studies have evaluated potential neurologic and vestibular effects of exposure to static magnetic fields up to 8-T range.

De Vocht et al. assessed health complaints and cognitive performance in workers with static magnetic fields ranging from 0.5 to 1.5 T from a MRI manufacturing company (De Vocht et al. 2006). They noted that vertigo, metallic taste, and concentration complaints were significantly and complaints were more frequently reported in the workers compared with the control. The workers moving rapidly through the static magnetic field reported more complaints than those who moved slower. In addition, they studied the effects of stray fields from a 7-T magnet on neurobehavioral performance, specifically, cognitive function of head movements in 27 healthy volunteers (average 25.0 years with 18–51 years; male 13; female 14) (De Vocht et al. 2007). The stray magnetic fields were designed to 1600, 800, and 2 mT (negligible exposure). The order of exposure was assigned at random and was masked by placing volunteers in a tent to hide their position relative to the magnet bore. Volunteers completed a test battery assessing auditory working memory, eye-hand coordination, and visual perception. During three sessions the volunteers were instructed to complete a series of standardized head movements to generate additional time-varying fields (~300 and ~150 mT/s_{rms}). No effects were observed on working memory. They suggest that there are effects on visual perception and eye-hand coordination, but these were weak. The magnitude of these effects may depend on the magnitude of time-varying fields and not so much on the static field.

The subjects and operators reported vertigo-like sensations or perception of movement in and around high field during MRI. For this effect, the change of firing rates of hair cells by induced currents, magnetohydrodynamics (MHD), and the forces induced by the static magnetic fields due to the magnetic susceptibility differences of tissues have been proposed as possible mechanisms. Glover et al. examined three possible mechanisms of vertigo and concluded that magnetic field–induced vertigo results from both magnetic susceptibility difference between vestibular organs and surrounding fluid, and induced currents acting on the vestibular hair cells (Glover et al. 2007). In this way, the mechanism depends on movement though magnetic fields or magnetic fields with high spatial gradients.

Heinrich et al. conducted a meta-analysis of studies published from 1992 to 2007 in order to evaluate whether cognitive processes, sensory perception, and vital signs might be influenced by static magnetic fields in MRI (Heinrich et al. 2011). Vital signs such as blood pressure and heart rate were not affected. With regard to effects on sensory perception, there was an increase of dizziness and vertigo caused by movement during gradient magnetic field exposures. They mentioned that the number of studies is very small and the experimental setup of some of analyzed studies makes it difficult to accurately determine the effects of static magnetic fields by themselves. Further, Heinrich et al. presented the results of a study on how cognitive functions in health subjects undergoing MRI are acutely impaired by static magnetic fields (Heinrich et al. 2013). The cognitive functions such as eye-hand coordination, attention, reaction time, and visual discrimination were not impaired by a static magnetic field in MRI systems (0, 1.5, 3.0, and 7.0 T). Dizziness, nystagmus, phosphenes, and head ringing were related to the strength of the static magnetic field, although there were no significant effects on cognitive function at any static magnetic strength. Sensory perceptions did vary according to field strength. They concluded that static magnetic fields as high as 7.0 T did not have a significant effect on cognition.

Van Nierop et al. showed that the neurocognitive functioning is modulated when human volunteers were only exposed to movement in stray field from a 7-T MRI scanner (Van Nierop et al. 2012). The healthy volunteers were tested in a sham (0 T), low (0.5 T), and high (1.0 T) exposure conditions. The volunteers are simultaneously exposed to movement-induced time-varying magnetic fields at each exposure conditions. The attention and concentration were negatively affected when exposed to time-varying magnetic field with static magnetic field of 7-T MRI scanners. The visuospatial orientation was also affected after exposure. Further, Van Nierop et al. studied vertigo of subjects exposed to only stray magnetic fields with movement-induced time-varying magnetic fields (Van Nierop et al. 2013). They evaluated the postural body sway of subjects sitting in front of a 7-T

MRI for 1 h followed by the movement of their heads for 16 sec. The postural body sway was expressed in sway path length, sway area, and sway velocity. The healthy volunteers performed two tasks, standing with eyes closed and feet in parallel and then in tandem position, after standardized head movements in a sham, low-exposure (stray magnetic fields: 0.24 T with time-varying magnetic field of 0.49 T/sec) and high-exposure conditions (0.37 T with 0.70 T/sec). The results show that sway path length, sway area, and the velocity of body movements were significantly higher with higher static magnetic fields. They commented that the investigation of the practical safety implications of this finding for surgeons and others working near MRI scanner is needed.

9.4 EXPOSURE GUIDELINES RELATED TO MRI

The static magnetic field, the gradient (time-varying) magnetic field, and the radio frequency electromagnetic field are used to operate MRI equipment. Here, relevant ICNIRP guidelines are introduced briefly (ICNIRP 1998, 2009a, 2010, 2014).

The ICNIRP is independent groups of experts. The responsibility of the ICNIRP is (1) to evaluate the science-based information on the effects of NIR and (2) to provide the recommendations, statements, and advice on the protection against harmful effects of the NIR. The ICNIRP develops basic guidelines and exposure limits to protect the exposure from NIR. These guidelines contain exposure limit both for workers and the general public. The ICNIRP publishes timey recommendations of exposure guidelines and statements, informing and advising us on the specific topics related to protection issues.

In 1998, the ICNIRP opened the guideline document for limiting exposure to time-varying electric, magnetic, and electromagnetic fields in the frequency range up to 300 GHz and in 2010 updated that document for the frequency range from 1 Hz to 100 kHz. The ICNIRP also published the updated guideline for the static magnetic field (ICNIRP 2009b). Guidelines are also established for movement-induced electric fields or time-varying magnetic fields below 1 Hz in order to prevent sensory effects such as vertigo associated with body movement (ICNIRP 2014). In addition, ICNIRP gave the statement on protection of patients during the MRI procedure (ICNIRP 2004, 2009c).

Guidelines of the ICNIRP are established for the protection of humans exposed to electric and magnetic fields in the low-frequency range, from 1 Hz to 100 kHz (ICNIRP 2010). This document is extended to 10 MHz, which covers the prevention of nervous system functions. The ICNIRP set the basic restrictions in terms of the induced electric field strength in the body to prevent perception of nerve stimulation.

Guidelines for exposure above 100 kHz are covered in another document (ICNIRP 1998). Between 100 kHz and 10 GHz, basic restrictions consider the SAR to prevent whole-body and localized tissue heating. For frequencies from 10 to 300 GHz, basic restrictions are set in terms of power density to prevent heating effects in tissue at or near the body surface (ICNIRP 1998).

As mentioned in the ICNIRP document, induced electric field strength and SAR cannot be measured directly. So the reference level was set by ICNIRP. For practical exposure assessment purposes, the ICNIRP has provided the reference level in terms of the strength of electric and magnetic fields. The reference levels are obtained from the basic restrictions by mathematical modeling.

9.4.1 Static Magnetic Field

The limits introduced by the ICNIRP for the general public and occupational exposure to static magnetic fields are shown in Table 9.1 (ICNIRP

TABLE 9.1   ICNIRP Guideline on the Limits of Exposure to Static Magnetic Fields[a]

Exposure Characteristics	Magnetic Flux Density
Occupational[b]	
Exposure of head and of trunk	2 T
Exposure of limbs[c]	8 T
General public[d]	
Exposure of any part of the body	400 mT

Source: Reproduced from ICNIRP, *Health Physics* **96**: 504–514, 2009.

[a] ICNIRP recommends that these limits should be viewed operationally as spatial peak exposure limits.

[b] For specific work applications, exposure up to 8 T can be justified if the environment is controlled and appropriate work practices are implemented to control movement-induced effects.

[c] Not enough information is available on which to base exposure limits beyond 8 T.

[d] Because of potential indirect adverse effects, ICNIRP recognizes that practical policies need to be implemented to prevent inadvertent harmful exposure of persons with implanted electronic medical devices and implants containing ferromagnetic material, and dangers from flying objects, which can lead to much lower restriction levels such as 0.5 mT.

2009b). The acute exposure of the general public should not exceed 400 mT (any part of the body). The term "general public" refers to the entire population including individuals of all ages and varying health status. For occupational exposure, the ICNIRP has a 2-T limit for the head and trunk and an 8-T limit for limbs in controlled situations.

9.4.2 Gradient (Time-Varying) Magnetic Field

The ICNIRP published a guideline for limiting exposure to time-varying electric and magnetic field in the frequencies range between 1 Hz and 100 kHz (ICNIRP 2010). This guideline has two separate guidances, one for occupational exposure and one for exposure of the general public. Occupational exposure refers to healthy adults exposed to time-varying electric and magnetic fields from 1 Hz to 10 MHz at their workplaces.

Basic restrictions for human exposure to time-varying electric and magnetic fields in the frequencies range between 1 Hz and 10 MHz are shown in Table 9.2. Basic restrictions are specified in terms of the induced

TABLE 9.2 Basic Restrictions for Human Exposure to Time-Varying Electric and Magnetic Fields from 1 Hz to 10 MHz

Exposure Characteristic	Frequency Range	Internal Electric Field $(V\ m^{-1})$
Occupational Exposure		
CNS tissue of the head	1–10 Hz	$0.5/f$
	10–25 Hz	0.05
	25–400 Hz	$2 \times 10^{-3}f$
	400 Hz–3 kHz	0.8
	3 kHz–10 MHz	$2.7 \times 10^{-4}f$
All tissues of head and body	1 Hz–3 kHz	0.8
	3 kHz–10 MHz	$2.7 \times 10^{-4}f$
General Public Exposure		
CNS tissue of the head	1–10 Hz	$0.1/f$
	10–25 Hz	0.01
	25–1000 Hz	$4 \times 10^{-4}f$
	1000 Hz–3 kHz	0.4
	3 kHz–10 MHz	$1.35 \times 10^{-4}f$
All tissues of head and body	1 Hz–3 kHz	0.4
	3 kHz–10 MHz	$1.35 \times 10^{-4}f$

Source: Reproduced from ICNIRP, *Health Physics* **99**: 818–836. Erratum. 2011. **100**: 112, 2010.

Note: f is the frequency in hertz. All values are rms. In the frequency range above 100 kHz, RF specific basic restrictions need to be considered additionally.

TABLE 9.3   Reference Levels for Occupational Exposure to Time-Varying Electric and Magnetic Fields (Unperturbed rms Values) from 1 Hz to 10 MHz

Frequency Range	E-Field Strength E (kV m^{-1})	Magnetic Field Strength H (A m^{-1})	Magnetic Flux Density B (T)
1–8 Hz	20	$1.63 \times 10^5/f^2$	$0.2/f^2$
8–25 Hz	20	$2 \times 10^4/f$	$2.5 \times 10^{-2}/f$
25–300 Hz	$5 \times 10^2/f$	8×10^2	1×10^{-3}
300 Hz–3 kHz	$5 \times 10^2/f$	$2.4 \times 10^5/f$	$0.3/f$
3 kHz–10 MHz	1.7×10^{-1}	80	1×10^{-4}

Source: Reproduced from ICNIRP, *Health Physics* 99: 818–836. Erratum. 2011. 100: 112, 2010.

Note: *f* in hertz.

electric field strength to prevent nervous system response including peripheral (PNS) and central nerve stimulation (CNS) and the induction of retinal phosphenes (magnetophosphenes).

Reference levels shown in Table 9.3 are obtained from the basic restrictions by mathematical modeling. Because of the physical quantity of the induced electric field (V/m) and SAR in the body cannot be measured directly, the reference levels up to 10 MHz are set using of the strength of electric and magnetic fields (unperturbed values) in the ICNIRP guideline (ICNIRP 2010). According to ICNIRP, occupational exposure below the reference levels assured that the basic restrictions are not exceeded.

9.4.3 Radio Frequency Electromagnetic Field

The revised guideline from ICNIRP covers only the exposure limit in the frequency range up to 100 kHz (ICNIRP 2010). This document replaces the low-frequency part of the 1998 guidelines of ICNIRP. To meet the responsibility of ICNIRP, the ICNIRP is now revising and drafting the guidelines for high-frequency range between 100 kHz and 300 GHz. So, before replacing the exposure limits in this range, one can use existing exposure limits of the 1998 guideline for the protection of human.

Here, the radio frequency electromagnetic field covers the frequency from 100 kHz to 300 GHz. From the 1998 ICNIRP guideline between 100 kHz and 10 GHz, basic restrictions on SAR are provided to prevent whole-body heat stress and excessive localized tissue heating. In the frequency range between 100 kHz and 10 MHz, basic restrictions are provided on both current density and SAR (Table 9.4). As shown in Table 9.5, basic restrictions between 10 and 300 GHz are provided on power density to prevent excessive heating in tissue at or near the body surface (ICNIRP 1998).

Reference levels for occupational exposure are shown in Table 9.6. In the frequency range between 100 kHz and 300 GHz, the exposure basic restriction and reference level are still based on the 1998 ICNIRP guidelines until there is a revision of the guidelines for the high-frequency portion of the electromagnetic spectrum between 100 kHz and 300 GHz.

According to guidelines (ICNIRP 1998, 2010), when the reference levels are exceeded, assessments with measurement and calculation are necessary to estimate whether the basic restrictions are exceeded.

9.4.4 Movement-Related Magnetic Field

In 2014, the ICNIRP published a document for the protection of workers moving in static magnetic fields or being exposed to magnetic fields with frequencies below 1 Hz (ICNIRP 2014). This document provides guidelines for the protection of workers against established adverse direct health effects and to avoid sensory effects that may be annoying and impair working ability.

The protection of patients during MRI examinations will be given in Section 9.4.5 following from the ICNIRP statements (ICNIRP 2004, 2009c). Table 9.7 presents the basic restrictions and reference levels of this guideline (ICNIRP 2014). The basic restrictions have been defined for the change in external magnetic flux density, ΔB, and for the induced internal electric field, V/m_{peak}. The induced internal electric field cannot be directly measured, so compliance can be demonstrated using reference level. In this guideline, the controlled and uncontrolled exposures are considered as two tiers. For controlled exposure, the basic restrictions are intended to be used in work environments. The work environment given by the ICNIRP is where access is restricted to workers who have been trained to understand the biological effects that may result from exposure and where the workers are able to control their movements to prevent annoying and disturbing sensory effects. For uncontrolled exposures, restrictions apply to all other occupational situations.

TABLE 9.4 Basic Restrictions for Time-Varying Electric, Magnetic, and Electromagnetic
Fields up to 10 GHz

Exposure Characteristics	Frequency Range	Current Density for Head and Trunk (mA m^{-2}) (rms)	Whole-Body Average SAR (W kg^{-1})	Localized SAR (Head and Trunk) (W kg^{-1})	Localized SAR (Limbs) (W kg^{-1})
Occupational exposure	Up to 1 Hz	40	–	–	–
	1–4 Hz	$40/f$	–	–	–
	4 Hz–1 kHz	10	–	–	–
	1–100 kHz	$f/100$	–	–	–
	100 kHz–10 MHz	$f/100$	0.4	10	20
	10 MHz–10 GHz	–	0.4	10	20
General public exposure	Up to 1 Hz	8	–	–	–
	1–4 Hz	$8/f$	–	–	–
	4 Hz–1 kHz	2	–	–	–
	1–100 kHz	$f/500$	–	–	–
	100 kHz–10 MHz	$f/500$	0.08	2	4
	10 MHz–10 GHz	–	0.08	2	4

Source: Reproduced from ICNIRP, *Health Physics* **74**(4): 494–522, 1998.

Note: 1. f is the frequency in hertz.

2. Because of electrical inhomogeneity of the body, current densities should be averaged over a cross section of 1 cm^2 perpendicular to the current direction.

3. For frequencies up to 100 kHz, peak current density values can be obtained by multiplying the rms value by $\sqrt{2}$ (~1.414). For pulses of duration t_p, the equivalent frequency to apply in the basic restrictions should be calculated as $f = 1/(2\, t_p)$.

4. For frequencies up to 100 kHz and for pulsed magnetic fields, the maximum current density associated with the pulses can be calculated from the rise/fall times and the maximum rate of change of magnetic flux density. The induced current density can then be compared with the appropriate basic restriction.

5. All SAR values are to be averaged over any 6-min period.

6. Localized SAR averaging mass is any 10 g of contiguous tissue; the maximum SAR so obtained should be the value used for the estimation of exposure.

7. For pulses of duration t_p, the equivalent frequency to apply in the basic restrictions should be calculated as $f = 1/(2\, t_p)$. Additionally, for pulsed exposures in the frequency range 0.3 to 10 GHz and for localized exposure of the head, in order to limit or avoid auditory effects caused by thermoelastic expansion, an additional basic restriction is recommended. This is that the SA should not exceed 10 mJ kg^{-1} for workers and 2 mJ kg^{-1} for the general public, averaged over 10 g tissue.

TABLE 9.5   Basic Restrictions for Power Density
for Frequencies between 10 and 300 GHz

Exposure Characteristics	Power Density (W m^{-2})
Occupational exposure	50
General public	10

Source: Reproduced from ICNIRP, *Health Physics* **74**(4): 494–522, 1998.

Note: 1. Power densities are to be averaged over any 20 cm^2 of exposed area and any $68/f^{1.05}$-min period (where f is in gigahertz) to compensate for progressively shorter penetration depth as the frequency increases.

2. Spatial maximum power densities, averaged over 1 cm^2, should not exceed 20 times the values above.

The ICNIRP recommended that ΔB should not exceed 2 T during any 3-sec period to prevent transient sensory effects such as vertigo and nausea from motion-induced electric field below few hertz. The ICNIRP also recommended that the induced electric field should not exceed the basic restriction of 1.1 V/m_{peak} to prevent stimulation of peripheral nerves in controlled exposure. In uncontrolled exposures, the basic restrictions are based on protection against magnetophosphenes and peripheral nerve stimulation. The reference level for peak dB/dt has been set to 2.7 T/sec as shown in Table 9.7.

The basic restrictions and reference levels above 1 Hz for occupational exposure are equal to the occupational guidelines of the ICNIRP (1998, 2010).

9.4.5 Guidelines on the Protection of Patients Undergoing MRI Examinations

The protection of patients undergoing MRI examinations is given in detail in the ICNIRP statements. In 2004, the ICNIRP opened the statement on the protection of patients undergoing medical MRI procedures (ICNIRP 2004). After the publication of this statement, ICNIRP has published revised guidelines on occupational and general public exposure to static magnetic fields (ICNIRP 2009b). In light of these updated guidelines, ICNIRP has decided to issue an amendment of the statement concerning patient exposure to static magnetic fields (ICNIRP 2009c).

The ICNIRP introduced three tiers of exposure: normal operating mode, controlled operating mode (under medical supervision), and experimental operating mode (with special ethical approval) in MRI statements

TABLE 9.6   Reference Levels for Occupational Exposure to Time-Varying Electric, Magnetic, and Electromagnetic Fields up to 300 GHz

Frequency Range	E-Field Strength (V m⁻¹)	H-Field Strength (A m⁻¹)	B-Field (µT)	Equivalent Plane Wave Power Density S_{eq} (W m⁻²)
Up to 1 Hz	–	1.63×10^5	2×10^5	–
1–8 Hz	20,000	$1.63 \times 10^5/f^2$	$2 \times 10^5/f^2$	–
8–25 Hz	20,000	$2 \times 10^4/f$	$2.5 \times 10^4/f$	–
0.025–0.82 kHz	$500/f$	$20/f$	$25/f$	–
0.82–65 kHz	610	24.4	30.7	–
0.065–1 MHz	610	$1.6/f$	$2.0/f$	–
1–10 MHz	$610/f$	$1.6/f$	$2.0/f$	–
10–400 MHz	61	0.16	0.2	10
400–2000 MHz	$3f^{1/2}$	$0.008f^{1/2}$	$0.01f^{1/2}$	$f/40$
2–300 GHz	137	0.36	0.45	50

Source: Reproduced from ICNIRP, *Health Physics* **74**(4): 494–522, 1998.

Note: 1. f as indicated in the frequency range column.
2. Provided that basic restrictions are met and adverse indirect effects can be excluded, field strength values can be exceeded.
3. For frequencies between 100 kHz and 10 GHz, S_{eq}, E^2, H^2, and B^2 are to be averaged over any 6-min period.
4. For peak values at frequencies up to 100 kHz, see Table 9.4, note 3. For peak values at frequencies exceeding 100 kHz. Between 100 kHz and 10 MHz, peak values for the field strengths are obtained by interpolation from the 1.5-fold peak at 100 kHz to the 32-fold peak at 10 MHz. For frequencies exceeding 10 MHz, it is suggested that the peak equivalent plane wave power density, as averaged over the pulse width, does not exceed 1000 times the S_{eq} restrictions or that the field strength does not exceed 32 times the field strength exposure levels given in the table.
5. For frequencies exceeding 10 GHz, S_{eq}, E^2, H^2, and B^2 are to be averaged over any $68/f^{1.05}$-min period (f in gigahertz).
6. No E-field value is provided for frequencies <1 Hz, which are effectively static electric fields. Electric shock from low impedance sources is prevented by established electrical safety procedures for such equipment.

(ICNIRP 2004). As a three-tier approach, the ICNIRP recommends the following:

1. Static magnetic field

   a. For normal operating mode, there should be an upper limit for whole-body exposure of 4 T including effects on fetuses and infants.

TABLE 9.7 Exposure Basic Restrictions for Controlling Movement in Static Magnetic Fields and Exposure to a Time-Varying Magnetic Field below 1 Hz

	Basic Restrictions				Reference Levels	
	ΔB (T)[a]	$B_{\text{peak to peak}}$ (T)	Internal Electric Field Strength [V m^{-1} (peak)]		dB/dt [T s^{-1} (peak)]	
Frequency f (Hz)						
Critical effect	Vertigo due to movement in static B-field	Vertigo due to time-varying B-field	PNS effects due to movement in static B-field and due to time-varying B-field	Phosphenes due to movement in static B-field and due to time-varying B-field	PNS effects due to movement in static B-field and due to time-varying B-field	Phosphenes due to movement in static B-field and due to time-varying B-field
Exposure condition[b]	Uncontrolled	Uncontrolled	Controlled	Uncontrolled	Controlled	Uncontrolled
0	2					
0–1		2				
0–0.66			1.1	1.1	2.7	2.7
0.66–1			1.1	0.7/f	2.7	1.8/f

Source: Reproduced from ICNIRP, *Health Physics* **106**(3): 418–425, 2014.

Note: For controlled exposure, the reference levels for a magnetic flux density may be converted to dB/dt by using $dB_0/dt = 2\pi f \sqrt{2} B_{\text{RMS}}$, where B_0 is the peak value of the sinusoidal magnetic flux density and B_{RMS} is the root-mean-square value.

[a] The maximum change of magnetic flux density ΔB is determined over any 3-sec period.

[b] For controlled exposure conditions, a ΔB of 2 T may be exceeded.

b. For controlled operating mode, there should be an upper limit for whole-body exposure of 8 T under medical supervision; and

c. For experimental operating mode above 8 T, a special ethical approval is required. No upper limit has been specified, but a progressively cautious approach is suggested for increasingly high magnetic flux densities due to uncertainties regarding possible effects of flow potentials on heart function. In light of these possible effects, it is determined that patients should only be exposed to such fields with appropriate clinical monitoring.

In addition, there is a need to ensure that patients should be moved slowly into the magnetic bore, in order to avoid the possibility of vertigo and nausea. Thresholds for motion-induced vertigo have been estimated to be around 1 T/sec for durations for greater than 1 sec, thus avoiding these sensations of induced electric fields and currents that arise as a consequence of motion in a static magnetic field.

2. Time-varying magnetic fields

a. For normal operating mode, it is recommended that the maximum exposure level should be a limit on the rate of change of the magnetic field (dB/dt) of 80 percent of the median perception threshold for peripheral nerve stimulation, as defined in the recommendations (ICNIRP 2004).

b. For controlled operating mode, there should be a limit on dB/dt of 100 percent of the median perception threshold.

c. For experimental operating mode, ICNIRP does not provide any explicit guidance for the procedures.

Here, the median perception threshold is defined by the following equation in the ICNIRP (2004): $dB/dt = 20(1 + 0.36/\tau)$ T/sec, where τ is the effective stimulus duration of the induced electrical stimulus in milliseconds. ICNIRP provided no explicit guideline for experimental operating mode.

3. Radio frequency electromagnetic fields

In order to avoid possible adverse thermal effects in patients during MRI examinations, ICNIRP has provided the limit for patients to protect against the rise of body and tissue temperature.

About the protection of patients undergoing MRI examinations, the recommended SAR limits for environmental temperature below 24°C and average time of 6 min are summarized in Table 9.8 (ICNIRP 2004). These SAR limits are for whole body and for the head, trunk, and extremities. These SAR levels should not be exceeded in order to limit temperature rise to the values given Table 9.9.

With respect to the application of the SAR levels defined in Table 9.8, the following points should be taken into account (ICNIRP 2004):

TABLE 9.8   Recommended SAR Level at Environmental Temperature below 24°C

		Averaging Time: 6 min					
	Whole Body SAR (W kg^{-1})	Partial-Body SAR (W kg^{-1})		Local SAR (Averaged over 10 g Tissue) (W kg^{-1})			
Body region → Operating mode ↓	Whole body	Any, except head	Head	Head	Trunk	Extremities
Normal	2	2–10[a]	3	10[b]	10	20
Controlled	4	4–10[a]	3	10[b]	10	20
Restricted	>4	(4–10)[a]	>3	10[b]	>10	>20
Short-term SAR	The SAR limit over any 10-sec period should not exceed three times the corresponding average SAR limit.					

Source: Reproduced from ICNIRP, *Health Physics* **106**(3): 418–425, 2014.
[a] Partial-body SARs scale dynamically with the ratio r between the patient mass exposed and the total patient mass:
  1. Normal operating mode: SAR = $(10-8 \cdot r)$ W kg^{-1}.
  2. Controlled operating mode: SAR = $(10-6 \cdot r)$ W kg^{-1}.
     The exposed patient mass and the actual SAR levels are calculated by the SAR monitor implemented in the MR system for each sequence and compared to the SAR limits.
[b] In cases where the eye is in the field of a small local coil used for RF transmission, care should be taken to ensure that the temperature rise is limited to 1°C.

TABLE 9.9   Basic Restrictions for Body Temperature Rise and Partial-Body Temperature

Operating Mode	Rise of Body Core Temperature (°C)	Spatially Localized Temperature Limits		
		Head (°C)	Trunk (°C)	Extremities (°C)
Normal	0.5	38	39	40
Controlled	1	38	39	40
Restricted	>1	>38	>39	>40

Source: Reproduced from ICNIRP, *Health Physics* **87**: 197–216, 2004.

- Partial-body SARs scale dynamically with the ratio r between the patient mass exposed and the total patient mass. For $r{\rightarrow}1$ they converge against the corresponding whole-body values. For $r{\rightarrow}0$, they converge against the localized SAR level of 10 W/kg defined by ICNIRP for occupational exposure of head and trunk (ICNIRP 1998).

- Recommended SAR restrictions do not relate to an individual MR sequence but rather to running SAR averages computed over each 6-min period, which is assumed to be a typical thermal equilibrium time of smaller masses of tissue (Brix et al. 2002).

- Whole-body SARs are valid at environmental temperatures below 24°C. At higher temperatures, they should be reduced, depending on actual environmental temperature and humidity.

It is recommended that the user requests detailed information on the energy deposition within the patient's body from the manufacturer of MRI devices. In addition the monitoring of the real-time temperature may be performed during MRI procedures in the controlled operating mode for patients and should be performed in all cases in the experimental operating mode.

Table 9.10 summarizes the ICNIRP recommended limits for patient exposure during MRI from guidelines (ICNIRP 2004, 2009c). These limits apply only to patients. The workers related to MRI examinations remain subject to the ICNIRP occupational guidelines with the protection of movement-induced adverse health effects arising from exposed to time-varying magnetic fields below 1 Hz (ICNIRP 2009b, 2010, 2014).

TABLE 9.10   Summary of Recommended Limits for Patient Exposure for Static, Gradient, and Radiofrequency during MRI Procedure

Operating Mode	Static Magnetic Field in Bore (T)	Gradient dB/dt as a Percentage of Median Perception Threshold (%)	Radiofrequency Field Maximum Core Temperature Rise (°C)	Temperature Limits (°C) Head	Trunk	Extremities
Normal	<4	80	0.5	38	39	40
Controlled	<8	100	1	38	39	40
Experimental	>8	100	>1	>38	>39	>40

9.5 OCCUPATIONAL EXPOSURE DURING MRI

The assessment of occupational exposure is an important issue and is needed for human health studies of MRI working staff. However, there is the difficulty in exposure assessment. In order to study the assessment of the occupational exposures, there are a number of approaches (Bongers et al. 2014; Bradley et al. 2007; De Vocht et al. 2009; Fuentes et al. 2008; Karpowicz et al. 2007; Karpowicz and Gryz 2006, 2013; Riches et al. 2007b; Schaap et al. 2013; Yamaguchi-Sekino et al. 2014).

The first approach is to record the movements of working staff during their activities in the MRI examinations and correlate these with the magnetic field maps obtained from the calculation and measurement around MRI scanners.

The second approach is to develop the magnetic field dosimeter and to record working staff's exposures with time elapse to static and time-varying magnetic fields with dosimeter during the working day.

The third approach is related more to the epidemiological study. One can identify and carefully characterize specific tasks of MRI working staff and the time spent on a worker's specific task and then measure the temporal and magnitude of the actual exposure of the worker. From the determinations of the specific task and actual exposure of the worker, one can formulate a job-exposure matrix. In an epidemiological study, this job-exposure matrix can be retrospectively used to estimate occupational exposures for different workers over a long period of time.

Using a questionnaire survey, Schaap et al. designed an assessment of the size and characteristics of the population who are occupationally exposed to electromagnetic fields from MRI scanners. In addition, they identify variability in job and workplace characteristics that determine the probability and type of exposure (Schaap et al. 2013). The survey results show that personal exposure measurements should be performed to characterize and quantify occupational exposure levels. Further, it is necessary to identify the role of exposure determinants in the MRI work environments.

Recently, the newly proposed Directive 2013/35/EU opened to the member states of European Union (EU Directive 2013/35/EU 2013). This Directive covers the limits of occupational exposure to the electromagnetic fields. This new Directive exempts MRI workers from electromagnetic field exposure limit.

9.5.1 Worker Exposure Assessment

In order to clarify health effects of MRI workers, the measurement and assessment of worker exposure level around MRI system and the evaluation of MRI-related static magnetic fields exposure are needed (Andreuccetti et al. 2013; Bongers et al. 2014; Börner et al. 2011; Bradley et al. 2007; De Vocht et al. 2009; Febles Santana et al. 2014; Fuentes et al. 2008; Hansson Mild et al. 2013; Kännälä et al. 2009; Karpowicz and Gryz 2006; Li et al. 2007; Liu and Crozier 2004; Riches et al. 2007a; Yamaguchi-Sekino et al. 2014; Wang et al. 2008).

The techniques have been used to estimate personal occupational exposure to MRI-related electromagnetic fields. There are four substantive studies where magnetometers were worn by MRI workers, specifically the radiologic technologist and engineering staff, during their routine duties. Bradley et al. investigated static magnetic field exposures in one open 0.6-T MRI magnet, four closed bore systems of 1.5-T MRI magnet, and one 3-T closed bore magnet (Bradley et al. 2007). The worker carried the dosimeter in the pocket. In this study, peak and 24 h time-averaged static magnetic fields were reported. Fuentes et al. measured and recorded the occupational exposure levels of static magnetic field, peak magnetic field (B), peak dB/dt, and average B in and around 1.5-, 2-, and 4-T MRI systems during their routine work (Fuentes et al. 2008).

Using a personal magnetic field dosimeter, De Vocht et al. monitored occupational exposure for MRI system engineering staff performing various tasks during routine work in 1-, 1.5-, and 3-T MRI systems (De Vocht et al. 2009). They reported task-based, time-weighted average exposures to static magnetic field as well as complementary task durations. Task level exposures (mT-min) were calculated based on the measurement data and were compared with task level exposures obtained from the historical exposure level and task durations.

On the other hand, Yamaguchi-Sekino et al. using a tri-axis Hall magnetometer, carried out occupational exposure monitoring of magnetic fields during routine examinations in 3-T MRI systems from two institutes (Yamaguchi-Sekino et al. 2014). A magnetometer was attached to a subject's chest during monitoring. Four radiologic technologists with magnetometers participated, and the exposure levels were recorded for 56 patients by workers and data acquisition of 103 samples. The relationship between the exposure levels and working duties was analyzed. The working duties are sorted into four categories: (1) MRI examination of the body and neck, (2) MRI examination of the body or extremities, (3) escort or assist patient, and (4) other work contents. The highest exposure was detected during MRI examination of the head or neck. The maximum stray fields were 628 ± 28 mT at the edge of the bed from one institute and 373 ± 2 mT at the edge of the bed from another institute. The reason why the stray magnetic field is different between the two institutes is that the sizes of the magnet and bore are different because they are from different suppliers.

Kännälä et al. examined the assessment of occupational exposure to gradient magnetic fields and time-varying magnetic fields generated by motion in nonhomogeneous static magnetic fields of two MRI scanners (1 T open and 3 T conventional) (Kännälä et al. 2009). These magnetic components can be measured simultaneously with an induction coil set up detecting the time rate of change of magnetic flux density (dB/dt). As a result, the highest motion-induced dB/dt was 0.7 T/sec for the 1-T MRI scanner and 3 T/s for the 3-T MRI scanner when only the static magnetic field was present.

Andreuccetti et al. proposed and tested the procedure for assessing occupational exposure due to MRI gradient magnetic fields and movement-induced effects in the static magnetic field (Andreuccetti et al. 2013). Their procedure considered two exposures of low-frequency switched gradient magnetic field and movement-induced effects in the static magnetic field. The radio frequency electromagnetic field was not taken into consideration. They tested in two 1.5-T whole-body MRI scanners and one 3-T head-only MRI scanner and evaluated exposure due to switched gradient magnetic fields in the location inside the magnet room where operators usually stay during those particular medical procedures. Movement-induced effects were evaluated considering the actual movements of volunteer operators during work activity by measuring the perceived time-varying magnetic field. The result was analyzed based on the ICNIRP 1998 and 2010 guidelines. Exposure to switched gradient magnetic fields in 1.5-T MRI scanner mostly resulted in noncompliance with the ICNIRP 1998 occupational levels, while at the same time, it showed as always compliant with the ICNIRP 2010 ones. Movement-induced effects resulted potentially noncompliant only in the case the operator moved the head inside the bore of 1.5-T MRI scanner.

There have been no epidemiological studies related to MRI operations. It is needed and very important from the point of long-term human health effects to conduct well-designed epidemiological studies of MRI workers.

In order to do this, Hansson Mild et al. proposed an exposure categorization for the different professions working with MRI equipment (Hansson Mild et al. 2013). As a basis for exposure assessment in epidemiological studies obtained from the estimation of exposure levels of electromagnetic field of about 100,000 workers, they define exposure in three categories, depending on whether people are exposed to only the static magnetic fields, to the static magnetic plus switched gradient magnetic, or to the static plus switched gradient magnetic plus radio frequency fields.

As a questionnaire-based descriptive pilot study, Wilén and De Vocht provided an indication of the self-reported prevalence of health complaints related to working with MRI systems (Wilén and de Vocht 2011). Fifty-nine nurses with on average 8 (±6) years' experience with MRI scanning procedures attended this study. In total, 9 nurses reported regularly experiencing at least one of the health complaints attributed to arise or be aggravated by their presence in the MRI scanning room. The results indicated that reporting of adverse symptoms was not related to the level of occupational workload/stress. However, reporting of health complaints was related to the strength of the magnet the nurses worked with, with 57 percent of symptoms reported by those nurses working with both 1.5- and 3-T MRI scanners.

9.5.2 European Electromagnetic Fields Directive

In relation to the exposure of workers, the European Directive 2004/40/EC "on the minimum health and safety requirements regarding the exposure of workers to the risks arising from physical agent (electromagnetic fields)" was published from the Council and Parliament of the European Union (EU Directive 2004/40/EC 2004). The aims of this Directive are to prevent occupational exposure to electromagnetic fields. This Directive set general rules for the obligation of employers.

The Directive was fundamentally based on the ICNIRP guidelines (ICNIRP 1998) and used the exposure limit values of ICNIRP. After the appearance of European Directive 2004/40/EC, the MRI community across Member States was concerned about the impact of implementation of this Directive on the use of MRI because of human risk assessments, manufacturing, and running costs (Keevil and Krestin 2010). So, the EU Directive 2004/40/EC was amended by EU Directive 2008/46/EC in April 2008. The EU Directive 2008/46/EC was adopted, delaying transposition deadline of the original Directive until 2012 (EU Directive 2008/46/EC).

The ICNIRP published two guidelines, one for static magnetic field and one for low-frequency electromagnetic fields (up to 100 kHz) (ICNIRP 2009b, 2010). Using these guidelines, the European Commission finally proposed a Directive replacement. This newly proposed EU Directive 2013/35/EU covers all known direct biophysical effects and other indirect effects caused by electromagnetic fields (up to 300 GHz) (EU Directive 2013/35/EU 2013). The Directive addresses short-term effects and does not cover suggested long-term effects. It does not cover the risks resulting from contact with live conductors. Finally, Member States are required to have transposed EU Directive 2013/35/EU into their respective national legislation framework by July 1, 2016.

The EU Directive 2013/35/EU covers exposure limit value and action levels of static electric, static, magnetic, and time-varying electric, magnetic, and electromagnetic fields with frequencies up to 300 GHz. The following discussions are focused on the issues related to MRI.

The basic restrictions in the ICNIRP become "exposure limit values" in the 2013 Directive and reference levels in the ICNIRP become "action levels." The Directive introduces two tiers of ELVs: sensory effects ELV and health effects ELV.

1. Exposure limit values (ELVs) and action levels (ALs) in the frequency range from 0 Hz to 10 MHz

   a. ELVs

      – ELVs below 1 Hz are limits for static magnetic field. For static magnetic field, the 2013 Directive sets exposure limit values identical to those in guidelines (ICNIRP 2009b).

      – ELVs for external magnetic flux density from 0 to 1 Hz

         – Sensory effects ELV is the ELV, 2 T for normal working conditions and is related to vertigo and other physiological effects related to disturbance of the human balance, resulting from the movement in a static magnetic field.

         – Health effects ELV is 8 T for controlled working conditions.

      – ELVs for frequencies between 1 Hz and 10 MHz are limits for induced electric fields in the body from exposure to time-varying electric and magnetic fields.

- Health effects ELVs for internal electric field strength from 1 Hz to 10 MHz are related to electric stimulation of all peripheral and central nervous system tissues in the body including the head.

- Sensory effects ELV for internal electric fields from 1 to 400 Hz are related to electric field effects on the central nervous system in the head, i.e., retinal phosphenes and minor transient changes in some brain functions.

b. ALs for exposure to magnetic fields shown in Table 9.11.

- Low action levels (low ALs) in the 2013 Directive for frequencies below 400 Hz are derived from the sensory effects ELV

TABLE 9.11   ALs for Exposure to Magnetic Fields from 1 Hz to 10 MHz

Frequency Range	Magnetic Flux Density Low ALs(B) [µT] (RMS)	Magnetic Flux Density High ALs(B) [µT] (RMS)	Magnetic Flux Density ALs for Exposure of Limbs to a Localized Magnetic Field [µT] (RMS)
$1 \leq f < 8$ Hz	$2.0 \times 10^5 / f^2$	$3.0 \times 10^5 / f$	$9.0 \times 10^5 / f$
$8 \leq f < 25$ Hz	$2.5 \times 10^4 / f$	$3.0 \times 10^5 / f$	$9.0 \times 10^5 / f$
$25 \leq f < 300$ Hz	1.0×10^3	$3.0 \times 10^5 / f$	$9.0 \times 10^5 / f$
300 Hz $\leq f < 3$ kHz	$3.0 \times 10^5 / f$	$3.0 \times 10^5 / f$	$9.0 \times 10^5 / f$
3 kHz $\leq f \leq 10$ MHz	1.0×10^2	1.0×10^2	3.0×10^2

Source: Reproduced from EU Directive 2013/35/EU, *Official Journal of the European Union* **56**: 1–21, L179, 2013.

Note: 1. f is the frequency expressed in hertz.

2. The low ALs and the high ALs are the RMS values that are equal to the peak values divided by $\sqrt{2}$ for sinusoidal fields. In the case of non-sinusoidal fields, the exposure evaluation carried out in accordance with Article 4 shall be based on the weighted peak method (filtering in the time domain), explained in practical guides referred to in Article 14; but other scientifically proven and validated exposure evaluation procedures can be applied, provided that they lead to approximately equivalent and comparable results.

3. ALs for exposure to magnetic fields represent maximum values at the workers' body position. This results in a conservative exposure assessment and automatic compliance with ELVs in all non-uniform exposure conditions. In order to simplify the assessment of compliance with ELVs, carried out in accordance with Article 4, in specific non-uniform conditions, criteria for the spatial averaging of measured fields based on established dosimetry will be laid down in the practical guides referred to in Article 14. In the case of a very localized source within a distance of a few centimeters from the body, the induced electric field shall be determined dosimetrically, case by case.

and for frequencies above 400 Hz, from the health effects ELVs for internal electric field.

- – High action levels (high ALs) are derived from the health effects ELV for internal electric fields related to electric stimulation of peripheral and autonomous nerve tissues in head and trunk.

2. ELVs and ALs in the frequency range from 100 kHz to 300 GHz.

   a. ELVs

   - – Health effects ELVs between 100 kHz and 6 GHz are limits for energy and power absorbed per unit mass of body tissue generated from exposure to electric and magnetic fields.

   - – Sensory effects ELVs from 0.3 to 6 GHz are limits on absorbed energy in a small mass tissue in the head from exposure to electromagnetic fields.

   - – Health effects ELVs between 6 and 300 GHz are limits for power density of an electromagnetic wave incident on the body surface.

   - – Sensory effect ELVs from 0.3 to 6 GHz are related to avoiding auditory effects caused by exposure of the head to pulsed microwave radiation.

   b. ALs for exposure to electric and magnetic fields shown in Table 9.12.

   - – ALs (B) are derived from the SAR or power density and ELVs based on the thresholds related to internal thermal effects caused by exposure to (external) electric and magnetic fields.

This EU Directive 2013/35/EU retains the derogation. Article 10 of the Directive is the scope of the derogations from the exposure limit values. It said that the exposure may exceed the ELVs if the exposure is related to the installation, testing, use, development, maintenance of, or research related to the MRI equipment for patients in the health sector, provided that certain conditions are met. These conditions are as follows:

1. Risk assessment has been carried out in accordance with Article 4 has demonstrated that the ELVs are exceeded.

TABLE 9.12   ALs for Exposure to Electric and Magnetic Fields from 100 kHz
to 300 GHz

Frequency Range	Electric Field Strength ALs(E) [V m^{-1}] (RMS)	Magnetic Flux Density ALs(B) [µT] (RMS)	Power Density ALs(S) [Wm^{-2}]
100 kHz $\leq f <$ 1 MHz	6.1×10^2	$2.0 \times 10^6/f$	–
$1 \leq f <$ 10 MHz	$6.1 \times 10^8/f$	$2.0 \times 10^6/f$	–
$10 \leq f <$ 400 MHz	61	0.2	–
400 MHz $\leq f <$ 2 GHz	$3 \times 10^{-3} f^{\frac{1}{2}}$	$1.0 \times 10^{-5} f^{\frac{1}{2}}$	–
$2 \leq f <$ 6 GHz	1.4×10^2	4.5×10^{-1}	–
$6 \leq f \leq$ 300 GHz	1.4×10^2	4.5×10^{-1}	50

Source: Reproduced from EU Directive 2013/35/EU, *Official Journal of the European
   Union* **56**: 1–21, L179, 2013.
Note: 1. f is the frequency expressed in hertz.
   2. [ALs(E)]^2 and [ALs(B)]^2 are to be averaged over a 6-min period. For RF pulses,
      the peak power density averaged over the pulse width shall not exceed
      1000 times the respective ALs(S) value. For multifrequency fields, the analysis
      shall be based on summation, as explained in the practical guides referred to in
      Article 14.
   3. ALs(E) and ALs(B) represent maximum calculated or measured values at the
      workers' body position. This results in a conservative exposure assessment and
      automatic compliance with ELVs in all non-uniform exposure conditions. In
      order to simplify the assessment of compliance with ELVs, carried out in accor-
      dance with Article 4, in specific non-uniform conditions, criteria for the spatial
      averaging of measured fields based on established dosimetry will be laid down in
      the practical guides referred to in Article 14. In the case of a very localized source
      within a distance of a few centimeters from the body, compliance with ELVs shall
      be determined dosimetrically, case by case.
   4. The power density shall be averaged over any 20 cm^2 of exposed area. Spatial
      maximum power densities averaged over 1 cm^2 should not exceed 20 times the
      value of 50 W m^{-2}. Power densities from 6 to 10 GHz are to be averaged over any
      6-min period. Above 10 GHz, the power density shall be averaged over any 68/
      $f^{1.05}$-min period (where f is the frequency in gigahertz) to compensate for pro-
      gressively shorter penetration depth as the frequency increases.

2. Given the state of the art, all technical and/or organizational measures have been applied.

3. Circumstances duly justify exceeding the ELVs.

4. Characteristics of the workplace, work equipment, or work practices
   have been taken into account.

5. Employer demonstrates that workers are still protected against
   adverse health effects and against safety risks.

On the basis of the above conditions, exposure related to MRI equipment for patients is not subject to the ELVs in the Annex.

Article 14 of this new Directive contains words, stating that the establishment of documented working procedures, as well as specific information and training measures for workers exposed to electromagnetic fields during MRI-related activities related to above five conditions (refer to Article 10 (1)a).

Overall, Stam summarized, in his review article, the origin of European Directive and its contents (Stam 2014). In addition, he compared magnetic field exposure level in high-risk workplace with the limits set in the revised Directive.

9.6 CONCLUSION

The MRI is the most successful technique in medicine, particularly, in clinical practices. In this chapter, we provided an overview and evaluation of the biological and health effects resulting from three kinds of electromagnetic fields (static magnetic field, gradient [time-varying] magnetic field, and radio frequency electromagnetic field) with the newest results, highlighting ICNIRP guidelines of occupational exposures, and European EMF Directives and its impact for MRI community. The electromagnetic fields associated with MRI have been used in medical technology without a concomitant risk of adverse health effects. Our scientific understanding of the biological and health effects of electromagnetic field has gradually improved from much research.

The WHO produced and planned three documents of EHC on the possible health effects of exposure to (1) static, (2) extremely low frequency (ELF), and (3) radio frequency electromagnetic fields. After publication of EHC documents, the WHO opened their research recommendations of MRI-related electromagnetic fields (WHO 2006b, 2010). The available data on possible effects of MRI examinations relevant to the safety of patients and worker are not sufficient to draw conclusions.

Since the introduction of MRI, patients, MRI staff, and other workers exposed to electromagnetic field associated with MRI have increased. Protection of workers requires occupational exposure assessment in MRI-working environments. In addition, in MRI environments, there has been no epidemiological research to assess the possible long-term health effects in patients, volunteers, and medical and MRI staffs. Because the health risk assessment for MRI procedures is incomplete, it is needed to address well-organized and high-quality research associated with exposure to MRI-related electromagnetic fields.

This chapter reviewed the current guidelines, their scope, and statement of protection of patients related to MRI, based on the origination from ICNIRP guidelines. The ICNIRP produced a guideline on time-varying electromagnetic field up to 300 GHz in 1998 and updated it for the frequency range 1 Hz to 100 kHz in 2010. The guideline on time-varying electromagnetic field below 1 Hz was published in 2014. The revision of the guideline for the frequency range from 100 kHz to 300 GHz is now under way. The ICNIRP produced a document on static magnetic field exposure in 1994, and it was updated in 2009. For the protection of patients or volunteers in MRI, statements from the ICNIRP appeared in 2004 and 2009. These documents are based on a review of exposures encountered during an MRI examination.

The European Directive 2013/35/EU was officially adopted to protect against effect of occupational exposure to electromagnetic field in the frequency range up to 300 GHz based on the ICNIRP. An important aspect of this new Directive and its impact to MRI community is shown in this chapter. The derogation for MRI in this new Directive will ensure that MRI scanners allow the use of diagnostic application. However, the guidelines of ICNIRP will hopefully apply toward the protection of MRI-related workers.

Finally, as a new activity, the WHO is considering developing an international standard for electromagnetic field exposure. The progress report 2013–2014 of international EMF project said "Member states are increasing interested in clear guidance based on harmonized standards and their application within a national framework of protection. The development of non-ionizing safety standards has been proposed by a Member State using the example of the International Ionizing radiation Basic Safety Standards developed as a collaborative approach between different UN organisations" (WHO 2014). The WHO International EMF project is set to develop the international harmonized global EMF exposure standard as "Safety standards."

ACKNOWLEDGMENTS

We thank Dr. Ikehata of Railway Technical Research Institute, Dr. Okano of Hakuju Institute for Health and Science, Dr. Sakurai of Gifu Pharmaceutical University, Dr. Yamazaki of Central Research Institute of Electric Power Industry, and Dr. Watanabe of NICT for their kind support on this manuscript.

REFERENCES

Abe, Z., Tanaka, K., Hotta, K., and Imai, M. 1974. Noninvasive measurements of biological information with application of nuclear magnetic resonance. In *Biological and Clinical Effects of Low Magnetic and Electric Fields*, Llaurado, J. G., Sances, A., and Battocletti, J. H., eds. Charles C. Thomas, Springfield, Ill.: 295–317.

Ahlbom, A., Green, A., Kheifets, L., Savitz, D., and Swerdlow, A. 2004. Epidemiology of health effects of radiofrequency exposure. *Environmental Health Perspectives* **112**: 1741–1754.

Andreuccetti, D., Contessa, G. M., Falsaperia, R., Lodato, R., Pinto, R., Zoppetti, N., and Rossi, P. 2013. Weighted-peak assessment of occupational exposure due to MRI gradient fields and movement in a nonhomogeneous static magnetic field. *Medical Physics* **40**(1): 011910.

Australian Radiation Protection and Nuclear Safety Agency (ARPANSA). 2014. Review of radiofrequency health effects research—Scientific literature 2000–2012. Technical Report Series No. **164**:1–68.

Baan, R., Gross, Y., Lauby-Secretan, B., El Ghissassi, F., Bouvard, V., Benbrahim-Tallaa Guha, N., Islami, F., Galicht, L., and Straif, K. 2011. Carcinogenicity of radio-frequency electromagnetic fields. *The Lancet Oncology* **12**(7): 624–626.

Barregard, L., Jarvholm, B., and Ungethum, E. 1985. Cancer among workers exposed to strong static magnetic fields. *The Lancet* **11**(8460): 892.

Bongers, S., Christopher, Y., Engels, H., Slottje, P., and Kromhout, H. 2014. Retrospective assessment of exposure to static magnetic fields during production and development of magnetic resonance imaging systems. *Annals of Occupational Hygiene* **58**(1): 85–102.

Börner, F., Brüggemeyer, H., Eggert, S., Fischer, M., Heinrich, H., Hentschel, K., and Neuschulz, H. 2011. Electromagnetic fields at workplaces: A new scientific approach to occupational health and safety. Bundesministeriums für Arbeit und Soziales.

Bradley, J. K., Nyekiova, M., Price, D. L., Lopez, L. D., and Grawley, T. 2007. Occupational exposure to static and time-varying gradient magnetic fields in MR units. *Journal of Magnetic Resonance Imaging* **26**(5): 1204–1209.

Brix, G., Seebass, M., Hellwig, G., and Griebel, J. 2002. Estimation of heat transfer and temperature rise in partial-body regions during MR procedures: An analytical approach with respect to safety considerations. *Magnetic Resonance Imaging* **20**: 65–76.

Cason, A. M., Kwon, B., Smith, J. C., and Houpt, T. A. 2009. Labyrinthectomy abolishes the behavioral and neural response of rats to a high-strength static magnetic field. *Physiology & Behavior* **97**: 36–43.

Damadian, R. 1971. Tumor detection by nuclear magnetic resonance. *Science* **171**: 1151–1153.

De Vocht, F., van Drooge, H., Engles, H., and Kromhout, H. 2006. Exposure, health complaints and cognitive performance among employees of an MRI scanners manufacturing department. *Journal of Magnetic Resonance Imaging* **23**: 197–204.

De Vocht, F., Stevens, T., Glover, P., Sunderland, A., Gowland, P., and Kromhout, H. 2007. Cognitive effects of head-movements in stray fields generated by a 7 Tesla whole-body MRI magnets. *Bioelectromagnetics* **28**(4): 247–255.

De Vocht, F., Muller, H., Engels, H., and Kromhout, H. 2009. Personal exposure to static and time-varying magnetic fields during MRI system test procedure. *Journal of Magnetic Resonance Imaging* **30**: 1223–1228.

EU Directive 2004/40/EC. 2004. EU Directive 2004/40/EC of the European Parliament and of the Council of 29 April 2004 on the minimum health and safety requirements regarding the exposure of workers to the risks arising from physical agents (electromagnetic fields). *Official Journal of the European Union* **47**: 1–9, L184.

EU Directive 2008/46/EC. 2008. EU Directive 2008/46/EC of the European Parliament and of the Council of 23 April 2008 amending Directive 2004/40/EC on the minimum health and safety requirements regarding the exposure of workers to the risks arising from physical agents (electromagnetic fields). *Official Journal of the European Union* **51**: 81–89, L114.

EU Directive 2013/35/EU. 2013. Directive 2013/35/EU of the European Parliament and of the Council of 29 June 2013 on the minimum health and safety requirements regarding the exposure of workers to the risks arising from physical agents (electromagnetic fields) (20th individual Directive within the meaning of Article 16(1) of Directive 89/391/EEC) and repealing Directive 2004/40/EC. *Official Journal of the European Union* **56**: 1–21, L179.

Febles Santana, V. M., Hernandez Armas, J. A., Martin Diaz, M. A., de Miguel Bibao, S., de Aldecoa Fernandez, J. C., and Ramos Gonzalez, V. 2014. Assessment of magnetic field in the surroundings of magnetic resonance systems: Risks for professional staff. *IFMBE Proceedings* **41**: 190–193.

Fuentes, M. A., Trikic, A., Wilson, S. J., and Crozier, S. 2008. Analysis and measurements of magnetic field exposure for healthcare workers in selected MR environment. *IEEE Transactions on Biomedical Engineering* **55**(4): 1355–1364.

Gaffey, C. T., and Tenforde, T. S., 1981. Alterations in the rat electrocardiogram induced by stationary magnetic fields. *Bioelectromagnetics* **2**: 357–370.

Glover, P. M., Cavini, I., Qian, W., Bowtell, R., and Gowland, P. A. 2007. Magnetic-field-induced vertigo: A theoretical and experimental investigation. *Bioelectromagnetics* **28**: 349–361.

Hansson Mild, K., Hand, J., Hietanen, M., Gowland, P., Karpowicz, J., Keevil, S., Van Rongen, E., Scarfi, M. R., and Wilen, J. 2013. Exposure classification of MRI workers in epidemiological studies. *Bioelectromagnetics* **34**(1): 81–84.

Health Protection Agency (HPA). 2008. Static magnetic fields. RCE-6 Report of the independent Advisory Group on Non-ionising Radiation. Chilton, UK.

Heinrich, A., Szostek, A., Nees, F., Meyer, P., Semmler, W., and Flor, H. 2011. Effects of static magnetic fields on cognition, vital signs, and sensory perception: A meta-analysis. *Journal of Magnetic Resonance Imaging* **34**(4): 758–763.

Heinrich, A., Szostek, A., Meyer, P., Nees, F., Rauschenberg, J., Groebner, J., Gilles, M., Paslakis, G., Deuschle, M., Semmler, W., and Flor, H. 2013. Cognition and sensation in very high static magnetic fields: A randomized case-crossover study with different field strengths. *Radiology* **266**(1): 236–245.

Houpt, T. A., and Houpt, C. E. 2010. Circular swimming in mice after exposure to a high magnetic field. *Physiology & Behavior* **100**: 284–290.

Houpt, T. A., Carella, L., Gonzalez, G., Janowitz, I., Mueller, A., Mueller, K., Neth, B., and Smith, J. C. 2011. Behavioral effects on rats of motion within a high static magnetic field. *Physiology & Behavior* **102**: 338–346.

HPA. 2012. Health effects from radiofrequency electromagnetic fields. RCE-20 Report of the independent Advisory Group on Non-ionising Radiation. Chilton, UK.

IARC. 2002. Non-ionizing radiation. Part 1: Static and extremely low frequency (ELF) electric and magnetic fields. IARC Monographs on the Evaluation of Carcinogenic Risks to Humans, Volume 80. International Agency for Research Cancer, Lyon.

IARC. 2013. Non-ionizing radiation. Part 2: Radiofrequency electromagnetic fields. IARC Monographs on the Evaluation of Carcinogenic Risks to Human, Volume 102. International Agency for Research Cancer, Lyon.

ICNIRP. 1998. Guidelines for limiting exposure to time-varying electric, magnetic, and electromagnetic fields (up to 300 GHz). *Health Physics* **74**(4): 494–522.

ICNIRP. 2003a. Exposure to static and low frequency electromagnetic fields, biological effects and health consequences (0-100kHz). Matthes, R, McKinlay, A. F., Bernhardt, J. H., Vecchia, O., and Veyret, B., eds. ICNIRP, Oberschleissheim, Germany.

ICNIRP. 2003b. Guidance on determining compliance of exposure to pulsed and complex non-sinusoidal waveforms below 100 kHz with ICNIRP guidelines. *Health Physics* **84**: 383–387.

ICNIRP. 2004. Medical magnetic resonance (MR) procedures: Protection of patients. *Health Physics* **87**:197–216.

ICNIRP. 2009a. ICNIRP statement on the "Guidelines for limiting exposure to time-varying electric, magnetic, and electromagnetic fields (up to 300 GHz)." *Health Physics* **93**: 257–258.

ICNIRP. 2009b. Guidelines on limits to exposure from static magnetic field. *Health Physics* **96**: 504–514.

ICNIRP. 2009c. ICNIRP statement on amendment to the ICNIRP "Statement on medical magnetic resonance procedures: Protection of patients." *Health Physics* **97**: 259–261.

ICNIRP. 2009d. Exposure to high frequency electromagnetic fields, biological effects and health consequences (100 kHz–300 GHz). Review of the Scientific Evidence and Health Consequences. ICNIRP, Oberschleissheim, Germany.

ICNIRP. 2010. Guidelines or limiting exposure to time-varying electric, magnetic, and electromagnetic fields (up to 100 kHz). *Health Physics* **99**: 818–836. (Erratum. 2011). **100**: 112.

ICNIRP. 2014. Guidelines for limiting exposure to electric fields induced by movement of the human body in a static magnetic fields and by time-varying magnetic fields below 1-Hz. *Health Physics* **106**(3): 418–425.

Ikehata, M., Koana, T., and Nakagawa, M. 1999. Effect of strong static magnetic fields on mutagenicity of chemical mutagens in bacterial mutation assay. In *Electricity and Magnetism in Biology and Medicine*, Bersani, F., eds. Plenum, New York: 683–686.

Kännälä, S., Toivo, T., Alanko, T., and Jokela, K. 2009. Occupational exposure measurements of static and pulsed gradient magnetic fields in the vicinity of MRI scanners. *Physics in Medicine and Biology* **54**(7): 2243–2257.

Karpowicz, J., and Gryz, K. 2006. Health risk assessment of occupational exposure to a magnetic field from magnetic resonance imaging devices. *International Journal of Occupational Safety and Ergonomics* **12**(2):155–167.

Karpowicz, J., and Gryz, K. 2013. The pattern of exposure to static magnetic field of nurses involved in activities related to contrast administration into patients diagnosed in 1.5 T MRI scanners. *Electromagnetic Biology and Medicine* **32**(2): 182–191.

Karpowicz, J, Hietanen, M., and Gryz, K. 2007. Occupational risk from static magnetic fields of MRI scanners. *Environmentalist* **27**: 533–538.

Keevil, S. F., and Krestin, G. P. 2010. EMF directive still poses a risk to MRI research in Europe. *The Lancet* **376**: 1124–1125.

Kinouchi, Y., Yamaguchi, H., and Tenforde, T. S. 1996. Threshold analysis of magnetic field interactions with aortic blood flow. *Bioelectromagnetics* **17**: 21–32.

Lauterbur, P. C. 1973. Image formation by induced local interactions: Examples employing nuclear magnetic resonance. *Nature* **242**: 90–191.

Lee, J. W., Kim, M. S., Kim, Y. J., Choi, Y. J., Lee, Y., and Chung, H. W. 2011. Genotoxic effects of 3 T magnetic resonance imaging in cultured human lymphocytes. *Bioelectromagnetics* **32**: 535–542.

Leitgeb, N. 2014. Childhood leukemia not linked with ELF magnetic fields. *Journal of Electromagnetic Analysis and Application* **6**: 174–183.

Li, Y., Hand, J. W., Wills, T., and Hajnal, J. V. 2007. Numerically simulated induced electric field and current density within a human model located close to a z-gradient coil. *Journal of Magnetic Resonance Imaging* **26**: 1286–1295.

Liu, F., and Crozier, F. 2004. A distributed equivalent magnetic current based FDTD method for the calculation of E-fields induced by gradient coils. *Journal of Magnetic Resonance Imaging* **169**: 323–327.

Mansfield, P., and Maudsley, A. A. 1977. Planar spin imaging by NMR. *Journal of Magnetic Resonance* **27**: 101–119.

Marsh, J. L., Armstrong, T. J., Jacobson, A. P., and Smith, R. G. 1982. Health effect of occupational exposure to steady magnetic fields. *American Industrial Hygiene Association Journal* **43**: 387–394.

Mevissen, M., Stamm, A., Buntenkoetter, S., Zwingleberg, R., Wahnschafe, U., and Loescher, W. 1993. Effects of magnetic fields on mammary tumour development induced by 7,12-dimethybenz(a)anthracene in rats. *Bioelectromagnetics* **14**: 131–143.

NIEHS. 1998. Assessment of health effects from exposure to power-line frequency electric and magnetic fields. *NIH Publ. 98-3981.*

NRPB (Advisory Group on Non-Ionising Radiation). 2003. Health effect from radiofrequency electromagnetic fields. Documents of the NRPB. 14.

Okano, H. 2008. Effects of static magnetic fields in biology: Role of free radicals. *Frontiers in Bioscience* **13**(16): 6106–6125.

Riches, S. F., Collins, D. J., Scuffham, J. W., and Leach, M. O. 2007a. EU Directive 2004/40: field measurements of a 1.5 T clinical MR scanner. *British Journal of Radiology* **80**(954): 483–487.

Riches, S. F., Collins, D. J., Charles-Edwards, G. D., Shafford, J. C., Cole, J, Keevil, S. F. et al. 2007b. Measurements of occupational exposure to switched gradient and spatially-varying magnetic fields in areas adjacent to 1.5 T clinical MRI systems. *Journal of Magnetic Resonance Imaging* **26**(5): 1346–1352.

Sakurai, T., Terashima, S., and Miyakoshi, J. 2009. Effects of strong magnetic fields used in magnetic resonance imaging on insulin-secreting cells. *Bioelectromagnetics* **30**: 1–8.

Sakurai, T., Hashimoto, A., Kiyokawa, T., Kikuchi, K., and Miyakoshi, J. 2012. Myotube orientation using strong static magnetic fields. *Bioelectromagnetics* **33**: 421–427.

Schaap, K., Christopher-De Vries, Y., Slottje, P., and Kromhout, H. 2013. Inventory of MRI application and workers exposed to MRI-related electromagnetic fields in the Netherlands. *European Journal of Radiology* **82**(12): 2279–2285.

Scientific Commission on Emerging and Newly Identified Health Risks (SCENIHR). 2009. Health effects of exposure to EMF. European Commission, Brussels http://ec.europa.eu/health/ph_risk/committees/04_scenihr/docs/scenihr_o_022.pdf.

Scientific Commission on Emerging and Newly Identified Health Risks (SCENIHR). 2013. Preliminary opinion on potential health effects of exposure to electromagnetic fields (EMF). European Commission, Brussels.

Shellock, F. G. 2000. Radiofrequency energy-induced heating during MR procedures: A review. *Journal of Magnetic Resonance Imaging* **12**(1): 30–36.

Sheppard, A. R., Swicord, M. L., and Balzano, Q. 2008. Quantitative evaluations of mechanisms of radiofrequency interactions with biological molecules and processes. *Health Physics* **95**(4): 365–396.

Stam, R. 2014. The revised electromagnetic fields directive and worker exposure in environments with high magnetic flux densities. *Annals of Occupational Hygiene*, in press.

Tenforde, T. S. 1983. Cardiovascular alterations in Macaca monkeys exposed to stationary magnetic fields: Experimental observation and theoretical analysis. *Bioelectromagnetics* **4**: 1–9.

Ueno, S., and Iwasaka, M. 1994a. Parting of water by magnetic fields. *IEEE Transactions on Magnetics* **30**: 4698.

Ueno, S., and Iwasaka, M. 1994b. Properties of diamagnetic fluid in high-gradient magnetic fields. *Journal of Applied Physics* **79**: 4705.

Ueno, S., and Shigemitsu, T. 2007. Biological effects of static magnetic fields. In *Handbook of Biological Effects of Electromagnetic Fields: Bioengineering and Biophysical Aspects of Electromagnetic Fields*, 3rd ed. Barnes, F. S., and Greenebaum, B., eds. CRC Press. Boca Raton, Fla.: 203

Ueno, S., and Okano, H. 2012. Static, low-frequency, and pulsed magnetic fields in biological systems. In *Electromagnetic Fields in Biological Systems*. Lin, J. C., ed. CRC Press, Boca Raton, Fla.: 115–196.

Van Nierop, L. E., Slottje, P., van Zandvoort, M. J., de Vocht, F., and Kromhout, H. 2012. Effects of magnetic stray fields from a 7 Tesla MRI scanner on neurocognition: A double-blind randomized crossover study. *Occupational & Environmental Medicine* **69**(10): 759–766.

Van Nierop, L. E., Slottje, P., Kingma, H., and Kromhout, H. 2013. MRI-related static magnetic stray fields and postural body sway: A double-blind randomized crossover study. *Magnetic Resonance in Medicine* **70**: 232–240.

Wang, H., Trakic, A., Liu, F., and Crozier, S. 2008. Numerical field evaluation of healthcare workers when bending towards high-field MRI magnets. *Magnetic Resonance in Medicine* **59**: 410–422.

WHO. 1987. Magnetic fields. *Environmental Health Criteria 69*. World Health Organization, Geneva.

WHO. 2006a. Static fields. *Environmental Health Criteria 232*. World Health Organization, Geneva.

WHO. 2006b. WHO research agenda for static fields. World Health Organization, Geneva.

WHO. 2007a. Extremely low frequency fields. *Environmental Health Criteria 238*. World Health Organization, Geneva.

WHO. 2007b. WHO Research agenda for extremely low frequency fields. World Health Organization, Geneva.

WHO. 2010. WHO research agenda for radiofrequency fields. World Health Organization, Geneva.

WHO. 2014. The International EMF project. Progress Report June 2013–2104. http://www.WHO-int/peh-emf/project/IAC_2014_Progress_Report .pdf?ua=1.

Wilén, J., and de Vocht, F. 2011. Health complaints among nurses working near MRI scanners- A descriptive pilot study. *European Journal of Radiology* **80**(2): 510–513.

Yamaguchi-Sekino, S., Sekino, M., and Ueno, S. 2011. Biological effects of electromagnetic fields and recently updated safety guidelines for strong static magnetic fields. *Magnetic Resonance in Medical Science* **10**(1): 1–10.

Yamaguchi-Sekino, S., Nakai, T., Imai, S., Izawa, S., and Okuno, T. 2014. Occupational exposure levels of static magnetic field during routine MRI examination in 3 T MR system. *Bioelectromagnetics* **35**(1): 70–75.

Zahedi, Y., Zaun, G., Maderwald, S., Orzada, S., Putter, C., Scherag, A., Winterhager, E., Ladd, M. E., and Grummer, R. 2014. Impact of repetitive exposure to strong static magnetic fields on pregnancy and embryonic development of mice. *Journal of Magnetic Resonance and Imaging* **39**(3): 691–699.

Zaun, G., Zahedi, Y., Maderwald, S., Orzada, S., Putter, C., Scherag, A., Winterhager, E., Ladd, M. E., and Grummer, R. 2014. Repetitive exposure of mice to strong static magnetic fields in utero does not impair fertility in adulthood but may affect placental weight of offspring. *Journal of Magnetic Resonance and Imaging* **39**(3): 683–690.

Zhao, G., Chen, S., Wang, L., Zhao, Y., Wang, J., Wang, X., Zhang, W., Wu, R., Lu, L., Wu, Y., and Xu, A. 2011. Cellular ATP content was decreased by a homogeneous 8.5 T static magnetic field exposure: Role of reactive oxygen species. *Bioelectromagnetics* **32**(2): 94–101.

CHAPTER 10

New Horizons in Biomagnetics and Bioimaging

Shoogo Ueno and Masaki Sekino

CONTENTS

Through the principles and applications of biomagnetic stimulation and imaging we discussed in the previous chapters, the field of biomagnetics is fertile ground for new research into the relationship between living organisms and magnetism. Applying an interdisciplinary approach, it covers a wide range of fields from medicine and biology to physics and bioengineering.

In this chapter, we discuss the recent advances in biomagnetics and bio-imaging to envision new horizons in the interdisciplinary fields.

10.1 ADVANCES IN BIOMAGNETICS

10.1.1 Deep Brain Stimulation for Therapeutic Application

A method of deep brain stimulation (DBS) with implanted electrodes has been used for the treatments of brain dysfunctions. The DBS has brought good news[1] for patients who are afflicted with, for example, unbearable chronic pain, resistant movement, and affective disorders including major depression. Although DBS is a powerful tool, it is an invasive method to insert and implant electrodes inside the deeper parts of the brain.

If an alternative noninvasive method for deep brain stimulation is introduced, new horizons will be opened in this area. The transcranial magnetic stimulation (TMS) is a promising potential tool for the realization of deep brain stimulation. In researching deep TMS (dTMS), several attempts have been reported. There are several coil configurations potentially suitable for dTMS; double cone, H-coil, and Halo coils.

The double-cone coil[2–4] operates on the same principle as the figure-eight coil configuration where a pair of opposing directed time-varying magnetic fields around a target increase induced electric fields in the targeted areas. The two circular coils have a fixed angle between them, and their diameter is larger than that of the figure- eight coil. The double-cone coil induces a greater electric field intensity to stimulate the deeper brain regions compared with other coils. However, the surrounding large areas in the brain are also stimulated.

Another coil configuration for dTMS is called the H-coil.[5,6] The H-coils have complex winding patterns based on numerical calculations of three-dimensional brain phantom models to achieve effective stimulation of deep brain structures and are used for the treatment of a variety of psychiatric and neurological disorders. Although the H-coils are used for clinical applications, nontargeted areas of the brain are also stimulated.

A family of dTMS coil designs, called the Halo coil, was proposed to increase the induced electric fields at depth in the brain.[7] The Halo coil system is composed of a small coil on the top of the head and a large circular coil around the head.

Computational studies were carried out in order to evaluate the focality of these three types of coil configurations.[8,9] Three-dimensional distributions of the induced electric fields in realistic head model by dTMS coils were calculated by an impedance method, and the results were compared with that of a standard figure-eight coil.[10,11] Simulation results show that double cone and H-coils have deep field penetration at the expense of induced higher and wider spread electric fields in superficial cortical regions. The combination of the Halo coil with a circular coil produce deeply penetrating electric fields the same as double cone and H-coils, but the stimulation in superficial brain tissues are much higher.[10,11] Further studies are needed for better dTMS with better focality.

The studies on TMS and reward circuits in the brain are important and interesting for the understanding of the neuronal connectivity in the brain as well as for the potential treatments of brain dysfunctions such as depression and Parkinson's disease. For example, Wassermann et al. have studied reward-related activity in the human motor cortex,[12] modulation of corticospinal excitability by reward,[13] reward processing abnormalities in Parkinson's disease,[14] and so on. The study concludes that TMS of the human primary motor cortex M1 may be useful as a quantitative measure of reward-related activity.[12]

It seems that repetitive transcranial magnetic stimulation (rTMS), transcranial direct current stimulation (tDCS), and other brain stimulation techniques are promising for treatments of depression and other psychiatric disorders.[15,16]

In 2011, Fitzgerald reviewed the current state of development and application of a wide range of brain stimulation approaches in the treatment of psychiatric disorders.[17] The brain stimulation methods that he reviewed include vagus nerve stimulation, rTMS, tDCS, electroconvulsive therapy, and magnetic seizure therapy. The review concludes that it appears likely that the range of psychiatric treatments available for patients will grow over the coming years to progressively include a number of novel brain stimulation techniques.[17]

In 2013, Fitzgerald et al. examined a double-blind, randomized trial of deep rTMS for autism spectrum disorder and came to the following conclusion: Deep rTMS to bilateral dorsomedial prefrontal cortex yielded a reduction in social relating impairment and socially related anxiety.[18]

Further studies are needed in the area of deep transcranial magnetic stimulation.

10.1.2 Combination of TMS with DTI and EEG

Many studies have been published on the topic of spatiotemporal patterns of brain electrical activities elicited by TMS in a high spatial and temporal resolution. These are combinations of TMS with diffusion tensor imaging (DTI), TMS with electroencephalographic (EEG) measurements, TMS with functional magnetic resonance imaging (fMRI), and so on.

The invention of DTI and *in vivo* fiber tractography[19,20] contribute to the studies on DTI-based neural trajectories in TMS. De Geeter et al.[21,22] studied a DTI-based model for TMS and effective electric fields along realistic neural trajectories for modeling the stimulation mechanisms of TMS. De Geeter et al. focused on the stimulation of the hand area of the left primary motor cortex, M1, with a figure-eight coil[10,11] to get deeper insights on the stimulation mechanisms. Including realistic neural geometry in the model demonstrates the strong and localized variations of the effective electric field between the tracts themselves and along them due to the interplay of factors such as the tract's position and orientation in relation to the TMS coil, the neural trajectory, and its course along the white and gray matter interface.[22] Thus, the studies using DTI and TMS are useful to visualize exciting fronts and neural trajectories and also to improve and integrate the early proposed models of nerve excitation elicited by magnetic stimulation.[23-30]

EEG measurements just after the onset of brain stimulation by TMS with a figure-eight coil[10,11] are important to study dynamic neuronal connectivity in the brain in a high temporal (msec) and high spatial (mm) resolution. Ilmoniemi et al. successfully measured neuronal electrical responses to magnetic brain stimulation.[31] The combination of TMS with simultaneous EEG has enabled us to study dynamic intra- and interhemispheric connectivity for the deeper understanding of cortico-cortical and interhemispheric interactions, cortical inhibitory processes, cortical plasticity and oscillations, and so on.[32-40]

The combination of TMS with fMRI or DTI is also interesting and important. It is rather difficult to combine TMS with simultaneous fMRI or DTI, but data of fMRI and DTI are used for studying the effects of TMS on the changes in functional organization of the brain.

Pascual-Leone et al. reviewed studies on measuring and manipulating brain connectivity with resting state functional connectivity MRI (fcMRI) and TMS, classifying many publications into three general network properties: anatomical connectivity, functional connectivity, and response to perturbation/stimulation.[41] TMS is a useful noninvasive method to assess brain connectivity in human subjects noninvasively for controlled, individualized neuronal network modulation.

10.1.3 Biomagnetic Approaches to Treatments of Cancer and Other Diseases

Magnetic control of cell growth and cancer therapy are important and challenging fields of research. Magnetic destruction of cancer cells[42] and magnetic suppression of cancer cell growth by electromagnetic pulses[43] are promising, and further *in vivo* studies using rodents are essential for future clinical use.

Studies on magnetic cell orientation and its use for bone growth acceleration[44] and nerve regeneration[45] are also very exciting and challenging subjects.

Studies on iron cage proteins (ferritins) exposed to microwaves are promising and exciting[46,47] for potential treatments of Alzheimer's and Parkinson's diseases in the future.

All topics discussed here require further studies for medical applications.

10.2 ADVANCES IN BIOIMAGING

10.2.1 MEG and Multimodal Integration

With its high temporal resolution, magnetoencephalography (MEG) has been a powerful tool to study and diagnose noninvasively the human brain function in the millisecond time scale. A variety of studies have been reported, for example, gamma phase locking,[48] resting cortical dynamics,[49] intra- and interfrequency brain network structure,[50] entrainment of alpha oscillations,[51] real-time MEG neurofeedback training,[52] visual object recognition related to feedback and feedforward inputs,[53] and language processing in atypical development.[54]

The combination of MEG and electroencephalographic (EEG) measurements can reduce ambiguities in data from usage of only one of the modalities.[55] The combination of MEG with fMRI[56,57] has accelerated functional brain studies such as the study on human emotion cognition[58] and gamma and alpha oscillations related to working memory performance.[59] Combining MEG with other noninvasive tools will further developments to neuroscience and clinical medicine.

10.2.2 Combination of SQUIDs and Ultra-Low Magnetic Field MRI

Clarke et al. and others showed that MRI at ultra-low magnetic fields of ~100 µT, Larmor frequencies of kHz, was possibly available by combining superconducting quantum interference device (SQUID) sensors.[60,61] Clarke's group reported the concept of SQUID-based MRI at ultra-low fields (ULFs) with readout magnetic fields as low as 132 μT and using pulsed pre-polarization.[60] In ULF-MRI, precession signals are detected at ~μT magnetic fields using highly sensitive SQUID sensors. Compared with conventional MRI with high magnetic fields of a few Teslas order, signal acquisition times or imaging times in ULF-MRI tend to be long. Many studies on SQUID-detected ULF-MRI have been reported in the 2000s and 2010s regarding distinct imaging with a shorter imaging time.[62-69]

10.2.3 Advances in MRI

In contrast with ULF-MRI, projects of ultra-high magnetic field MRI (UHF-MRI) are also very important to get ultra-clear and distinct imaging. Big projects for UHF-MRI have been promoted and are now being promoted by different institutions, for example, by a group at the University of Minnesota in the United States conducted by Ugurbil[70] and by a group at NeuroSpin in France conducted by Le Bihan.[71] Larmor frequency, or resonant frequency (RF), proportionally increases with increase in magnetic field. Many important issues need to be resolved in UHF-MRI such as safety, superconducting coils, gradient coils, and RF coils related to the inhomogeneity in RF field distributions and coil design, etc. The issues on parallel transmission and RF coil design were studied[72-74] to overcome the problem in RF field inhomogeneity.

Diffusion tensor MRI (DTI) and *in vivo* fiber tractography invented by Basser et al.[19,20] have become powerful tools in the imaging of ultra-fine anatomical structures of neuronal fibers in the human brain. Functional MRI (fMRI) based on blood oxygenation level dependent (BOLD) effects invented by Ogawa et al.[56,57] has become a powerful tool to reveal the functional organization of the human brain. Data obtained by these two tools have enabled us to study anatomical and functional connectivity in the brain. Polonara et al. studied the topography and connectivity of the corpus callosum in patients with partial callosal resection using both fMRI and DTI.[75]

Thomas et al. evaluated the inherent limit for anatomical accuracy of brain connections derived from DTI tractography by comparing the DTI data with the data obtained from tracer studies in the macaque.[76] Anatomical accuracy is highly dependent upon parameters of the tractography algorithm, with different optimal values for mapping different pathways. The results by Thomas et al. suggest there is an inherent limitation in determining long-range anatomical projections based on boxel averaged estimates of local fiber orientation obtained from DTI data that is unlikely to be overcome by improvements in data acquisition and analysis alone.[76]

To further understanding of diffusion coefficients in the brain, the relationship between membrane permeability and extra- and intra-cellular diffusion coefficients was studied using rat brain and human subjects.[77-79]

Studies on imaging electrical impedance and neuronal currents based on MRI have been reported. Further studies are needed in the areas of MR-based neuronal current imaging or MR neuroimaging.

10.2.4 Magnetic Particle Imaging and Microwave Imaging

Recent advances in magnetic nanotechnology and biochemistry have accelerated studies on applications of magnetic nanoparticles for medical diagnoses and treatments such as hyperthermia, drug delivery system, and magnetic particle imaging (MPI). This newly developing field, which we call nanobiomagnetics or biomedical nanomagnetics, covers a variety of fields from diagnostics to therapy that includes development, functionalizing and optimizing magnetic nanoparticles for targeting, drug and gene delivery, image enhancements and hyperthermia, for example, like Krishnan's group[80] at the University of Washington, USA is promoting.

Other groups have studied magnetic sensing and imaging of sentinel lymph nodes in patients with lung or breast cancer.[81-85] Detection of sentinel lymph nodes is crucially important for early detection of cancer before metastasis.

MPI methods have difficulty in detecting deeply seated target cells from the surface of the body because magnetic fields from the magnetic particles combined with cancer cells attenuate rapidly at the surface of the body. Empuke et al.[86] developed highly sensitive magnetic nanoparticle imaging using cooled-Cu/HTS (high-critical-temperature superconductor) pickup coils. Empuke et al. demonstrated the detection of 100 μg of magnetic nanoparticles and obtained a clear contour map of the magnetic fields from the particles at the distance between the target and pickup coil of 100 mm.

Persson et al.[87] developed two different brain diagnostic devices based on microwave technology, and they associated two first proof of principle measurements that the systems can differentiate hemorrhagic from ischemic stroke in acute stroke patients, as well differentiate hemorrhagic patients from healthy volunteers. The system was based on microwave scattering measurements with an antenna system worn on the head. This system will certainly contribute to quick diagnosis so that ischemic stroke patients may receive acute thrombolytic treatment at hospitals, dramatically reducing or eliminating symptoms.[87]

10.2.5 Molecular Imaging

Studies of molecular imaging have been rapidly accelerated for medical applications in the 2000s through the 2010s. Brain research through advancing innovative neurotechnologies (BRAIN) initiative was introduced by President Obama in the United States, on April 2, 2013, and a National Institutes of Health (NIH) Workshop on clinical translation of molecular imaging probes and technology was held on August 2, 2013, in Bethesda, Maryland.[88,89] Pettigrew, director of the national institute of biomedical imaging and bioengineering (NIBIE) at the NIH, stated that "This high-priority BRAIN initiative on next-generation human imaging technologies is a striking call for teams of our best and brightest minds from multiple disciplines to think big, aim high, and reach far beyond previous boundaries in human imaging science."[89]

Studies on *in vivo* imaging using rodents are promising and closest to human imaging, and new strategies for fluorescent probe design in medical diagnostic imaging were reviewed by Kobayashi et al.[90] Among them, Urano et al., for example, developed a method of selective molecular imaging of viable cancer cells with a pH-activatable fluorescence probe.[91] Urano's group also developed a unique method of rapid cancer detection by topically spraying a γ-glutamyltranspeptidase (GGT)-activated fluorescent probe.[92] This probe is practical for clinical application during surgical or endoscopic procedures because of its rapid and strong activation upon contact with GGT on the surface of cancer cells.

The future of science and medicine will be rapidly progressing with the integration of advanced technologies, including biomagnetics, bioimaging, drug delivery system, and nanobioscience.[93]

Biomagnetics and bioimaging are thus leading medicine and biology into new horizons through their novel applications of magnetism and imaging technology. With the increasing integration of medicine and engineering, biomagnetics and bioimaging are merging into a new science that encompasses a wide range of fields including their diverse cultures to create a new and original discipline.

REFERENCES

1. Kringelbach, M. L., Jenkinson, N., Owen, S. L. F., and Aziz, T. Z. 2007. Translational principles of deep brain stimulation. *Nature Reviews Neuroscience* **8**(8): 623–635.
2. Ugawa, Y., Uesaka, Y., Terao, Y., Hanajima, R., and Kanazawa, I. 1995. Magnetic stimulation over the cerebellum in humans. *Annals Neurology* **37**: 703–713.

3. Roth, Y., Zangen, A., and Hallett, M. 2002. A coil design for transcranial magnetic stimulation of deep brain regions. *Journal of Clinical Neurophysiology* **19**: 361–370.
4. Lontis, E. R., Voigt, M., and Struijk, J. J. 2006. Focality assessment in transcranial magnetic stimulation with double and cone coils. *Journal of Clinical Neurophysiology* **23**: 462–471.
5. Zangen, A., Roth, Y., Voller, B., and Hallett, M. 2005. Transcranial magnetic stimulation of deep brain regions; evidence for efficacy of the H-coil. *Clinical Neurophysiology* **116**:775–779.
6. Roth, Y., Amir, A., Levkovitz, Y., and Zangen, A. 2007. Three-dimensional distributions of the electric fields induced in the brain by transcranial magnetic stimulation using figure-8 and deep H-coils. *Journal of Clinical Neurophysiology* **24**: 31–38.
7. Crowther, L. J., Marketos, P., Williams, P. I., Melikhov, Y., and Jiles, D. C. 2011. Transcranial magnetic stimulation: Improved coil design for deep brain investigation. *Journal of Applied Physics* **109**: 07B314.
8. Lu, M., and Ueno, S. 2013. Calculating the electric fields in the human brain by deep transcranial magnetic stimulation. *Proceedings of the IEEE Engineering in Medicine and Biology Society*, 376–379.
9. Lu, M., and Ueno, S. 2013. Numerical simulation of deep brain transcranial magnetic stimulation. In *Transcranial Magnetic Stimulation*, Lucia Alba-Ferrara, L., ed. Nova Science Publishers, Hauppauge, New York: 41–60.
10. Ueno, S., Tashiro, T., and Harada, S. 1988. Localized stimulation of neural tissues in the brain by means of a paired configuration of time-varying magnetic fields. *Journal of Applied Physics* **64**: 5862–5864.
11. Ueno, S., Matsuda, T., and Fujiki, M. 1990. Functional mapping of the human motor cortex obtained by focal and vectorial magnetic stimulation of the brain. *IEEE Transactions on Magnetics* **26**: 1539–1544.
12. Kapogiannis, D., Campion, P., Grafman, J., and Wassermann, E. M. 2008. Reward-related activity in the human motor cortex. *European Journal of Neuroscience* **27**: 1836–1842.
13. Mooshagian, E., Keisler, A., Zimmermann, T., Schweickert, J. M., and Wassermann, E. M. 2015. Modulation of corticospinal excitability by reward depends on task framing. *Neuropsychologia* **68**: 31–37.
14. Kapogiannis, D., Mooshagian, E., Campion, P., Grafman, J., Zimmermann, T. J., Ladt, K. C., and Wassermann, E. M. 2011. Reward processing abnormalities in Parkinson's disease. *Movement Disorders* **26**: 1451–1457.
15. Fitzgerald, P. B., and Daskalakis, Z. J. 2012. A practical guide to the use of repetitive transcranial magnetic stimulation in the treatment of depression. *Brain Stimulation* **5**: 287–296.
16. Fitzgerald, P. B., Hoy, K. E., Singh, A., Gunewardene, R., Slack, C., Ibrahim, S., Hall, P. J., and Daskalakis, Z. J. 2013. Equivalent beneficial effects of unilateral and bilateral prefrontal cortex transcranial magnetic stimulation in a large randomized trial in treatment-resistant major depression. *International Journal of Neuropsychopharmacology* **16**: 1975–1984.

17. Fitzgerald, P. B. 2011. The emerging use of brain stimulation treatments for psychiatric disorders. *Australian and New Zealand Journal of Psychiatry* **45**: 923–938.

18. Enticott, P. G., Fitzgibbon, B. M., Kennedy, H. A., Arnold, S. L., Elliot, D., Peachey, A., Zangen, A., and Fitzgerald, P. B. 2013. A double-blind, randomized trial of deep repetitive transcranial magnetic stimulation (fTMS) for autism spectrum disorder. *Brain Stimulation* **6**: 1–6.

19. Basser, P. J., Mattiello, J., and Le Bihan, D. 1994. Estimation of the effective self-diffusion tensor from the NMR spin echo. *Journal of Magnetic Resonance B* **103**: 247–254.

20. Basser, P. J., Pajevic, S., Pierpaoli, C., Duda, J., and Aldroubi, A. 2000. In vivo fiber tractography using DT-MRI data. *Magnetic Resonance in Medicine* **44**: 625–632.

21. De Geeter, N., Crevecoeur, G., Dupre, L., Van Hecke, W., and Leemans, A. 2012. A DTI-based model for TMS using the independent impedance method with frequency-dependent tissue parameters. *Physics in Medicine and Biology* **57**: 2169–2188.

22. De Geeter, N., Crevecoeur, G., Leemans, A., and Dupre, L. 2015. Effective electric fields along realistic DTI-based neural trajectories for modelling the stimulation mechanisms of TMS. *Physics in Medicine and Biology* **60**: 453–471.

23. Roth, B. J., and Basser, P. J. 1990. A model of the stimulation of a nerve fiber by electromagnetic induction. *IEEE Transactions on Biomedical Engineering* **37**: 588–597.

24. Ueno, S., Matsuda, T., and Hiwaki, O. 1991. Estimation of structures of neural fibers in the human brain by vectorial magnetic stimulation. *IEEE Transactions on Magnetics* **27**: 5387–5389.

25. Basser, P., Wijesinghe, R. S., and Roth, B. J. 1992. The activating function for magnetic stimulation derived from a three-dimensional volume conductor model. *IEEE Transactions on Biomedical Engineering* **39**: 1207–1211.

26. Nagarajan, S. S., Dutand, M., and Warman, E. N. 1993. Effects of induced electric fields on finite neuronal structures: A simulation study. *IEEE Transactions on Biomedical Engineering* **40**: 1175–1188.

27. Hyodo, A., and Ueno, S. 1996. Nerve excitation model for localized magnetic stimulation of finite neuronal structures. *IEEE Transactions on Magnetics* **32**: 5112–5114.

28. Liu, R., and Ueno, S. 2000. Calculating the activating function of nerve excitation in inhomogeneous volume conductor during magnetic stimulation using finite element method. *IEEE Transactions on Magnetics* **36**: 1796–1799.

29. Lu, M., Ueno, S., Thorlin, T., and Persson, M. 2008. Calculating the activating function in the human brain by transcranial magnetic stimulation. *IEEE Transactions on Magnetics* **44**: 1438–1441.

30. Hiwaki, O., and Inoue, T. 2009. A method for estimation of stimulated brain sites based on columnar structure of cerebral cortex in transcranial magnetic stimulation. *Journal of Applied Physics* **105**: 07B303.

31. Ilmoniemi, R. J., Virtanen, J., Karhu, J., Aronen, H. J., Naatanen, R., and Katila, T. 1997. Neuronal responses to magnetic stimulation reveal cortical reactivity and connectivity. *Neuroreport* **8**: 3537–3540.

32. Ilmoniemi, R. J., and Karhu, J. 2008. TMS and electroencephalography: Methods and current advances. In *The Oxford Handbook of Transcranial Stimulation*, Wasserman E. M., Epstein, C. M., Ziemann, U., Paus, T., and Lisanby, S. H., eds. Oxford University Press, New York: 593–608.

33. Raij, T., Karhu, J., Kicic, D., Lioumis, P., Julkunen, P., Lin, F. H., Ahveninen, J., Ilmoniemi, R. J., Makela, J. P., Hamalainen, M., Rosen, B. R., and Belliveau, J. W. 2008. Parallel input makes the brain run faster. *NeuroImage* **40**: 1792–1797.

34. Ilmoniemi, R. J., and Dubravko, K. 2010. Methodology for combined TMS and EEG. *Brain Topography* **22**: 233–248.

35. Maki, H., and Ilmoniemi, R. J. 2010. EEG oscillations and magnetically evoked motor potentials reflect motor system excitability in overlapping neuronal populations. *Clinical Neurophysiology* **121**: 492–501.

36. Iramina, K., Maeno, T., Kowatari, Y., and Ueno, S. 2002. Effects of transcranial magnetic stimulation on EEG activity. *IEEE Transactions on Magnetics* **38**: 3347–3349.

37. Iramina, K., Maeno, T., Nonaka, Y., and Ueno, S. 2003. Measurement of evoked electroencephalography induced by transcranial magnetic stimulation. *Journal of Applied Physics* **93**: 6718–6720.

38. Iwahashi, M., Koyama, Y., Hyodo, A., Hayami, T., Ueno, S., and Iramina, K. 2009. Measurements of evoked electroencephalograph by transcranial magnetic stimulation applied to motor cortex and posterior parietal cortex. *Journal of Applied Physics* **105**: 07B321.

39. Torii, T., Sato, A., Nakahara, Y., Iwahashi, M., Itoh, Y., and Iramina, K. 2012. Frequency-dependent effects of repetitive transcranial magnetic stimulation on the human brain. *Neuroreport* **19**: 1065–1070.

40. Marshall, T. R., O'Shea, J., Jensen, O., and Bergmann, T. O. 2015. Frontal eye fields control attentional modulation of alpha and gamma oscillations in contralateral occipitoparietal cortex. *Journal of Neuroscience* **35**: 1–10.

41. Fox, M. D., Halko, M. A., Eldaief, M. C., and Pascual-Leone, A. 2012. Measuring and manipulating brain connectivity with resting state functional connectivity magnetic resonance imaging (fcMRI) and transcranial magnetic stimulation (TMS). *Neuroimaging* **62**: 2232–2243.

42. Ogiue-Ikeda, M., Sato, Y., and Ueno, S. 2003. A new method to destruct targeted cells using magnetizable beads and pulsed magnetic force. *IEEE Transactions on Nanobioscience* **2**: 262–265.

43. Yamaguchi, S., Ogiue-Ikeda, M., Sekino, M., and Ueno, S. 2006. The effect of repetitive magnetic stimulation on tumor and immune functions in mice. *Bioelectromagnetics* **27**: 64–72.

44. Kotani, H., Kawaguchi, H., Shimoaka, T., Iwasaka, M., Ueno, S., Ozawa, H., Nakamura, K., and Hoshi, K. 2002. Strong static magnetic field stimulates bone formation to a definite orientation *in vivo* and *in vitro*. *Journal of Bone and Mineral Research* **17**: 1814–1821.

45. Eguchi, Y., Ohtori, S., and Ueno, S. 2015. The effectiveness of magnetically aligned collagen for neural regeneration *in vitro* and *in vivo*. *Bioelectromagnetics* **36**: 233–243.

46. Cespedes, O., and Ueno, S. 2009. Effects of radio frequency magnetic fields on iron release from cage proteins. *Bioelectromagnetics* **30**: 336–342.

47. Cespedes, O., Inmoto, O., Kai, S., Nibu, Y., Yamaguchi, T., Sakamoto, N., Akune, T., Inoue, M., Kiss, T., and Ueno, S. 2010. Radio frequency magnetic field effects on molecular dynamics and iron uptake in cage proteins. *Bioelectromagnetics* **31**: 311–317.

48. Han, J., Mody, M., and Ahlfors, S. P. 2012. Gamma phase locking modulated by phonological contrast during auditory comprehension in reading disability. *Cognitive neuroscience and neuropsychology* **24**: 851–856.

49. Shriki, O., Alstott, J., Carver, F., Holroyd, T., Henson, R. N. A., Smith, M. L., Coppola, R., Bullmore, E., and Plenz, D. Neuronal avalanches in the resting MEG of the human brain. *Journal of Neuroscience* **33**: 7079–7090.

50. Siebenhuner, F., Weiss, S. A., Coppola, R., Weinberger, D. R., and Bassett, D. S. 2013. Intra- and inter-frequency brain network structure in health and schizophrenia. *PLoS One* **8**: 1–13.

51. Spaak, E., de Lange, F. P., and Jensen, O. 2014. Local entrainment of alpha oscillations by visual stimuli causes cyclic modulation of perception. *Journal of Neuroscience* **34**: 3536–3544.

52. Okazaki, Y., Horschig, J. M., Luther, L., Oostenveld, R., Murakami, I., and Jensen, O. 2015. *NeuroImage* **107**: 323–332.

53. Ahlfors, S. P., Jones, S. R., Ahveninen, J., Hamalainen, M. S., Belliveau, J. W., and Bar, M. 2015. Direction of magnetoenchephalography source associated with feedback and feedforward contributions in a visual object recognition task. *Neuroscience Letters* **585**: 149–154.

54. Mody, M. 2014. Language processing in atypical development: looking below the surface with MEG. In *Magnetoencephalography*, Supek, S., and Aine, C. J., eds. Springer-Verlag, Berlin: 579–597.

55. Ahlfors, S. P. 2014. MEG and multimodal integration. In *Magnetoencephalography*, Supek, S., and Aine, C. J., eds. Springer-Verlag, Berlin: 183–198.

56. Ogawa, S., Lee, T. M., Nayak, A. S., and Glynn, P. 1990. Oxygenation-sensitive contrast in magnetic resonance image of rodent brain at high magnetic field. *Magnetic Resonance in Medicine* **14**: 68–78.

57. Ogawa, S., Lee, T. M., Kay, A. R., and Tank, D. W. 1990. Brain magnetic resonance imaging with contract dependent on blood oxygenation. *Proceedings of the National Academy of Sciences of the United States of America* **87**: 9868–9872.

58. Jabbi, M., Kohn, P. D., Nash, T., Ianni, A., Coutlee, C., Holroyd, R., Carver, F. W., Chen, Q., Cropp, B., Kippenhan, J. S., Robinson, S. E., Coppola, R., and Berman, K. F. 2014. Convergent BOLD and beta-band activity in superior temporal sulcus and frontolimbic circuitry underpins human emotion cognition. *Cerebral Cortex*, in press.

59. Lozano-Soldevilla, D., ter Huurne, N. Cools, R., and Jensen, O. 2014. GABAergic modulation of visual gamma and alpha oscillations and its consequences for working memory performance. *Current Biology* **24**: 2876–2887.

60. Clarke, J., Hatridge, M., and Mossle. M. 2007. SQUID detected magnetic resonance in microtesla fields. *Annual Reviews of Biomedical Engineering* 9: 389–413.
61. Mossle, M., Han, S. I., Myers, W. R., Lee, S. K., Kelso, N., Pines, A., and Clarke, J. 2006. SQUID-detected microtesla MRI in the presence of metal. *Journal of Magnetic Resonance* **179**: 146–151.
62. Matlachov, A. N., Volegov, P. L., Espy, M. A., George, J. S., and Kraus Jr., R. H. 2004. SQUID detected NMR in microtesla magnetic fields. *Journal of Magnetic Resonance* **170**: 1–7.
63. Espy, M., Matlashov, A., and Volegov, P. 2013. SQUID-detected ultra-low field MRI. *Journal of Magnetic Resonance* **229**: 127–141.
64. Nieminen, J. O., Burghoff, M., Trahms, L., and Ilmoniemi, R. J. 2010. Polarization encoding as a novel approach to MRI. *Journal of Magnetic Resonance* **202**: 211–216.
65. Vesanen, P. T., Nieminen, J. O., Zevenhoven, K. C., Dabek, J., Parkkonen, L. T., Zhdanov, A. V., Luomahaara, J., Hassel, J., Penttila, J., Simonla, J., Ahonen, A. I., Makela, J. P., and Ilmoniemi, R. J. 2013. Hybrid ultra-low-field MRI and magnetoenchephalography system based on a commercial whole-head neuromagnetometer. *Magnetic Resonance in Medicine* **69**: 1795–1804.
66. Nieminen, J. O., Zevenhoven, K. C. J., Vesanen, P. T., Hsu, Y. C., and Ilmoniemi, R. J. 2014. Current-density imaging using ultra-low-field MRI with adiabatic pulses. *Magnetic Resonance Imaging* **32**: 54–59.
67. Vesanen, P. T., Nieminen, J. O., Zevenhoven, K. C. J., Hsu, Y. C., and Ilmoniemi, R. J. 2014. Current-density imaging using ultra-low-field MRI with zero-field encoding. *Magnetic Resonance Imaging* **32**: 1–5.
68. Espy, M. A., Magnelind, P. E., Matlashov, A. N., Newman, S. G., Sandin, H. J., Schultz, L. J., Sedillo, R., Urbaitis, A. V., and Volegov, P. L. 2015. Progress toward a deployable SQUID-based ultra-low field MRI system for anatomical imaging. *IEEE Transactions on Applied Superconductivity* **25**: 1601705.
69. Woo, T., Nagase, M., Ohashi, H., and Sekino, M. 2015. Development of a SQUID system for ultralow-field MRI measurement. *International Journal of Applied Electromagnetics and Mechanics*, in press.
70. Ugurbil, K. 2014. Magnetic resonance imaging at ultrahigh fields. *IEEE Transactions on Biomedical Engineering* **61**: 1364–1379.
71. Le Bihan, D. 2014. Diffusion MRI: What water tells us about the brain. *EMBO Molecular Medicine* **6**: 569–573.
72. Pruessmann, K. P. 2006. Encoding and reconstruction in parallel NMR. *NMR in Biomedicine* **19**: 288–299.
73. Katscher, U., and Bornert, P. 2006. Parallel RF transmission in MRI. *NMR in Biomedicine* **19**: 393–400.
74. Sekino, M., Boulant, N., Luong, M., Amadon, A., Ohsaki, H., and Le Bihan, D. 2009. Effects of relaxation during RF pulses on the homogeneity of signal intensity in parallel transmission. *Proceedings of the ISMRM 17th Scientific Meeting and Exhibition* 2568.

75. Polonara, G., Mascioll, G., Foschi, N., Salvolini, U., Pierpali, C., Manzoni, T., Fabri, M., and Barbaresi, P. 2014. Further evidence for the topography and connectivity of the corpus callosum: An fMRI study of patients with partial callosal resection. *Journal of Neuroimaging*, in press.

76. Thomas, C., Ye, F. Q., Irfanoglu, M. O., Madi, P., Saleen, K. S., Leopold, D. A., and Pierpaoli, C. 2015. Anatomical accuracy of brain connections derived from diffusion MRI tractography is inherently limited. *Proceedings of the National Academy of Sciences of the United States of America* **111**: 16,574–16,579.

77. Sekino, M., Sano, M., Ogiue-Ikeda, M., and Ueno, S. 2006. Estimation of membrane permeability and intracellular diffusion coefficient. *Magnetic Resonance in Medical Sciences* **5**: 1–6.

78. Imae, T., Shinohara, H., Sekino, M., Ueno, S., Ohsaki, H., Mima, K., and Ohtomo, K. 2008. Estimation of cell membrane permeability of the rat brain using diffusion magnetic resonance imaging. *Journal of Applied Physics* **103**: 07A311.

79. Imae, T., Shinohara, T., Sekino, M., Ueno, S., Ohsaki, H., Mima, K., and Ohtomo, K. 2009. Estimation of cell membrane permeability and intracellular diffusion coefficient of the human gray matter. *Magnetic Resonance in Medical Science* **8**: 1–7.

80. Krishnan, K. M. 2010. Biomedical nanomagnetics: A spin through new possibilities in mapping, diagnostics and therapy. *IEEE Transactions on Magnetics Advances in Magnetics* **46**: 2523–2558.

81. Shiozawa, M., Kobayashi, S., Sato, Y., Maeshima, H., Hozumi, Y., Lefor, A. T., Kurihara, K., Sata, N., and Yasuda, Y. 2012. Magnetic resonance lymphography of sentinel lymph nodes in patients with breast cancer using superparamagnetic iron oxide: A feasibility study. *Breast Cancer* **21**(4): 394–401.

82. Minamiya, Y., Ito, M., Katayose, Y., Saito, H., Imai, K., Sato, Y., and Ogawa, J. 2006. Intraoperative sentinel lymph node mapping using a new sterilizable magnetometer in patients with nonsmall cell lung cancer. *Annals of Thoracic Surgery* **81**: 327–330.

83. Ookubo, T., Inoue, Y., Kim, D., Ohsaki, H., Masahiko, Y., Kusakabe, M., and Sekino, M. 2013. Characteristics of magnetic probes for identifying sentinel lymph nodes. *Proceedings of the IEEE Engineering in Medicine and Biology Society* **2013**: 5485–5488.

84. Gleich, B., and Weizenecker, J. 2005. Tomographic imaging using the non-linear response of magnetic particles. *Nature* **435**: 1214–1217.

85. Sattel, T. F., Knopp, T., Biederer, S., Gleich, B., Weizenecker, J., Borgert, J., and Buzug, T. M. 2009. Single-sided device for magnetic particle imaging. *Journal of Physics D: Applied Physics* **42**: 022001.

86. Morishige, T., Mihaya, T., Bai, S., Miyazaki, T., Yoshida, T., Matsuo, M., and Empuku, K. 2014. Highly sensitive magnetic nanoparticle imaging using cool-Cu/HTS-superconductor pickup coils. *IEEE Transactions on Applied Superconductivity* **24**: 1800105.

87. Persson, M., Fhager, A., Dobsicek, H., Yu, Y., McKelvey, T., Pegenius, G., Karlsson, J. E., and Elam, M. 2014. Microwave-based stroke diagnosis making global pre-hospital thrombolytic treatment possible. *IEEE Transactions on Biomedical Engineering* **61**: 2806–2817.
88. Liu, C. H., Sastre, A., Conroy, R., Seto, B., and Pettigrew, R. I. 2014. NIH workshop on clinical translation of molecular imaging probes and technology—Meeting report. *Molecular Imaging and Biology* **16**: 595–604.
89. Pettigrew, R. I. 2014. BRAIN initiative to transform human imaging. *Neuroscience* **6**: 1–2.
90. Kobayashi, H., Ogawa, M., Alford, R., Choyke, P. L., and Urano, Y. 2010. New strategies for fluorescent probe design in medical diagnostic imaging. *Chemical Reviews* **110**: 2620–2640.
91. Urano, Y., Asanuma, D., Hama, Y., Koyama, Y., Barrett, T., Kamiya, M., Nagano, T., Watanabe, T., Hasegawa, A., Choyke, P. L., and Kobayashi, H. 2009. Selective molecular imaging of viable cancer cells with pH-activatable fluorescence probes. *Nature Medicine* **15**: 104–109.
92. Urano, Y., Sakabe, M., Kosaka, N., Ogawa, M., Mitsunaga, M., Asanuma, D., Kamiya, M., Young, M. R., Nagano, T., Choyke, P. L., and Kobayashi, H. 2011. Rapid cancer detection by topically spraying a γ-glutamyltranspeptidase-activated fluorescent probe. *Science Translational Medicine* **3**: 1–10.
93. Ueno, S., Ando, J., Sugawara, T., Jimbo, Y., Itaka, K, Kataoka, K., and Ushida, T. 2006. The state of the art of nanobioscience in Japan. *IEEE Transactions on Nanobioscience* **5**: 1–12.

Index

Page numbers followed by f and t indicate figures and tables, respectively.